Canadian Mathematical Society
Société mathématique du Canada

Roberto Lucchetti

Convexity and Well-Posed Problems

With 46 Figures

 Springer

Roberto Lucchetti
Dipto. Matematica
Politecnico di Milano
Milano, 20133
Italy
rel@como.polimi.it

Editors-in-Chief ₁
Rédacteurs-en-chef
Jonathan Borwein
Karl Dilcher
Department of Mathematics and Statistics
Dalhousie University
Halifax, Nova Scotia B3H 3J5
Canada
cbs-editors@cms.math.ca

Mathematics Subject Classification (2000): 49-01, 46N10, 26B25, 90-01

ISBN   978-1-4419-2111-6   e-ISBN   978-0-387-31082-4

Printed on acid-free paper.

9 8 7 6 5 4 3 2 1

springeronline.com

Dedicated to my family, pets included.

Contents

Preface

This book deals mainly with the study of convex functions and their behavior from the point of view of stability with respect to perturbations. We shall consider convex functions from the most modern point of view: a function is defined to be *convex* whenever its epigraph, the set of the points lying above the graph, is a convex set. Thus many of its properties can be seen also as properties of a certain convex set related to it. Moreover, we shall consider extended real valued functions, i.e., functions taking possibly the values $-\infty$ and $+\infty$. The reason for considering the value $+\infty$ is the powerful device of including the constraint set of a constrained minimum problem into the objective function itself (by redefining it as $+\infty$ outside the constraint set). Except for trivial cases, the minimum value must be taken at a point where the function is not $+\infty$, hence at a point in the constraint set. And the value $-\infty$ is allowed because useful operations, such as the inf-convolution, can give rise to functions valued $-\infty$ even when the primitive objects are real valued.

Observe that defining the objective function to be $+\infty$ outside the closed constraint set preserves lower semicontinuity, which is the pivotal and minimal continuity assumption one needs when dealing with minimum problems. Variational calculus is usually based on derivatives. In the convex case too, of course, the study of the derivative is of the utmost importance in the analysis of the problems. But another concept naturally arises, which is a very important tool for the analysis. This is the *subdifferential* of a function at a given point x, which, as opposed to the derivative, does not require the function to be finite on a whole ball around x. It also exists when the graph of the function has angles, and preserves many important properties of the derivatives. Thus a chapter is dedicated to the study of some properties of the subdifferential: its connections with the directional derivatives and the Gâteaux and Fréchet differentials whenever they exist, and its behavior as a multifunction. The following chapter, after introducing the most fundamental existence theorem in minimum problems, the Weierstrass theorem, is dedicated to the Ekeland variational principle which, among other things, establishes, for a very general class $\mathcal{F}$ of functions (lower semicontinuous, lower bounded) defined on a complete metric space X, an existence theorem on a dense (for a natural topology on $\mathcal{F}$) set. This gives a way around lack of a topology on X, and allows for application of the Weierstrass theorem. We also analyze in some detail some of the very interesting consequences of the principle, mainly in the convex setting.

Next, we introduce the fundamental operation of Fenchel conjugation. This is the basis of all the duality theory which we develop, essentially following the approach of Ekeland–Temam (see [ET]). We then give a representative number of examples of its applications, including zero sum games, including the beautiful proof of the famous von Neumann theorem on the existence of an equilibrium in mixed strategies for finite games. This also allows us to get interesting results for linear programming. I want to stress at this point that, notwithstanding that the minimization of a scalar convex function is the primary subject of study of this book, the basic underlying concept that motivated me to write it is "optimization". For this reason, I include in it some game theory, one of the most modern and challenging aspects of optimization, with a glance as well to vector optimization. My hope is that readers will be stimulated and encouraged to bring the ideas, developed here for the convex, extended real valued functions, (mainly stability and well-posedness) to these domains too. To this end I must however say that some research is already in progress in this direction, although it is not so well established as to have a place in this book.

Coming back to the content of the book, I have to mention that my primary goal is to illustrate the ideas of stability and well-posedness, mainly in the convex case. Stability means that the basic parameters of a minimum problem, the infimal value and the set of the minimizers, do not vary much if we slightly change the initial data, the objective function and the constraint set. On the other hand, well-posedness means that points with values close to the value of the problem must be close to actual solutions. In studying this, one is naturally led to consider perturbations of functions and of sets. But it turns out that neither traditional convergences of functions, pointwise convergence, compact-open topology, nor classical convergence of sets, Hausdorff and Vietoris, are well suited to our setting. The stability issue explains why scholars of optimization have devoted so much time to defining and studying various convergence structures on the space of closed subsets of a metric space. Moreover, this approach perfectly fits with the idea of regarding functions as sets. Thus beginning with Chapter 8, the second part of the book starts with an introduction to the basic material concerning convergence of the closed subsets of a metric space X, and the topological nature of these convergences. These topologies are usually called hypertopologies, in the sense that the space X can be embedded in the hyperspace (whose points are closed sets), and the topology in the hyperspace respects the topology of X. A sequence $\{x_n\}$ in X converges in X if and only if the sequence of sets $\{\{x_n\}\}$ converges in the hyperspace. Since this topic appears to be interesting in itself, Appendix B is dedicated to exploring in more detail some basic ideas underlying the construction and study of these topologies/convergences, but it is not necessary to the comprehension of the rest of the book.

Using these topologies requires also knowing the continuity of basic operations involving them. For instance, when identifying functions with sets, it is not clear (nor even true) whether the sum of two convergent (in some particular sense) sequences converges to the sum of the limits. Yet having this property is very fundamental, for instance to ensure a good Lagrange multipliers rule in constrained problems. Thus, Chapter 9 is dedicated to this issue.

We then turn our attention to the study of well-posed problems, and the connection between stability and well-posedness. In doing this, we give some emphasis to a very recent and fruitful new well-posedness concept, which in some sense contains at the same time the two classical notions of stability and Tykhonov well-posedness.

Since there are many important classes of minimization problems for which existence cannot be guaranteed universally for all elements of the class, it is interesting to know "how many" of these problems will have solutions and also enjoy the property of being well-posed. This is the subject of Chapter 11. We consider here the idea of "many" from the point of view of the Baire category, and in the sense of σ-porosity, a recent and interesting notion which provides more refined results than the Baire approach. This part contains the most recent results in the book, and is mainly based on some papers by Ioffe, Revalski and myself.

The book ends with some appendices, entitled "Functional analysis" (a quick review of the Hahn–Banach theorem and the Banach–Dieudonné–Krein–Smulian theorem), "Topology" (the theorem of Baire, and a deeper insight to hypertopologies) and "More game theory".

A few words on the structure of the book. The part on convexity is standard, and much of the inspiration is taken from the classical and beautiful books cited in the References, such as those by Ekeland–Temam, Rockafellar, Phelps, and Lemaréchal–Hiriart-Urruty. I also quote more recent and equally interesting books, such as those of Borwein–Lewis and of Zalinescu. The study of hypertopologies is instead a less classical issue, the only book available is the one by G. Beer [Be]. However my point of view here is different from his and I hope that, though very condensed, this section will help people unfamiliar with hypertopologies to learn how to use them in the context of optimization problems. Finally, the sections related to stability have roots in the book by Dontchev–Zolezzi, but here we focus mainly on convexity.

About the (short) bibliography, I should emphasize that, as far as the first part is concerned, I do not quote references to original papers, since most of the results which are presented are now classical; thus I only mention the most important books in the area, and I refer the reader to them for a more complete bibliography. The references for hypertopologies and classical notions of well-posedness are the books by [Be],[DZ] respectively. When dealing with more recent results, which are not yet available in a book, I quote the original papers. Finally, the section concerning game theory developed in the duality chapter is inspired by [Ow].

The book contains more than 120 exercises, and some 45 figures. The exercises, which are an essential part of this work, are not all of the same level of difficulty. Some are suitable for students, while others are statements one can find in recent papers. This does not mean that I consider these results to be straightforward. I have merely used the exercise form to establish some interesting facts worth mentioning but whose proof was inessential to a reading of the book. I have chosen to start each chapter with one of my favorite quotations, with no attempt to tie the quote directly to the chapter.

Since this is my first and last book of this type, I would like to make several acknowledgements. First of all, I want to thank all my coauthors. I have learned much from all of them, in particular, A. Ioffe and J. Revalski. Most of the material concerning the genericity results is taken from some of their most recent papers with me. More importantly, I am very happy to share with them a friendship going far beyond the pleasure of writing papers together. For several years these notes were used to teach a class at the Department of Mathematics and Physics at the Catholic University of Brescia, and a graduate class held at the Faculty of Economics at the University of Pavia. I would like to thank my colleagues M. Degiovanni and A. Guerraggio for inviting me to teach these classes, and all students (in particular I want to mention Alessandro Giacomini) who patiently helped me in greatly improving the material, and correcting misprints. I also wish to thank some colleagues whom I asked to comment on parts of the book, in particular G. Beer, who provided me with some excellent remarks on the chapters dedicated to hyper-topologies. Also, comments by the series editors J. Borwein and K. Dilcher to improve the final version of the book were greatly appreciated. I owe thanks to Mary Peverelli and Elisa Zanellati for undertaking the big task of outlining figures copied from my horrible and incomprehensible drawings. Last but not least, I would like to express my appreciation for an invitation from CNRS to spend three months at the University of Limoges, attached to LACO. The nice, quiet and friendly atmosphere of the department allowed me to complete the revision of all material. In particular, I thank my host M. Théra, and the director of the LACO, A. Movahhedi.

While going over the book for the last time, I learned of the passing away of my friend and colleague Jan Pelant. A great man and a great mathematician, his loss hurts me and all who had the good fortune to meet and know him.

1

Convex sets and convex functions: the fundamentals

*Nobody realizes that some people expend
a tremendous amount of energy
merely to be normal.*
(A. Camus)

In this first chapter we introduce the basic objects of this book: convex sets and convex functions. For sets, we provide the notions of convex set, convex cone, the convex, conic and affine hulls of a set, and the recession cone. All these objects are very useful in highlighting interesting properties of convex sets. For instance, we see that a closed convex set, in finite dimensions, is the closure of its relative interior, and we provide a sufficient condition in order that the sum of two closed convex sets be closed, without using any compactness assumption. To conclude the introduction of these basic geometric objects of the convex analysis, we take a look at the important theorems by Carathéodory, Radon and Helly.

We then introduce the idea of extended real valued convex function, mainly from a geometric point of view. We provide several important examples of convex functions and see what type of operations between functions preserve convexity. We also introduce the very important operation of inf-convolution.

In this introductory chapter we mainly focus on the geometry of convexity, while in the second chapter we shall begin to consider the continuity properties of the extended real valued convex functions.

1.1 Convex sets: basic definitions and properties

Let X be a linear space and C a subset of X.

Definition 1.1.1 C is said to be *convex* provided

$$x, y \in C, \lambda \in (0, 1) \text{ imply } \lambda x + (1 - \lambda)y \in C.$$

The empty set is assumed to be convex by definition. C is a *cone* if $x \in C$, $\lambda \geq 0$ imply $\lambda x \in C$.

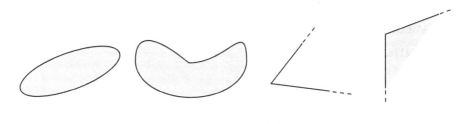

Convex set.   Nonconvex set.   Cone.   Convex cone.

Figure 1.1.

Exercise 1.1.2 A cone is convex if and only if $x, y \in C$ implies $x + y \in C$.

For sets A, C and for $t \in \mathbb{R}$, we set

$$A + C := \{a + c : a \in A, c \in C\}, \quad tA := \{ta : a \in A, t \in \mathbb{R}\}.$$

Exercise 1.1.3 Let A, C be convex (cones). Then $A + C$ and tA are convex (cones). Also, if C_α is an arbitrary family of convex sets (convex cones), then $\bigcap_\alpha C_\alpha$ is a convex set (convex cone). If X, Y are linear spaces, $L: X \to Y$ a linear operator, and C is a convex set (cone), then $L(C)$ is a convex set (cone). The same holds for inverse images.

Definition 1.1.4 We shall call a *convex combination* of elements $x_1, \ldots, x_n$ any vector x of the form

$$x = \lambda_1 x_1 + \cdots + \lambda_n x_n,$$

with $\lambda_1 \geq 0, \ldots, \lambda_n \geq 0$ and $\sum_{i=1}^n \lambda_i = 1$.

We now see that a set C is convex if and only if it contains any convex combination of elements belonging to it.

Proposition 1.1.5 *A set C is convex if and only if for every $\lambda_1 \geq 0, \ldots, \lambda_n \geq 0$ such that $\sum_{i=1}^n \lambda_i = 1$, for every $c_1, \ldots, c_n \in C$, for all n, then $\sum_{i=1}^n \lambda_i c_i \in C$.*

Proof. Let

$$A = \left\{ \sum_{i=1}^n \lambda_i c_i \; : \; \lambda_i \geq 0, \sum_i \lambda_i = 1, c_i \in C \; \forall i, n \in \mathbb{R} \right\}.$$

We must prove that $A = C$ if and only if C is convex. Observe that A contains C. Next, A is convex. This is very easy to see, and tedious to write, and so we omit it. Thus the proof will be concluded once we show that $A \subset C$ provided C is convex. Take an element $x \in A$. Then

$$x = \sum_{i=1}^{n} \lambda_i c_i,$$

with $\lambda_i \geq 0, \sum_i \lambda_i = 1, c_i \in C$. If $n = 2$, then $x \in C$ just by definition of convexity. Suppose now $n > 2$ and that the statement is true for any convex combination of (at most) $n - 1$ elements. Then

$$x = \lambda_1 c_1 + \cdots + \lambda_n c_n = \lambda_1 c_1 + (1 - \lambda_1)y,$$

where

$$y = \frac{\lambda_2}{1 - \lambda_1} c_2 + \cdots + \frac{\lambda_n}{1 - \lambda_1} c_n.$$

Now observe that y is a convex combination of $n - 1$ elements of C and thus, by inductive assumption, it belongs to C. Then $x \in C$ as it is a convex combination of two elements.   □

If C is not convex, then there is a smallest convex set (convex cone) containing C: it is the intersection of all convex sets (convex cones) containing C.

Definition 1.1.6 The *convex hull* of a set C, denoted by $\operatorname{co} C$, is defined as

$$\operatorname{co} C := \bigcap \{A : C \subset A, \ A \text{ is convex}\}.$$

The *conic hull* denoted by $\operatorname{cone} C$, is

$$\operatorname{cone} C := \bigcap \{A : C \subset A, \ A \text{ is a convex cone}\}.$$

Proposition 1.1.7 *Given a set C,*

$$\operatorname{co} C = \left\{ \sum_{i=1}^{n} \lambda_i c_i : \lambda_i \geq 0, \sum_{i=1}^{n} \lambda_i = 1, c_i \in C \ \forall i, n \in \mathbb{R} \right\}.$$

Proof. It easily follows from Proposition 1.1.5.   □

Definition 1.1.8 Let A be a convex set. A point $x \in A$ is said to be an *extreme point* of A if it is not the middle point of a segment contained in A. A *simplex* S is the convex hull of a finite number of points $x_1, \ldots, x_k$.

Exercise 1.1.9 Given a simplex S as in the above definition, show that the extreme points of S are a subset of $\{x_1, \ldots, x_k\}$.

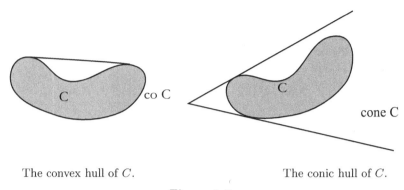

The convex hull of C.   The conic hull of C.

Figure 1.2.

Suppose now X is a Euclidean space.

Definition 1.1.10 A nonempty set $A \subset X$ is said to be *affine* provided

$$x, y \in A \implies ax + by \in A \ \forall a, b : a + b = 1.$$

Given a nonempty convex set C, we define aff C as the smallest affine set containing C, and we denote it by aff C.

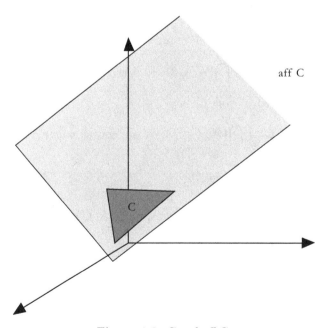

Figure 1.3. C and aff C.

Clearly, an affine set is convex since, if it contains two points, it also contains the whole line joining the points. Moreover, if it contains the zero element, it is a subspace. This is easy to see. First of all, if it contains x, then it contains $ax = ax + (1 - a)0$ for all a. Then if it contains x, y, it contains $2[(1/2)x + (1/2)y]$; it is then closed with respect to the two operations of sum and multiplication by a scalar. Moreover, it is easy to see that the following formula holds:

$$\text{aff } C = \{ax + by : x, y \in C \; a, b : \; a + b = 1\},$$

and that, for $x \in X$ and C convex,

$$x + \text{aff } C = \text{aff}(x + C).$$

Suppose now X is a normed space. One very important property of a closed convex set C with a nonempty interior is that

$$C = \text{cl int } C,$$

where, for a set A, int A denotes the set of its interior points, while cl A denotes its closure (sometimes we shall also use the notation $\overline{A}$ to indicate cl A). This, of course, does not usually hold for an arbitrary set.

Figure 1.4. $C \neq \text{cl int } C$.

However, if we think, for instance, of a triangle embedded in three dimensional Euclidean space, it is clear that even though the set does not have internal points in the topology of the space, we perfectly understand the meaning of the words "internal points of the triangle". To make this idea more precise, we now introduce the useful concept of a relative interior.

Definition 1.1.11 Given a nonempty convex set C, the *relative interior* of C, denoted by ri C, is the set of the interior points of C, considered as a subset of aff C, endowed with the relative topology inherited by X.

It is clear that

$$\text{ri } C = \{x \in X : \exists \varepsilon > 0 \; B(x; \varepsilon) \cap \text{aff } C \subset C\},$$

where $B(x;\varepsilon)$ denotes the open ball centered at x with radius ε (we shall use the notation $B[x;\varepsilon]$ for the corresponding closed ball).

The relative interior of a point is the point itself, the relative interior of a segment contains all points except the endpoints. Clearly, $\operatorname{ri} C$ is an open set as a subset of $\operatorname{aff} C$.

Proposition 1.1.12 *The following formula holds:*

$$\operatorname{ri}(x + C) = x + \operatorname{ri} C.$$

Proof. $z \in x + \operatorname{ri} C$ if and only if there exists $\varepsilon > 0$ such that

$$B(z;\varepsilon) \cap \operatorname{aff}(x + C) \subset x + C,$$

if and only if

$$B(z - x;\varepsilon) \cap \operatorname{aff} C \subset C,$$

if and only if

$$z - x \in \operatorname{ri} C \Longleftrightarrow z \in x + \operatorname{ri} C.$$

$\square$

Now, we prove the following important result.

Proposition 1.1.13 *Let $C \subset \mathbb{R}^n$ be nonempty and convex. Then $\operatorname{ri} C$ is nonempty.*

Proof. From the formula $\operatorname{ri}(x + C) = x + \operatorname{ri} C$, we can assume, without loss of generality, that $0 \in \operatorname{ri} C$. Then $A = \operatorname{aff} C$ is a subspace. If it is $\{0\}$, there is nothing to prove. Otherwise, consider a maximal subset $\{e_1, \ldots, e_j\}$ of linearly independent vectors in C. Clearly, $\{e_1, \ldots, e_j\}$ is a basis for A. Moreover, as $0 \in C$, then

$$\lambda_1 e_1 + \cdots + \lambda_j e_j \in C \quad \text{if } \lambda_i \geq 0 \ \forall i \text{ and } \sum_i \lambda_i \leq 1. \tag{1.1}$$

Consider the element $z = \frac{1}{j+1} e_1 + \cdots + \frac{1}{j+1} e_j$. In view of (1.1), we have that $z + \alpha e_i \in C$ for all $i = 1, \ldots, j$ and for $|\alpha| \leq \frac{1}{j+1}$. This means that z belongs to the interior of a full dimensional box, relative to A, which is contained in C, and so $z \in \operatorname{ri} C$. $\square$

The following proposition highlights some properties of the relative interior of a set.

Proposition 1.1.14 *Let C be a nonempty convex set. Then*

(i) $\operatorname{ri} C$ *is a convex set;*
(ii) $x \in \operatorname{ri} C$, $y \in C$ *imply* $\lambda x + (1 - \lambda)y \in \operatorname{ri} C$ *for all* $0 < \lambda \leq 1$;
(iii) $\operatorname{cl} \operatorname{ri} C = \operatorname{cl} C$;

(iv) ri cl C = ri C;
(v)   *if* int $C \neq \emptyset$, *then* cl int C = cl C.

Proof. (Outline). (ii) clearly implies (i). Now suppose $y = 0$, $B(x; \varepsilon) \cap$ aff $C \subset C$ and let $0 < \lambda < 1$. Suppose $\|z - \lambda x\| < \lambda \varepsilon$. Then $\frac{z}{\lambda} \in B(x; \varepsilon)$ and $z = \lambda(\frac{z}{\lambda}) + (1 - \lambda)0 \in C$. The same idea, with a few more technicalities, works if we suppose, more generally, $y \in$ cl C. This helps proving (iii) and (iv).   □

Though the concept of relative interior can be given in any normed space, it is particularly interesting in finite dimensions, because of Proposition 1.1.13. In infinite dimensions, it can happen that ri C is empty. A typical example of this is a dense hyperplane with no interior points (see for instance Example A.1.4).

We now introduce another important geometrical object related to a convex set C.

Definition 1.1.15 Let C be a nonempty convex set. The *recession cone* to C, denoted by $0^+(C)$ is the following set:

$$0^+(C) := \{x : x + c \in C, \forall c \in C\}.$$

Figure 1.5. C and the recession cone of C: two examples.

Proposition 1.1.16 *Let C be a nonempty closed convex set. Then $0^+(C)$ is a closed convex cone. If C is bounded, then $0^+(C) = \{0\}$; the converse is true in finite dimensions.*

Proof. If $x \in 0^+(C)$ then it is obvious that $nx \in 0^+(C)$. Now, fix $a \geq 0$, $c \in C$ and take $n \geq a$. Then

$$ax + c = \frac{a}{n}(nx + c) + \left(1 - \frac{a}{n}\right)c \in C,$$

and this shows that $0^+(C)$ is a cone. As $x, y \in 0^+(C)$ clearly implies $x + y \in 0^+(C)$, then $0^+(C)$ is a convex cone. It is easily seen that it is closed. Finally, suppose dim $X < \infty$ and let C be unbounded. Take $\{c_n\} \subset C$ such that

$\|c_n\| \to \infty$. Then, up to a subsequence, $\{c_n/\|c_n\|\} \to x$ (and $\|x\| = 1$). Fix $c \in C$. Then

$$C \ni \left(1 - \frac{1}{\|c_n\|}\right)c + \frac{1}{\|c_n\|}c_n \to c + x.$$

This shows that $x \in 0^+(C)$, so that $0^+(C)$ contains at least a nonzero element.

□

Proposition 1.1.17 *Let C be a closed convex set, let $x \in X$. Then*

(i) $0^+(C) = 0^+(x + C)$;

(ii) $0^+(C) = \{z : \exists c \in C, c + tz \in C, \forall t > 0\}$.

Proof. We prove only the second claim. Suppose $c + tz \in C$ for some c, let $x \in C$ and prove $x + z \in C$. Write

$$\lambda x + (1 - \lambda)c + z = \lambda x + (1 - \lambda)\left(c + \frac{z}{1 - \lambda}\right).$$

Then $\lambda x + (1 - \lambda)c + z \in C$ for all $\lambda \in [0, 1)$. The conclusion now follows from the fact that C is closed.

□

Let us recall some simple topological facts to motivate the introduction of the recession cone of a convex set. First, it is easy to see that in a normed space X, if C is any set and A is an open set, then $C + A$ is an open set. The situation changes if we consider the sum of two closed sets. In this case, even if we assume C and A to be convex, the sum $A + C$ need not be closed.

Figure 1.6.

On the other hand, it is an easy exercise to see that if one of the two sets is compact, then the sum is closed (with no need of convexity). The next result shows how the idea of the recession cone allows us to generalize this result.

Proposition 1.1.18 *Suppose A, C be nonempty closed convex subsets of a Euclidean space. Suppose moreover*

$$0^+(A) \cap -0^+(C) = \{0\}.$$

Then $A + C$ is a closed (convex) set.

Proof. Let $\{a_n\} \subset A$, $\{c_n\} \subset C$ be such that $a_n + c_n \to z$. We need to prove that $z \in A + C$. Suppose $\|a_n\| \to \infty$. Then, for a subsequence, $\{a_n/\|a_n\|\} \to a \neq 0$. It is easy to see that $a \in 0^+(A)$. As

$$\frac{a_n}{\|a_n\|} + \frac{c_n}{\|a_n\|} \to 0,$$

then $c_n/\|a_n\| \to -a$, and this implies $-a \in 0^+(C)$. This is impossible, and thus $\{a_n\}$ must be bounded. The conclusion easily follows. □

Remark 1.1.19 In the proof above we have shown that given a closed convex set A and an unbounded sequence $\{a_n\}$ in it, any norm one element a which is a limit point of $\{\frac{a_n}{\|a_n\|}\}$ is in the recession cone of A. Such a unit vector is often called a *recession direction* for A.

A similar result holds for the L-image of a closed convex set, where L is a linear operator.

Proposition 1.1.20 *Let X be a Euclidean space, Y a normed space, $C \subset X$ a closed convex set, and finally let $L: X \to Y$ be a linear operator. Denoting by N the kernel of L, suppose moreover*

$$N \cap 0^+(C) = \{0\}.$$

Then $L(C)$ is a closed (convex) set.

Proof. Let $\{y_n\} \subset Y$ be such that $y_n \in L(C)$ for all n and $y_n \to y$. There is $c_n \in C$ such that $y_n = Lc_n$ for all n. Write $c_n = z_n + x_n$, with $x_n \in N$, $z_n \in N^\perp$. As $z_n \in N^\perp$ and $\{Lz_n\}$ is bounded, it follows that $\{z_n\}$ is bounded (see Exercise 1.1.21 below). Now suppose $\|x_n\| \to \infty$ (up to a subsequence). Then there is a norm one limit point x of $\{\frac{x_n}{\|x_n\|}\}$, and $x \in N$. Fix $c \in C$. Then

$$C \ni \left(1 - \frac{1}{\|x_n\|}\right) c + \frac{1}{\|x_n\|} x_n \to c + x.$$

It follows that $x \in 0^+(C) \cap N$, which is impossible. Thus $\{x_n\}$ is bounded and this yields the result. □

Exercise 1.1.21 Let $L: X \to Y$ be a linear operator. Suppose moreover L is $1 - 1$. Then there is $a > 0$ such that $\|Lx\| \geq a\|x\|$ for all x.

Hint. Suppose there exists $\{x_n\}$ such that $\|x_n\| = 1$ for all n and $Lx_n \to 0$. Then $\{x_n\}$ has a limit point.

Proposition 1.1.22 *Given points $x_1, \ldots, x_n$ the conic hull of $\{x_1, \ldots, x_n\}$ is given by*

$$\text{cone}\{x_1, \ldots, x_n\} = \mathbb{R}_+ x_1 + \cdots + \mathbb{R}_+ x_n,$$

(where $\mathbb{R}_+ x := \{y : y = tx\}$, for some $t \geq 0$) and it is a closed set.

Proof. (We take $n = 2$, the general case being similar.). The set

$$C := \mathbb{R}_+ x_1 + \mathbb{R}_+ x_2$$

is clearly a cone, it is convex, and contains both x_1 and x_2. Thus it contains cone$\{x_1, x_2\}$. On the other hand, let $ax_1 + bx_2$ be an element of C. Then $\frac{x_1}{b}$ and $\frac{x_2}{a}$ both belong to cone$\{x_1, x_2\}$. Thus their sum $\frac{x_1}{b} + \frac{x_2}{a}$ also belongs to it. To conclude, $ax_1 + bx_2 = ab(\frac{x_1}{b} + \frac{x_2}{a}) \in$ cone$\{x_1, x_2\}$. Now, observe that $0^+ \mathbb{R}_+ x = \mathbb{R}_+ x$ for all x, and appeal to Proposition 1.1.18 to conclude.   □

We end this section by proposing three beautiful and famous results on convex sets as guided exercises.

Exercise 1.1.23 (Carathéodory's Theorem.) Let $C \subset \mathbb{R}^n$ be a convex set, and let $c \in C$. Then c can be written as a convex combination of *at most* $n+1$ elements of C.

Hint. Suppose that for any representation of x as a convex combination of elements of C, i.e.,

$$x = l_1 x_1 + \cdots + l_k x_k, \quad x_i \in C,$$

k must be greater than $n + 1$, and suppose that the above is a representation of x with a minimal set of elements. Consider the following linear system, with unknown $\lambda_1, \ldots, \lambda_k$:

$$\lambda_1 x_1 + \cdots + \lambda_k x_k = 0, \quad \lambda_1 + \cdots + \lambda_k = 0. \qquad (1.2)$$

Observe that this homogeneous system has $n + 1$ equations, and more unknowns than equations, so it must have a nontrivial solution $(\lambda_1, \ldots, \lambda_k)$. At least one component must be positive. Set

$$\sigma = \min \left\{ \frac{l_i}{\lambda_i} : \lambda_i > 0 \right\}.$$

Let $m_i = l_i - \sigma \lambda_i$. Show that the convex combination made by the m_i's again gives x, that at least one of the m_i's is zero, and that this is a contradiction.

Exercise 1.1.24 (Radon's Theorem.) Any collection of $j > n + 1$ (distinct) points in $\mathbb{R}^n$ can be partitioned into two subsets such that the intersection of the two convex hulls is nonempty.

Hint. Let $x_1, \ldots, x_k$ be a collection of $k > n + 1$ points in $\mathbb{R}^n$. Consider a nontrivial solution of (1.2), as in Exercise 1.1.23. Let $I^+(I^-)$ be the set of indices corresponding to nonnegative λ's (negative λ's). Set $\lambda = \sum_{i \in I^+} \lambda_i$, show that the element

$$\sum_{i \in I^+} \lambda_i x_i$$

belongs to co$\{x_i : i \in I^-\}$, and observe that this shows that $\{x_i : i \in I^-\}$, $\{x_i : i \in I^+\}$ is a required partition.

Exercise 1.1.25 (Helly's Theorem.) Let $C_1, \dots, C_k$ be convex subsets of $\mathbb{R}^n$ such that the intersection of any $n+1$ of them is nonempty. Then

$$\bigcap_{i=1}^{k} C_i \neq \emptyset.$$

Hint. Suppose the property holds for every collection of k sets, with $k > n$, and prove the statement for a collection of $k+1$ sets. Let $C_1, \dots, C_{k+1}$ be such a collection. Let $c_j \in \bigcap_{i \neq j} C_i$. If two of the c_j do coincide, the statement is proved. Otherwise, we have $k+1 > n+1$ distinct points in $\mathbb{R}^n$ and thus, by the Radon Theorem, they can be partitioned in such a way that the two partitions have a common point c. Prove that $c \in \bigcap_{i=1}^{k} C_i$.

1.2 Convex functions: basic definitions and properties

Let $f \colon X \to [-\infty, \infty]$ be a given, extended real valued function. Let us define

$$\text{epi}\, f := \{(x, r) \in X \times \mathbb{R} : f(x) \leq r\}, \text{ the } \textit{epigraph} \text{ of } f,$$
$$\text{s-epi}\, f := \{(x, r) \in X \times \mathbb{R} : f(x) < r\}, \text{ the } \textit{strict epigraph} \text{ of } f,$$
$$\text{dom}\, f := \{x \in X : f(x) < \infty\}, \text{ its } \textit{(effective) domain}$$

and

$$f^a := \{x \in X : f(x) \leq a\}, \text{ its } \textit{level set} \text{ at height } a \in \mathbb{R}$$

(which could be empty for some a). Observe that $\text{dom}\, f$ is the projection of $\text{epi}\, f$ on the space X.

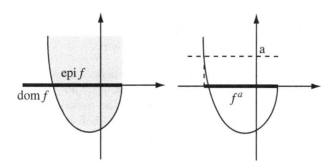

Figure 1.7.

Now let X be a linear space.

Definition 1.2.1 We shall say that $f \colon X \to [-\infty, \infty]$ is *convex* provided $\text{epi}\, f$ is a convex set. $f \colon X \to [-\infty, \infty]$ is said to be *concave* provided $-f$ is convex.

Exercise 1.2.2 Verify that f is convex if and only if its strict epigraph is convex; verify that, if f is convex, then dom f is a convex subset of X, and so are all level sets f^a. On the other hand, if all the level sets f^a are convex, then f need not be convex.

The classical definition of a convex function is a bit different:

Definition 1.2.3 Let $C \subset X$ be a convex set and let $f\colon C \to \mathbb{R}$ be a given function. We say that f is *convex, in the classical sense,* if $\forall x, y \in C, \forall \lambda \in (0,1)$,

$$f(\lambda x + (1 - \lambda)y) \leq \lambda f(x) + (1 - \lambda)f(y).$$

Remark 1.2.4 There is a clear connection between the two definitions of a convex function given above. If $f\colon X \to (-\infty, \infty]$ is a convex function in the geometric sense of Definition 1.2.1, then as $C = \operatorname{dom} f$ is a convex set, it is possible to consider the restriction of f to the set C; it is easy to verify that $f\colon C \to \mathbb{R}$ is convex in the analytic sense described by Definition 1.2.3. Conversely, given $f\colon C \to \mathbb{R}$, convex in the sense of Definition 1.2.3, if we define it also outside C, by simply assigning to it the value ∞ there, its extension is convex in the sense of Definition 1.2.1.

Exercise 1.2.5 Verify that $f\colon X \to [-\infty, \infty]$ is convex if and only if

$$\forall x, y \in X, \forall \lambda \in (0,1), f(\lambda x + (1 - \lambda)y) \leq \lambda f(x) + (1 - \lambda)f(y),$$

with the agreement that $-\infty + \infty = +\infty$.

Remark 1.2.6 Suppose that a given convex function assumes the value $-\infty$ at a given point x, and consider any half line originating from x. Only the following cases can occur: f is $-\infty$ on the whole half line, or it is valued ∞ on the whole line except x, or else it has value $-\infty$ in an interval $[x, y)$, $f(y)$ is arbitrary, and the function has value ∞ elsewhere.

It makes sense to consider convex functions possibly assuming the value $-\infty$ because important operations between functions do not guarantee a priori that the result is not $-\infty$ at some point, even if the resulting function is still convex; however, when using such operations, we shall usually try to prove that we do not fall in such an essentially degenerated case. An example of such a situation is given later in this section, when we define the inf-convolution operation.

Example 1.2.7 The following are convex functions:

- $X = \mathbb{R}$, $f(x) = |x|^a$, $a > 1$;
- $X = \mathbb{R}$, $f(x) = \begin{cases} \infty & \text{if } x \leq 0, \\ -\ln x & \text{if } x > 0; \end{cases}$
- $X = \mathbb{R}^2$, $f(x, y) = ax^2 + 2bxy + cy^2$, provided $a > 0$, $ac - b^2 > 0$;
- $X = \mathbb{R}^2$, $f(x, y) = |x| + |y|$, $f_p(x, y) = \sqrt[p]{x^p + y^p}$, $f_\infty(x, y) = \max\{|x|, |y|\}$;

- X a linear space, $f(x) = l(x)$, $f(x) = |l(x)|$, where $l \colon X \to \mathbb{R}$ a linear functional.

Exercise 1.2.8 (Young's inequality.) Prove that for $x, y > 0$,

$$xy \le \frac{x^p}{p} + \frac{y^q}{q},$$

where $p, q \in (1, \infty)$ are such that $\frac{1}{p} + \frac{1}{q} = 1$.

Hint. Write $xy = e^{\ln xy}$ and use convexity of the exponential function.

Exercise 1.2.9 Let C be a given nonempty set and let $I_C(x)$ be the following function:

$$I_C(x) = \begin{cases} 0 & \text{if } x \in C, \\ \infty & \text{elsewhere.} \end{cases}$$

Then the function I_C is convex if and only if C is a convex set. The function I_C is called the *indicator function* of the set C.

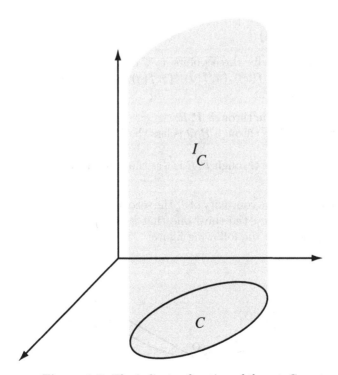

Figure 1.8. The indicator function of the set C.

Exercise 1.2.10 A *norm* on a linear space X is a real valued function $\| \cdot \| \colon X \to \mathbb{R}$ such that

(i) $\|x\| \geq 0$ for all x and $\|x\| = 0$ if and only if $x = 0$;
(ii) $\|ax\| = |a| \|x\|$ for every $a \in \mathbb{R}$ and $x \in X$;
(iii) $\|x + y\| \leq \|x\| + \|y\|$ for every $x, y \in X$.

Prove that $f(x) = \|x\|$ is a convex function; prove that f_1, f_p, f_∞ in Example 1.2.7 are convex functions and make a picture of the corresponding unit balls; prove that for $p = 3, 4, \dots$ in the boundary of the unit ball there are no points with both coordinates rational. (Do not be worried if you do not succeed, and be worried if you think you succeeded!)

Several fundamental properties of the convex functions, even when defined in infinite-dimensional spaces, rely for their behavior on the one-dimensional spaces. The next proposition is a basic example of this claim.

Proposition 1.2.11 *Let $I \subset \mathbb{R}$ be a nonempty interval and let $f \colon I \to \mathbb{R}$ be a given function. Then f is convex if and only if, $\forall x_0 \in I$, the function*

$$x \mapsto s_f(x; x_0) := \frac{f(x) - f(x_0)}{x - x_0}, \quad x \neq x_0$$

is increasing in $I \setminus \{x_0\}$.

Proof. It is enough to fix three points $x < u < y$ and, calling respectively P, Q, R the points $(x, f(x)), (u, f(u)), (y, f(y))$, to show that the following conditions are equivalent:

- Q lies below the line through P, R;
- the slope of the line through P, Q is less than the slope of the line through P, R;
- the slope of the line through P, R is less than the slope of the line through Q, R.

The first condition is the convexity of f, the second one says that $x \mapsto s_f(x; x_0)$ is increasing for $x > x_0$, the third one that $x \mapsto s_f(x; x_0)$ is increasing for $x < x_0$. The proof is in the following figure: □

Figure 1.9. A figure can be more convincing than a page of calculations ....

The following is sometimes a useful criterion to see if a given function is convex.

Proposition 1.2.12 *Let* $f\colon [0,\infty] \to \mathbb{R}$ *and let* $g\colon [0,\infty] \to \mathbb{R}$ *be defined as*

$$g(x) := xf\left(\frac{1}{x}\right).$$

Then f *is convex if and only if* g *is convex.*

Proof. Let $x_0 > 0$. Then, with the notation of the Proposition 1.2.11,

$$s_g(x;x_0) = f\left(\frac{1}{x_0}\right) - \frac{1}{x_0}s_f\left(\frac{1}{x};\frac{1}{x_0}\right).$$

Moreover $f(x) = xg(\frac{1}{x})$. □

Example 1.2.13 $x\ln x$, $x\exp^{\frac{1}{x}}$ are convex functions.

A particular and important class of a convex function is the family of the subadditive and positively homogeneous (for short, sublinear) functions.

Definition 1.2.14 $f\colon X \to (-\infty,\infty]$ *is said to be* sublinear *if the following hold* $\forall x,y \in X, \forall a > 0$:

(i) $f(x+y) \le f(x) + f(y)$;
(ii) $f(ax) = af(x)$.

The sublinear functions are precisely those convex functions that are also positively homogeneous; hence their epigraph is a convex cone.

Exercise 1.2.15 Let C be a convex subset of a Banach space X containing 0 and let $m_C(x)$ be the following function:

$$m_C(x) := \inf\{\lambda > 0 : \frac{x}{\lambda} \in C\}.$$

Show that m_C is a sublinear function, which is finite when C is *absorbing*, which means that $X = \bigcup_{\lambda>0} \lambda C$, and $m_C(x) \le 1$ for all $x \in C$. When C is absorbing, $m_C(x) < 1$ if and only if $x \in \operatorname{int} C$. Then m_C is called the *Minkowski functional* of C.

Hint. When C is absorbing, then $\operatorname{int} C \ne \emptyset$, as a consequence of the Baire's theorem (see B.1.2), and also $0 \in \operatorname{int} C$. If $m_C(x) < 1$, then there is $a > 1$ such that $ax \in C$. Now conclude with the help of Figure 1.10.

We saw above that convex functions assuming the value $-\infty$ have a particular shape. We shall see later that if we also impose some weak form of continuity, then they cannot assume any real value. So, it is often useful to exclude these pathological behaviors. This explains why we shall concentrate on a particular subset of convex functions, identified by the following definition.

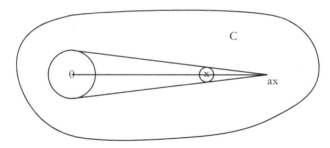

Figure 1.10.

Definition 1.2.16 $f : X \to [-\infty, \infty]$ is said to be *proper* if it never assumes the value $-\infty$ and it is not identically ∞. We shall indicate by

$$\mathcal{F}(X) := \{f : X \to [-\infty, \infty] : f \text{ is proper and convex}\}.$$

Inside $\mathcal{F}(X)$ we find all the extended real valued functions whose epigraph is nonempty, convex and does not contain vertical lines.

Proposition 1.2.17 $f \in \mathcal{F}(X)$ *if and only if* f *is proper and* $\forall x_1, \ldots, x_n \in X, \forall \lambda_1, \ldots, \lambda_n$ *such that* $\lambda_i > 0, \forall i = 1, \ldots, n, \sum_{i=1}^{n} \lambda_i = 1,$

$$f\left(\sum_{i=1}^{n} \lambda_i x_i\right) \leq \sum_{i=1}^{n} \lambda_i f(x_i).$$

Proof. The case $n = 2$ is just the definition. The general case is easily deduced from this one by using finite induction. For, suppose the statement is true for any convex combination of $n - 1$ elements. Given $x_1, \ldots, x_n \in X, \lambda_1, \ldots, \lambda_n$ such that $\lambda_i > 0, \sum_{i=1}^{n} \lambda_i = 1$, write

$$\lambda_1 x_1 + \cdots + \lambda_n x_n = \lambda_1 x_1 + (1 - \lambda_1)y,$$

where

$$y = \frac{\lambda_2}{1 - \lambda_1} x_2 + \cdots + \frac{\lambda_n}{1 - \lambda_1} x_n.$$

Now observe that y is a convex combination of $n-1$ elements, use the analytic definition of convexity to get that

$$f(\lambda_1 x_1 + \cdots + \lambda_n x_n) \leq \lambda_1 f(x_1) + (1 - \lambda_1)f(y)$$

and conclude by applying the inductive assumption. $\qquad \square$

Proposition 1.2.18 *Let* $f_i \in \mathcal{F}(X) \ \forall i = 1, \ldots, n$ *and let* $t_1, \ldots, t_n > 0$. *If there exists* $x_0 \in X$ *such that* $f_i(x_0) < \infty \ \forall i$, *then* $(\sum_{i=1}^{n} t_i f_i) \in \mathcal{F}(X)$.

Proof. Use the characterization given in Proposition 1.2.17. $\qquad \square$

Proposition 1.2.19 *Let $f_i \in \mathcal{F}(X)$ $\forall i \in J$, where J is an arbitrary index set. If there exists $x_0 \in X$ such that $\sup_{i \in J} f_i(x_0) < \infty$, then $(\sup_{i \in J} f_i) \in \mathcal{F}(X)$.*

Proof. epi$(\sup_{i \in J} f_i) = \bigcap_{i \in J}$ epi f_i. $\square$

If we make the pointwise infimum of convex functions, in general we do not get a convex function. Thus it is useful to define another inf operation, more complicated from the analytical point of view, but with a clear geometrical meaning.

Definition 1.2.20 *Let $f, g \in \mathcal{F}(X)$. We define the inf-convolution, or epi-sum, between f and g to be the function*

$$(f \nabla g)(x) := \inf\{f(x_1) + g(x_2) : x_1 + x_2 = x\} = \inf\{f(y) + g(x - y) : y \in X\}.$$

The inf-convolution is said to be *exact* at $x \in \text{dom}(f \nabla g)$, provided the inf appearing in the definition is attained.

The following exercise highlights the geometrical meaning of the inf-convolution operation, and explains the more modern name of epi-sum.

Exercise 1.2.21 Verify that

$$\text{s-epi}(f \nabla g) = \text{s-epi}\, f + \text{s-epi}\, g.$$

Hint. Let $(x, r) \in \text{s-epi}(f \nabla g)$. Then there are x_1, x_2 such that $x_1 + x_2 = x$ and $f(x_1) + g(x_2) < r$. Now choose suitable $a > f(x_1)$, $b > g(x_2)$ such that $a + b = r$. From this we conclude that $\text{s-epi}(f \nabla g) \subset \text{s-epi}\, f + \text{s-epi}\, g$, etc.

Then the epi-sum operation provides a convex function which, however, need not be proper. Here is a simple situation when it is.

Proposition 1.2.22 *Let $f, g \in \mathcal{F}(X)$. Suppose there are a linear functional l on X and $a \in \mathbb{R}$ such that $f(x) \geq l(x) - a$, $g(x) \geq l(x) - a$ $\forall x \in X$. Then $(f \nabla g) \in \mathcal{F}(X)$.*

Proof. From Exercise 1.2.21 we get that $(f \nabla g)$ is a convex function. Moreover, the common lower bound by the affine function $l(x) - a$, gives $(f \nabla g)(x) \geq l(x) - 2a$ $\forall x \in X$. Since the sum of two nonempty sets is obviously nonempty, then $(f \nabla g) \in \mathcal{F}(X)$. $\square$

Proposition 1.2.23 *Let C be a nonempty convex set. Let*

$$d(x, C) := \inf_{c \in C} \|x - c\|.$$

Then $d(\cdot, C)$ is a convex function.

Proof. It is enough to observe that

$$d(\cdot, C) = (\| \, \| \nabla I_C)(\cdot).$$

                    $\square$

The next exercise familiarizes the reader with the inf-convolution operation.

Exercise 1.2.24 Evaluate $(f \nabla g)$, and make a picture whenever possible,
when f and g are

- $f(x) = I_C(x)$, $g(x) = I_D(x)$, with C and D two convex sets;
- $f(x) = x$, $g(x) = 2x$;
- $f: X \to \mathbb{R}$, $g(x) = \begin{cases} 0 & \text{if } x = x_0, \\ \infty & \text{elsewhere;} \end{cases}$
- $f: X \to (-\infty, \infty]$, $g(x) = \begin{cases} r & \text{if } x = 0, \\ \infty & \text{elsewhere;} \end{cases}$
- $f(x) = \frac{1}{2}|x|^2$, $g(x) = I_{[0,1]}$;

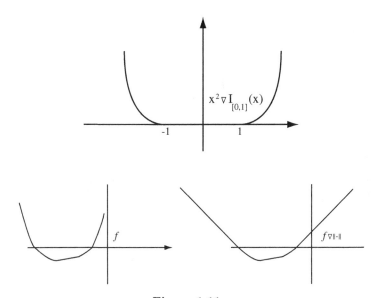

Figure 1.11.

- $f \in \mathcal{F}(X)$, $g(x) = \begin{cases} 0 & \text{if } |x| \le r, \\ \infty & \text{elsewhere;} \end{cases}$
- $f(x) = \frac{1}{2}x^2$, $g(x) = x$;
- $f(x) = \frac{1}{2}\|x\|^2$, $g(x) = \|x\|$;
- $f \in \mathcal{F}(X)$, $g(x) = k\|x\|$.

Exercise 1.2.25 Prove or disprove that the pointwise limit of convex functions is a convex function.

Exercise 1.2.26 Let $f: X \to \mathbb{R}$ be convex, let $S = \text{co}\{x_1, \ldots, x_n\}$. Then there is j such that

$$\max_{x \in S} f(x) = f(x_j).$$

2

Continuity and $\Gamma(X)$

Par délicatesse j'ai perdu ma vie
(A. Rimbaud, "Chanson de la plus haute tour")

Continuity, Lipschitz behavior, existence of directional derivatives, and differentiability are, of course, topics of the utmost importance in analysis. Thus the next two chapters will be dedicated to a description of the special features of convex functions from this point of view. Specifically, in this chapter we analyze the continuity of the convex functions and their Lipschitz behavior.

The first results show that a convex function which is bounded above around a point is continuous at that point, and that if it is at the same time lower and upper bounded on a ball centered at some point x, then it is Lipschitz in every smaller ball centered at x. The above continuity result entails also that a convex function is continuous at the interior points of its effective domain. It follows, in particular, that a convex, real valued function defined on a Euclidean space is everywhere continuous. This is no longer true in infinite dimensions.

We then introduce the notion of lower semicontinuity, and we see that if we require this additional property, then a real valued convex function is everywhere continuous in general Banach spaces. Lower semicontinuity, on the other hand, has a nice geometrical meaning, since it is equivalent to requiring that the epigraph of f, and all its level sets, are closed sets: one more time we relate an analytical property to a geometrical one. It is then very natural to introduce, for a Banach space X, the fundamental class $\Gamma(X)$ of convex, lower semicontinuous functions whose epigraph is nonempty (closed, convex) and does not contain vertical lines.

The chapter ends with a very fundamental characterization of a function in $\Gamma(X)$: it is the pointwise supremum of all affine functions minorizing it. Its proof relies, quite naturally, on the Hahn-Banach separation theorems recalled in Appendix A.

2.1 Continuity and Lipschitz behavior

Henceforth, as we shall deal with topological issues, every linear space will be endowed with a norm.

Convex functions have remarkable continuity properties. A key result is the following lemma, asserting that continuity at a point is implied by upper boundedness in a neighborhood of the point.

Lemma 2.1.1 *Let $f\colon X \to [-\infty, \infty]$ be convex, let $x_0 \in X$. Suppose there are a neighborhood V of x_0 and a real number a such that $f(x) \le a \ \forall x \in V$. Then f is continuous at x_0.*

Proof. We show the case when $f(x_0) \in \mathbb{R}$. By a translation of coordinates, which obviously does not affect continuity, we can suppose $x_0 = 0 = f(0)$. We can also suppose that V is a symmetric neighborhood of the origin. Suppose $x \in \varepsilon V$. Then $\frac{x}{\varepsilon} \in V$ and we get

$$f(x) \le (1 - \varepsilon)f(0) + \varepsilon f(\frac{x}{\varepsilon}) \le \varepsilon a.$$

Now, write $0 = \frac{\varepsilon}{1+\varepsilon}(-\frac{x}{\varepsilon}) + \frac{1}{1+\varepsilon}x$ to get

$$0 \le \frac{\varepsilon}{1 + \varepsilon}f(-\frac{x}{\varepsilon}) + \frac{1}{1 + \varepsilon}f(x),$$

whence

$$f(x) \ge -\varepsilon f(-\frac{x}{\varepsilon}) \ge -\varepsilon a.$$

$\square$

From the previous result, it is easy to get the fundamental

Theorem 2.1.2 *Let $f \in \mathcal{F}(X)$. The following are equivalent:*

(i)   *There are a nonempty open set O and a real number a such that $f(x) \le a$ $\forall x \in O$;*

(ii)   *int dom $f \ne \emptyset$, and f is continuous at all points of int dom f.*

Proof. The only nontrivial thing to show is that, whenever (i) holds, f is continuous at each point $x \in$ int dom f. We shall exploit boundedness of f in O to find a nonempty open set I containing x where f is upper bounded. Suppose $f(z) \le a \ \forall z \in O$ and, without loss of generality, that $x = 0$. Fix a point $v \in O$. There exists $t > 0$ such that $-tv \in$ int dom f. Now, let $h(y) := \frac{t+1}{t}y + v$. Then $h(0) = v$ and $I = h^{-1}(O)$ is a neighborhood of $x = 0$. Let $y \in I$. Then $y = \frac{t}{t+1}h(y) + \frac{1}{t+1}(-tv)$ and

$$f(y) \le \frac{t}{t + 1}a + \frac{1}{t + 1}f(-tv) \le a + f(-tv).$$

We found an upper bound for f in I, and this concludes the proof.

$\square$

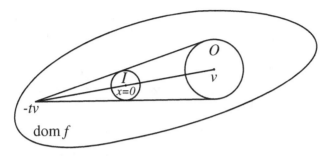

Figure 2.1.

Corollary 2.1.3 *Let* $f \in \mathcal{F}(\mathbb{R}^n)$. *Then* f *is continuous at each point of* int dom f. *In particular, if* f *is real valued, then it is everywhere continuous.*

Proof. If $x \in$ int dom f, to show that f is upper bounded in a neighborhood of x, it is enough to observe that x can be put in the interior of a simplex, where f is bounded above by the maximum value assumed by f on the vertices of the simplex (see Exercise 1.2.26). □

Remark 2.1.4 The continuity of f at the boundary points of dom f is a more delicate issue. For instance, the function

$$f(x) = \begin{cases} 0 & \text{if } |x| < 1, \\ 1 & \text{if } |x| = 1, \\ \infty & \text{if } |x| > 1, \end{cases}$$

is convex and at the boundary points does not fulfill any continuity condition.

The next exercise characterizes the continuity of a sublinear function.

Exercise 2.1.5 Show the following:

Proposition 2.1.6 *Let* $h \colon X \to (-\infty, \infty]$ *be a sublinear function. Then the following are equivalent:*
(i) h *is finite at a point* $x_0 \neq 0$ *and continuous at* $-x_0$;
(ii) h *is upper bounded on a neighborhood of zero;*
(iii) h *is continuous at zero;*
(iv) h *is everywhere continuous.*

Hint. To show that (i) implies (ii), observe that $h(x_0) < \infty$ and $h(x) \leq h(x - x_0) + h(x_0)$. Moreover, observe that (iii) implies that h is everywhere real valued.

Exercise 2.1.7 Referring to Exercise 1.2.15, show that the Minkowski functional is continuous if and only if C is an absorbing set.

We saw that upper boundedness around a point guarantees continuity; the next lemma shows that a convex function is Lipschitz around a point if it is upper and lower bounded near that point.

Lemma 2.1.8 *Let $f \in \mathcal{F}(X)$, and let $x_0 \in X, R > 0, m, M \in \mathbb{R}$. Suppose $m \leq f(x) \leq M, \forall x \in B(x_0; R)$. Then f is Lipschitz on $B(x_0; r)$, for all $r < R$, with Lipschitz constant $\frac{M-m}{R-r}$.*

Proof. Let $x, y \in B(x_0; r)$ and let $z = y + \frac{R-r}{\|y-x\|}(y-x)$. Then $z \in B(x_0; R)$, hence $f(z) \leq M$. Moreover y is a convex combination of x and z:

$$y = \frac{\|y-x\|}{R-r+\|y-x\|}z + \frac{R-r}{R-r+\|y-x\|}x.$$

Hence

$$f(y) - f(x) \leq \frac{\|y-x\|}{R-r+\|y-x\|}M - \frac{\|y-x\|}{R-r+\|y-x\|}m \leq \frac{M-m}{R-r}\|y-x\|.$$

By interchanging the roles of x and y we get the result.   □

2.2 Lower semicontinuity and $\Gamma(X)$

Let X be a topological space. Let $f\colon X \to (-\infty, \infty]$, $x \in X$, and denote by $\mathcal{N}$ the family of all neighborhoods of x. Remember that

$$\liminf_{y \to x} f(y) = \sup_{W \in \mathcal{N}} \inf_{y \in W \setminus \{x\}} f(y).$$

Definition 2.2.1 Let $f\colon X \to (-\infty, \infty]$. f is said to be *lower semicontinuous* if epi f is a closed subset of $X \times \mathbb{R}$. Given $x \in X$, f is said to be *lower semicontinuous at x* if

$$\liminf_{y \to x} f(y) \geq f(x).$$

Exercise 2.2.2 A subset E of $X \times \mathbb{R}$ is an epigraph if and only if $(x, a) \in E$ implies $(x, b) \in E$ for all $b \geq a$. If E is an epigraph, then $\operatorname{cl} E = \operatorname{epi} f$ with $f(x) = \inf\{a : (x, a) \in E\}$, and f is lower semicontinuous.

Definition 2.2.3 Let $f\colon X \to (-\infty, \infty]$. The *lower semicontinuous regularization* of f is the function $\bar{f}$ such that

$$\operatorname{epi} \bar{f} := \operatorname{cl} \operatorname{epi} f.$$

The definition above is consistent because $\operatorname{cl} \operatorname{epi}(f)$ is an epigraph, as is easy to prove (see Exercise 2.2.2). Moreover, it is obvious that $\bar{f}$ is the greatest lower semicontinuous function minorizing f: if $g \leq f$ and g is lower semicontinuous, then $g \leq \bar{f}$. Namely, epi g is a closed set containing epi f, and thus it contains its closure too.

Exercise 2.2.4 Show that f is lower semicontinuous if and only if it is lower semicontinuous at $x, \forall x \in X$. Show that f is lower semicontinuous at x if and only if $f(x) = \bar{f}(x)$.

Hint. Let $l = \liminf_{y \to x} f(y)$. Show that $(x, l) \in \mathrm{cl}\,\mathrm{epi}\, f$. If f is everywhere lower semicontinuous, show that if $(x, r) \in \mathrm{cl}\,\mathrm{epi}\, f$, $\forall \varepsilon > 0$, $\forall W$ neighborhood of x, there is $y \in W$ such that $f(y) < r + \varepsilon$. Next, suppose f lower semicontinuous at x, observe that $(x, \bar{f}(x)) \in \mathrm{cl}\,\mathrm{epi}\, f$ and see that this implies $f(x) \leq \bar{f}(x)$. Finally, to see that $f(x) = \bar{f}(x)$ implies f lower semicontinuous at x, observe that $f(y) \geq \bar{f}(y)$ $\forall y \in X$ and use the definition.

Proposition 2.2.5 *Let* $f \colon X \to (-\infty, \infty]$. *Then* f *is lower semicontinuous if and only if* f^a *is a closed set* $\forall a \in \mathbb{R}$.

Proof. Let $x_0 \notin f^a$. Then $(x_0, a) \notin \mathrm{epi}\, f$. Thus there is an open set W containing x_0 such that $f(x) > a$ $\forall x \in W$. This shows that $(f^a)^c$ is open. Suppose, by way of contradiction, f^a closed for all a, and let $(x, b) \notin \mathrm{epi}\, f$. Then there is $\varepsilon > 0$ such that $f(x) > b + \varepsilon$, so that $x \notin f^{b+\varepsilon}$. Then there exists an open set W containing x such that $\forall y \in W$ $f(y) \geq b + \varepsilon$. Thus $W \times (-\infty, b+\varepsilon) s \cap \mathrm{epi}\, f = \emptyset$, which means that $(\mathrm{epi}\, f)^c$ is open and this ends the proof. $\square$

When X is first countable, for instance a metric space, then lower semicontinuity of f at x can be given in terms of sequences: f is lower semicontinuous at x if and only if $\forall x_n \to x$,

$$\liminf_{n \to \infty} f(x_n) \geq f(x).$$

Example 2.2.6 I_C is lower semicontinuous if and only if C is a closed set.

Remark 2.2.7 Let $f \colon \mathbb{R} \to (-\infty, \infty]$ be convex. Then $\mathrm{dom}\, f$ is an interval, possibly containing its endpoints. If f is lower semicontinuous, then f restricted to $\mathrm{cl}\,\mathrm{dom}\, f$ is continuous.

We saw in Corollary 2.1.3 that a real valued convex function defined on a finite-dimensional space is everywhere continuous. The result fails in infinite dimensions. To see this, it is enough to consider a linear functional which is not continuous. However continuity can be recovered by assuming that f is lower semicontinuous. The following result holds:

Theorem 2.2.8 *Let* X *be a Banach space and let* $f \colon X \to (-\infty, \infty]$ *be a convex and lower semicontinuous function. Then* f *is continuous at the points of* $\mathrm{int}\,\mathrm{dom}\, f$.

Proof. Suppose $0 \in \mathrm{int}\,\mathrm{dom}\, f$, let $a > f(0)$ and let V be the closure of an open neighborhood of the origin which is contained in $\mathrm{dom}\, f$. Let us see that the closed convex set $f^a \cap V$ is absorbing (in V). Let $x \in V$. Then $g(t) := f(tx)$ defines a convex function on the real line. We have that $[-b, b] \in \mathrm{dom}\, g$ for some $b > 0$. Then g is continuous at $t = 0$, and thus it follows that there is

$\bar{t} > 0$ such that $\bar{t}x \in f^a$. By convexity and since $0 \in f^a$, we then have that $x \in nf^a$, for some large n. Thus

$$V = \bigcup_{n=1}^{\infty} n(f^a \cap V).$$

As a consequence of Baire's theorem (see Proposition B.1.1), $f^a \cap V$ is a neighborhood of the origin, (in V, and so in X), where f is upper bounded. Then f is continuous at the points of int dom f, see Theorem 2.1.2.   □

The family of convex, lower semicontinuous functions plays a key role in optimization, so that now we shall focus our attention on this class. For a Banach space X, we denote by $\Gamma(X)$ the set

$$\Gamma(X) := \{f \in \mathcal{F}(X) : f \text{ is lower semicontinuous}\}.$$

In other words, $\Gamma(X)$ is the subset of $\mathcal{F}(X)$ of the functions with a nonempty closed convex epigraph not containing vertical lines.

Example 2.2.9 $I_C \in \Gamma(X)$ if and only if C is a nonempty closed convex set.

Exercise 2.2.10 Verify that

$$f(x,y) := \begin{cases} \frac{y^2}{x} & \text{if } x > 0, y > 0, \\ 0 & \text{if } x \geq 0, y = 0, \\ \infty & \text{otherwise} \end{cases}$$

belongs to $\Gamma(\mathbb{R}^2)$. Verify also that f does not assume a maximum on the (compact, convex) set $C = \{(x,y) \in \mathbb{R}^2 : 0 \leq x, y \leq 1, \ y \leq \sqrt{x} - x^2\}$.

Hint. Consider the sequence $\{(1/n, (1/\sqrt{n} - 1/n^2))\}$.

The example above shows that $f(\text{dom } f \cap C))$ need not be closed, even if C is compact. The next exercise highlights the structure of the image of a convex set by a function in $\Gamma(X)$.

Exercise 2.2.11 Prove that for $f \in \Gamma(X)$, $f(\text{dom } f \cap C))$ is an interval for every (closed) convex set C.

Hint. Let $a, b \in f(\text{dom } f \cap C)$. Then there exist $x \in C$, $y \in C$ such that $f(x) = a$, $f(y) = b$. Now consider $g(t) = f(tx + (1-t)y)$, $t \in [0,1]$.

We see now that $\Gamma(X)$ is an (essentially) stable family with respect to some operations.

Proposition 2.2.12 *Let $f_i \in \Gamma(X), \forall i = 1,\ldots,n$ and let $t_1,\ldots,t_n > 0$. If for some $x_0 \in X$ $f_i(x_0) < \infty \ \forall i$, then $(\sum_{i=1}^{n} t_i f_i) \in \Gamma(X)$.*

Proof. From Proposition 1.2.18 and because for $a, b > 0$, $f, g \in \Gamma(X)$, $x \in X$, W a neighborhood of x,

$$\inf_{y \in W \setminus \{x\}} f(y) + g(y) \geq \inf_{y \in W \setminus \{x\}} f(y) + \inf_{y \in W \setminus \{x\}} g(y).$$

Thus

$$\sup_{W} \inf_{y \in W \setminus \{x\}} af(y) + bg(y) \geq \sup_{W} \left(a \inf_{y \in W \setminus \{x\}} f(y) + b \inf_{y \in W \setminus \{x\}} g(y) \right)$$

$$= a \sup_{W} \inf_{y \in W \setminus \{x\}} f(y) + b \sup_{W} \inf_{y \in W \setminus \{x\}} g(y).$$

$\square$

Proposition 2.2.13 *Let $f_i \in \Gamma(X), \forall i \in J$, where J is an arbitrary index set. If for some $x_0 \in X$ $\sup_{i \in J} f_i(x_0) < \infty$, then $(\sup_{i \in J} f_i) \in \Gamma(X)$.*

Proof. $\mathrm{epi}(\sup_{i \in J} f_i) = \bigcap_{i \in J} \mathrm{epi}\, f_i$. $\square$

The following Example shows that $\Gamma(X)$ is not closed with respect to the inf-convolution operation.

Example 2.2.14 Let C_1, C_2 be closed convex sets. Then $I_{C_1} \nabla I_{C_2} = I_{C_1 + C_2}$ (see Exercise 1.2.24). On the other hand, the function I_C is lower semicontinuous if and only if C is a closed convex set. Taking

$$C_1 := \{(x, y) \in \mathbb{R}^2 : x \leq 0 \text{ and } y \geq 0\}$$

and

$$C_2 := \{(x, y) \in \mathbb{R}^2 : x \geq 0 \text{ and } y \geq \frac{1}{x}\},$$

since $C_1 + C_2$ is not a closed set, then $I_{C_1} \nabla I_{C_2} \notin \Gamma(X)$.

Remark 2.2.15 An example as above cannot be constructed for functions defined on the real line. Actually, in this case the inf-convolution of two convex lower semicontinuous functions is lower semicontinuous. It is enough to observe that the effective domain of $(f \nabla g)$ is an interval. Let us consider, for instance, its right endpoint b, assuming that $(f \nabla g)(b) \in \mathbb{R}$ (the other case is left for the reader). Then if b_1 is the right endpoint of $\mathrm{dom}\, f$ and b_2 is the right endpoint of $\mathrm{dom}\, g$, it follows that

$$(f \nabla g)(b) = f(b_1) + g(b_2),$$

and if $x_k \to b^-$, taking x_k^1, x_k^2 with $x_k^1 + x_k^2 = x_k$ and $f(x_k^1) + g(x_k^2) \leq (f \nabla g)(x_k) + \frac{1}{k}$, then $x_k^1 \to b_1^-$, $x_k^2 \to b_2^-$ and

$$(f \nabla g)(b) = f(b_1) + g(b_2) \leq \liminf(f(x_k^1) + g(x_k^2))$$

$$\leq \liminf((f \nabla g)(x_k) + \frac{1}{k}) = \liminf(f \nabla g)(x_k).$$

We intend now to prove a fundamental result for functions in $\Gamma(X)$. We start with some preliminary facts. Let X be a Banach space and denote by X^* its topological dual space, the space of all real valued linear continuous functionals defined on X. Then X^* is a Banach space, when endowed with the canonical norm $\|x^*\|_* = \sup\{\langle x^*, x\rangle : \|x\| = 1\}$.

Lemma 2.2.16 Let $f \in \Gamma(X)$, $x_0 \in \operatorname{dom} f$ and $k < f(x_0)$. Then there are $y^* \in X^*$ and $q \in \mathbb{R}$ such that the affine function $l(x) = \langle y^*, x\rangle + q$ fulfills

$$f(x) \geq l(x), \forall x \in X, l(x_0) > k.$$

Proof. In $X \times \mathbb{R}$, let us consider the closed convex set epi f and the point (x_0, k). They can be separated by a closed hyperplane (Theorem A.1.6): there are $x^* \in X^*$, $r, c \in \mathbb{R}$ such that

$$\langle x^*, x\rangle + rb > c > \langle x^*, x_0\rangle + rk, \forall x \in \operatorname{dom} f, \forall b \geq f(x).$$

With the choice of $x = x_0, b = f(x_0)$ in the left part of the above formula, we get $r(f(x_0) - k) > 0$, and so $r > 0$. Let us consider the affine function $l(x) = \langle y^*, x\rangle + q$, with $y^* = \frac{-x^*}{r}$, $q = \frac{c}{r}$. It is then easy to see that $l(x) \leq f(x) \forall x \in X$ and that $l(x_0) > k$. □

Corollary 2.2.17 Let $f \in \Gamma(X)$. Then there exists an affine function minorizing f.

Corollary 2.2.18 Let $f \in \Gamma(X)$. Then f is lower bounded on bounded sets.

Corollary 2.2.19 Let $f \in \Gamma(X)$ be upper bounded on a neighborhood of $x \in X$. Then f is locally Lipschitz around x.

Proof. From the previous Corollary and Lemma 2.1.8. □

Remark 2.2.20 The conclusion of Corollary 2.2.19 can be strengthened if X is finite-dimensional and f is real valued. In this case f is Lipschitz on all bounded sets. This is no longer true in infinite dimensions, because then it can happen that f is not upper bounded on all bounded sets, as the following example shows. Consider a separable Hilbert space X, and let $\{e_n\}$ be an orthonormal basis. Consider the function

$$f(x) = \sum_{n=1}^{\infty} n(x, e_n)^{2n}.$$

Then f is not upper bounded on the unit ball.

Theorem 2.2.21 Let $f\colon X \to (-\infty, \infty]$ be not identically ∞. Then $f \in \Gamma(X)$ if and only if, $\forall x \in X$

$$f(x) = \sup\{\langle x^*, x\rangle + a : x^* \in X^*, a \in \mathbb{R}, f(x) \geq \langle x^*, x\rangle + a\}.$$

Proof. Denote by $h(x)$ the function $h(x) = \sup\{\langle x^*, x \rangle + a : x^* \in X^*, a \in \mathbb{R}, f(x) \geq \langle x^*, x \rangle + a\}$. Then $h(x) \leq f(x)$ and, being the pointwise supremum of affine functions, $h \in \Gamma(X)$ (see Proposition 2.2.13); this provides one of the implications. As far as the other one is concerned, let us consider $x_0 \in X$, $k < f(x_0)$ and prove that $h(x_0) > k$. Lemma 2.2.16 shows that $h(x_0) > k$ if $x_0 \in \text{dom } f$. We then consider the case $f(x_0) = \infty$.

Recalling the proof of Lemma 2.2.16, we can claim existence of $x^* \in X^*$, $r, c \in \mathbb{R}$ such that

$$\langle x^*, x \rangle + rb > c > \langle x^*, x_0 \rangle + rk, \forall x \in \text{dom } f, \forall b \geq f(x).$$

If $r \neq 0$, we conclude as in Lemma 2.2.16. If $r = 0$, which geometrically means that the hyperplane separating epi f and (x_0, k) is vertical, then

$$\langle x^*, x \rangle > c > \langle x^*, x_0 \rangle, \forall x \in \text{dom } f.$$

Calling $l(x) = \langle -x^*, x \rangle + c$, we have $l(x_0) > 0$ and $l(x) < 0, \forall x \in \text{dom } f$. From Corollary 2.2.17, there exists an affine function $m(x) := \langle y^*, x \rangle + q$ with the property that $f(x) \geq m(x), \forall x \in X$. Hence, $\forall h > 0$, $m(x) + hl(x) \leq f(x), \forall x \in \text{dom } f$, whence $m(x) + hl(x) \leq f(x), \forall x \in X$. On the other hand, as $l(x_0) > 0$, for a sufficiently large h, $(m + hl)(x_0) > k$, and this concludes the proof. □

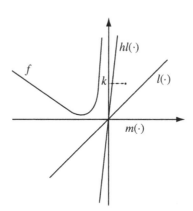

Figure 2.2.

The previous theorem can be refined if f is also a positively homogeneous function.

Corollary 2.2.22 *Let $h \in \Gamma(X)$ be sublinear. Then*

$$h(x) = \sup\{\langle x^*, x \rangle : x^* \in X^*, h(x) \geq \langle x^*, x \rangle\}.$$

Proof. It is enough to show that if the *affine* function $\langle x^*, \cdot \rangle + c$ minorizes h, then the *linear* function $\langle x^*, \cdot \rangle$ minorizes h. Now, since h is positively homogeneous, $\forall x \in X$, $\forall t > 0$,

$$\langle x^*, \frac{x}{t} \rangle + \frac{c}{t} \le h\left(\frac{x}{t}\right),$$

i.e.,

$$\langle x^*, y \rangle + \frac{c}{t} \le h(y),$$

$\forall y \in X$. We conclude now by letting t go to ∞. $\qquad \square$

Exercise 2.2.23 Let C be a nonempty closed convex set. Let $d(\cdot, C)$ be the distance function from C: $d(x, C) = \inf_{c \in C} \|x - c\|$. Then d is 1-Lipschitz.

3

The derivatives and the subdifferential

Something must still happen, but my strength is over,
my fingers empty gloves,
nothing extraordinary in my eyes,
nothing driving me.
(M. Atwood, *Surfacing*)

In the previous chapter we have seen that convex functions enjoy nice properties from the point of view of continuity. Here we see that the same happens with directional derivatives. The limit involved in the definition of directional derivative always exists, and thus in order to claim the existence of the directional derivative at a given point and along a fixed direction, it is enough to check that such a limit is a real number. Moreover, the directional derivative at a given point is a sublinear function, i.e., a very particular convex function, with respect to the direction.

We then introduce and study the very important concept of gradient. Remember that we are considering extended real valued functions. Thus it can happen that the interior of the effective domain of a function is empty. This would mean that a concept of derivative would be useless in this case. However, we know that a convex function which is differentiable at a given point enjoys the property that its graph lies above that tangent line at that point, a remarkable *global* property. This simple remark led to the very useful idea of subgradient for a convex function at a given point. The definition does not require that the function be real valued at a neighborhood of the point, keeps most of the important properties of the derivative (in particular, if zero belongs to the subdifferential of f at a given point x, then x is a global minimizer for f), and if f is smooth, then it reduces to the classical derivative of f. The subdifferential of f at a given point, i.e., the set of its subgradients at that point, is also related to its directional derivatives.

Clearly, an object such as the subdifferential is more complicated to handle than a derivative. For instance, the simple formula that the derivative of the sum of two functions f and g is the sum of the derivatives of f and g must be rewritten here, and its proof is not obvious at all. Moreover, studying continuity of the derivative here requires concepts of continuity for multivalued functions, which we briefly introduce. We also briefly analyze concepts of twice differentiability for convex functions, to see that the theory can be extended beyond the smooth case. Thus, the subdifferential calculus introduced and analyzed in this chapter is of the utmost importance in the study of convex functions.

3.1 Properties of the directional derivatives

We shall now see that the same happens with directional derivatives. In particular, the limit in the definition of the directional derivative at a given point and for a fixed direction always exists. Thus, to claim existence of a directional derivative it is enough to check that such a limit is a real number.

Definition 3.1.1 Let $f \in \Gamma(X), x, d \in X$. The *directional derivative* of f at x along the vector d, denoted by $f'(x; d)$, is the following limit:

$$f'(x; d) = \lim_{t \to 0^+} \frac{f(x + td) - f(x)}{t},$$

whenever it is finite.

Proposition 3.1.2 *Let $f \in \Gamma(X), x, d \in X$. The directional derivative of f at x along the vector d exists if and only if the quotient*

$$\frac{f(x + td) - f(x)}{t}$$

is finite for some $\bar{t} > 0$ and is lower bounded in $(0, \infty)$.

Proof. Let $x, d \in X$. We know from Proposition 1.2.11 that the function

$$0 < t \mapsto g(t; d) := \frac{f(x + td) - f(x)}{t},$$

is increasing. This implies that $\lim_{t \to 0^+} g(t; d)$ always exists and

$$\lim_{t \to 0^+} g(t; d) = \inf_{t > 0} g(t).$$

If there is $\bar{t} > 0$ such that $g(\bar{t}) \in \mathbb{R}$ and if g is lower bounded, then the limit must be finite. □

Of course, $\lim_{t \to 0^+} \frac{f(x+td)-f(x)}{t} = \infty$ if and only if $f(x + td) = \infty$ for all $t > 0$. Note that we shall use the word directional derivative, even if d is not a unit vector.

The next estimate for the directional derivative is immediate.

Proposition 3.1.3 *Let $f \in \Gamma(X)$ be Lipschitz with constant k in a neighborhood V of x. Then*

$$|f'(x; d)| \le k, \ \forall d \in X : \|d\| = 1.$$

Proposition 3.1.4 *Let $f \in \Gamma(X)$, and let $x \in \mathrm{dom}\, f$. Then $X \ni d \mapsto \lim_{t \to 0+} \frac{f(x+td)-f(x)}{t}$ is a sublinear function.*

Proof. We shall prove that $X \ni d \mapsto g(t; d)$ is convex and positively homogeneous.

$$
\begin{aligned}
f(x + t(\lambda d_1 + (1-\lambda)d_2)) &= f(\lambda(x + td_1) + (1-\lambda)(x + td_2)) \\
&\le \lambda f(x + td_1) + (1-\lambda) f(x + td_2),
\end{aligned}
$$

providing convexity of $d \mapsto \lim_{t \to 0+} \frac{f(x+td)-f(x)}{t}$. It is immediate to verify that it is positively homogeneous. □

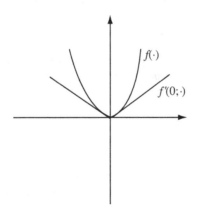

Figure 3.1.

The following example shows that the limit in the definition of the directional derivative can assume value $-\infty$.

$$
f(x) = \begin{cases} -\sqrt{x} & \text{if } x \ge 0, \\ \infty & \text{elsewhere.} \end{cases}
$$

If there exists d such that the limit in the definition is $-\infty$, as $f'(x; 0) = 0$, then $d \mapsto \lim_{t \to 0+} \frac{f(x+td)-f(x)}{t}$ is never lower semicontinuous, because a convex lower semicontinuous function assuming value $-\infty$ never assumes a real value (prove it, remembering Remark 1.2.6).

The next theorem provides a condition under which $d \mapsto f'(x; d) \in \Gamma(X)$.

Theorem 3.1.5 *Let $f \in \Gamma(X)$. Let $x_0 \in \operatorname{dom} f$. Suppose moreover,*

$$F := \mathbb{R}_+(\operatorname{dom} f - x_0)$$

is a closed vector space of X. Then $d \mapsto f'(x_0; d) \in \Gamma(X)$.

Proof. By translation, we can suppose that $x_0 = 0$. It is easy to show that

$$F = \bigcup_{n=1}^{\infty} nf^n.$$

As nf^n is a closed set for each $n \in \mathbb{R}$, and since F is a complete metric space, it follows from Baire's theorem that there exists $\bar{n}$ such that $\operatorname{int}_{|F} \bar{n}f^{\bar{n}}$ (hence $\operatorname{int}_{|F} f^{\bar{n}}) \neq \emptyset$. Thus f, restricted to F, is upper bounded on a neighborhood of a point $\bar{x}$. As $-t\bar{x} \in \operatorname{dom} f$ for some $t > 0$, it follows that $f_{|F}$ is upper bounded on a neighborhood of 0 (see the proof of Theorem 2.1.2), whence continuous and locally Lipschitz (Corollary 2.2.19) on a neighborhood of 0. It follows that $F \ni d \mapsto f'(0; d)$ is upper bounded on a neighborhood of zero and, by Proposition 2.1.5, is everywhere continuous. As $f'(0; d) = \infty$ if $d \notin F$ and F is a closed set, we conclude that $d \mapsto f'(x_0; d) \in \Gamma(X)$. □

Corollary 3.1.6 *Let $f \in \Gamma(X)$. Let $x_0 \in \operatorname{int} \operatorname{dom} f$. Then $d \mapsto f'(x_0; d)$ is a convex, positively homogeneous and everywhere continuous function.*

3.2 The subgradient

We now introduce the notion of subgradient of a function at a given point. It is a generalization of the idea of derivative, and it has several nice properties. It is a useful notion, both from a theoretical and a computational point of view.

Definition 3.2.1 Let $f: X \to (-\infty, \infty]$. $x^* \in X^*$ is said to be a *subgradient* of f at the point x_0 if $x_0 \in \operatorname{dom} f$ and $\forall x \in X$,

$$f(x) \geq f(x_0) + \langle x^*, x - x_0 \rangle.$$

The *subdifferential* of f at the point x_0, denoted by $\partial f(x_0)$, is the possibly empty set of all subgradients of f at the point x_0.

The above definition makes sense for any function f. However, a definition of derivative, as above, requiring a *global* property, is useful mainly in the convex case.

Definition 3.2.2 Let $A \subset X$ and $x \in A$. We say that $0^* \neq x^* \in X^*$ *supports* A *at* x if

$$\langle x^*, x \rangle \geq \langle x^*, a \rangle, \ \forall a \in A.$$

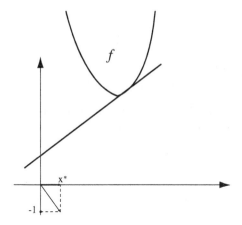

Figure 3.2. x^* is a subgradient of f at the point x_0.

Remark 3.2.3 $x^* \in \partial f(x_0)$ if and only if the pair $(x^*, -1)$ supports epi f at the point $(x_0, f(x_0))$. For, $\forall x \in X$

$$\langle x^*, x_0 \rangle - f(x_0) \geq \langle x^*, x \rangle - r, \forall r \geq f(x) \iff f(x) \geq f(x_0) + \langle x^*, x - x_0 \rangle.$$

Example 3.2.4 Here are some examples of subgradients:

- $f(x) = |x|$. Then $\partial f(x) = \{\frac{x}{|x|}\}$ if $x \neq 0$, $\partial f(0) = [-1, 1]$ (try to extend this result to the function $f(x) = \|x\|$ defined on a Hilbert space X);
- $f \colon \mathbb{R} \to [0, \infty]$, $f(x) = I_{\{0\}}(x)$. Then $\partial f(0) = (-\infty, \infty)$;
- Let C be a closed convex set. $x^* \in \partial I_C(x) \iff x \in C$ and $\langle x^*, c \rangle \leq \langle x^*, x \rangle, \forall c \in C$. That is, if $x^* \neq 0^*$, then $x^* \in \partial I_C(x)$ if and only if x^* supports C at x; $\partial I_C(x)$ is said to be the *normal cone* of C at $\dot{x}$ and it is sometimes indicated also by $N_C(x)$.

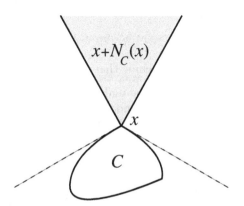

Figure 3.3. The normal cone to C at x.

- Let

$$f(x) = \begin{cases} -\sqrt{x} & \text{if } x \geq 0, \\ \infty & \text{otherwise .} \end{cases}$$

Then $\partial f(0) = \emptyset$, $\partial f(x) = \{-\frac{1}{2\sqrt{x}}\}$ if $x > 0$.

Exercise 3.2.5 Let $f: \mathbb{R}^2 \to \mathbb{R}$ be the following function: $f(x,y) = \max\{|x|, |y|\}$. Find the subdifferential of f at the points $(0,0)$, $(0,1)$, $(1,1)$.

Hint. $\{(x^*, y^*) : |x^*| + |y^*| \leq 1\}, \{(0,1)\}, \{(x^*, y^*) : x^* \geq 0, y^* \geq 0, x^* + y^* = 1\}$ respectively.

Definition 3.2.6 Given a Banach space X, the *duality mapping* $\delta: X \to X^*$ is defined as

$$\delta(x) := \{x^* \in X^* : \|x^*\|_* = 1 \text{ and } \langle x^*, x \rangle = \|x\|\}.$$

It is well known that $\delta(x) \neq \emptyset$ for all $x \in X$. The proof of this relies on the fact that the function $x^* \mapsto \langle x^*, x \rangle$ is weak* continuous.

Example 3.2.7 Let X be a Banach space, let $f(x) = \|x\|$. Then, for all $x \neq 0$,

$$(\partial \| \cdot \|)(x) = \delta(x).$$

We leave as an exercise the proof that $\delta(x) \subset (\partial \| \cdot \|)(x)$. To show the opposite inclusion, let $x^* \in \partial(\|x\|)$. Then, for all y,

$$\|y\| \geq \|x\| + \langle x^*, y - x \rangle. \tag{3.1}$$

The choices of $y = 0$ and $y = 2x$ show that

$$\langle x^*, x \rangle = \|x\|. \tag{3.2}$$

From (3.1) and (3.2) we get that

$$\|y\| \geq \langle x^*, y \rangle, \forall y \in X.$$

Combining this with (3.2), we conclude that $\|x^*\|_* = 1$ and so $x^* \in \delta(x)$.

Exercise 3.2.5 shows that δ can be multivalued at some point. Those Banach spaces having a norm which is smooth outside the origin (in this case δ must be single valued) are important. We shall discuss this later.

Example 3.2.8 Let $X = l^2$, with $\{e_n\}_{n \in \mathbb{N}}$ the canonical basis, and $C = \{x \in l^2, x = (x_1, x_2, \ldots, x_n, \ldots) : |x_n| \leq 2^{-n}\}$. Let

$$f(x) = \begin{cases} -\sum_{n=1}^{\infty} \sqrt{2^{-n} + x_n} & \text{if } x \in C, \\ \infty & \text{elsewhere.} \end{cases}$$

Then f is convex and its restriction to the set C is a continuous function. An easy calculation shows that $f'(0; e_n) = -2^{\frac{n-2}{2}}$. Now suppose $x^* \in \partial f(0)$. Then

$$f(2^{-n} e_n) \geq f(0) + \langle x^*, 2^{-n} e_n \rangle, \ \forall n \in \mathbb{N},$$

whence

$$(1 - \sqrt{2}) 2^{\frac{n}{2}} \geq \langle x^*, e_n \rangle, \ \forall n \in \mathbb{N}.$$

Thus f has all directional derivatives at 0, but $\partial f(0) = \emptyset$. Observe that this cannot happen in finite dimensions, as Exercise 3.2.13 below shows.

Remark 3.2.9 Let $x \in \operatorname{dom} f$, $x^* \in \partial f(x)$, u^* in the normal cone to $\operatorname{dom} f$ at x ($\langle u^*, x - u \rangle \leq 0, \forall u \in \operatorname{dom} f$). Then $x^* + u^* \in \partial f(x)$. This does not provide any information if $x \in \operatorname{int} \operatorname{dom} f$, for instance if f is continuous at x, as the normal cone to $\operatorname{dom} f$ at x reduces to 0^*. However this information is interesting if $x \notin \operatorname{int} \operatorname{dom} f$. In many situations, for instance if X is finite-dimensional or if $\operatorname{dom} f$ has interior points, there exists at least a $0^* \neq u^*$ belonging to the normal cone at x, which thus is an unbounded set (the existence of such a $0^* \neq u^*$ in the normal cone follows from the fact that there is a hyperplane supporting $\operatorname{dom} f$ at x. The complete argument is suggested in Exercise 3.2.10). Hence, in the boundary points of $\operatorname{dom} f$ it can happen that the subdifferential of f is either empty or an unbounded set.

Exercise 3.2.10 Let X be a Banach space and let $\operatorname{int} \operatorname{dom} f \neq \emptyset$. Let $x \in \operatorname{dom} f \setminus \operatorname{int} \operatorname{dom} f$. Prove that the normal cone to $\operatorname{dom} f$ at the point x is unbounded.

Hint. Use Theorem A.1.5 by separating x from $\operatorname{int} \operatorname{dom} f$.

We now see how to evaluate the subdifferential of the inf convolution, at least in a particular case.

Proposition 3.2.11 *Let X be a Banach space, let $f, g \in \Gamma(X)$, let $x \in X$ and let u, v be such that*

$$u + v = x \ and \ (f \nabla g)(x) = f(u) + g(v).$$

Then

$$\partial (f \nabla g)(x) = \partial f(u) \cap \partial g(v).$$

Proof. Let $x^* \in \partial f(u) \cap \partial g(v)$. Thus, for all $y \in X$ and $z \in X$

$$f(y) \geq f(u) + \langle x^*, y - u \rangle, \tag{3.3}$$

$$g(z) \geq g(v) + \langle x^*, z - v \rangle. \tag{3.4}$$

Let $w \in X$ and let $y, z \in X$ be such that $y + z = w$. Summing up (3.3) and (3.4) we get

$$f(y) + g(z) \geq (f \nabla g)(x) + \langle x^*, w - x \rangle. \tag{3.5}$$

By taking, in the left side of (3.5), the infimum over all y, z such that $y+z = w$, we can conclude that $x^* \in \partial(f \nabla g)(x)$. Conversely, suppose for all $y \in X$,

$$(f \nabla g)(y) \geq f(u) + g(v) + \langle x^*, y - (u + v) \rangle. \tag{3.6}$$

Then, given any $z \in X$, put $y = z + v$ in (3.6). We get

$$f(z) + g(v) \geq f(u) + g(v) + \langle x^*, z - v \rangle,$$

showing that $x^* \in \partial f(u)$. The same argument applied to $y = z + u$ shows that $x^* \in \partial g(v)$ and this ends the proof.   □

The above formula applies to points where the inf-convolution is exact. A much more involved formula, involving approximate subdifferentials, can be shown to hold at any point. We shall use the above formula to calculate, in a Euclidean space, the subdifferential of the function $d(\cdot, C)$, where C is a closed convex set.

In the next few results we investigate the connections between the subdifferential of a function at a given point and its directional derivatives at that point.

Proposition 3.2.12 *Let* $f \in \Gamma(X)$ *and* $x \in \operatorname{dom} f$. *Then*

$$\partial f(x) = \{x^* \in X^* : \langle x^*, d \rangle \leq f'(x; d), \forall d \in X\}.$$

Proof. $x^* \in \partial f(x)$ if and only if

$$\frac{f(x + td) - f(x)}{t} \geq \langle x^*, d \rangle, \forall d \in X, \forall t > 0,$$

if and only if, taking the inf for $t > 0$ in the left side of the above inequality,

$$f'(x; d) \geq \langle x^*, d \rangle, \forall d \in X.$$

□

Exercise 3.2.13 If $f \in \Gamma(\mathbb{R}^n)$, if $f'(x; d)$ exists and is finite for all d, then $\partial f(x) \neq \emptyset$.

Hint. $f'(x; d)$ is sublinear and continuous. Now apply a corollary to the Hahn–Banach theorem (Corollary A.1.2) and Proposition 3.2.12.

Theorem 3.2.14 *Let* $f \in \Gamma(X)$ *and* $x \in \operatorname{dom} f$. *If*

$$F := \mathbb{R}_+(\operatorname{dom} f - x)$$

is a closed vector space, then

$$d \mapsto f'(x; d) = \sup\{\langle x^*, d \rangle : x^* \in \partial f(x)\}.$$

Proof. The function $d \mapsto f'(x;d)$ is sublinear (Proposition 3.1.4). From Theorem 3.1.5 $d \mapsto f'(x;d) \in \Gamma(X)$. Hence $d \mapsto f'(x;d)$ is the pointwise supremum of all linear functionals minorizing it (Corollary 2.2.22):

$$d \mapsto f'(x;d) = \sup\{\langle x^*, d \rangle : \langle x^*, d \rangle \le f'(x;d), \forall d \in X\}.$$

We conclude by Proposition 3.2.12, since $\langle x^*, d \rangle \le f'(x;d), \forall d \in X$ if and only if $x^* \in \partial f(x)$. $\square$

The next theorem shows that the subdifferential is nonempty at "many" points.

Theorem 3.2.15 *Let $f \in \Gamma(X)$. Then $\partial f(x) \ne \emptyset, \forall x \in \text{int dom } f$.*

Proof. If $x \in \text{int dom } f$, then $\mathbb{R}_+(\text{dom } f - x) = X$. Now apply Theorem 3.2.14. $\square$

If X is finite dimensional, the previous result can be refined (same proof) since $\partial f(x) \ne \emptyset \; \forall x \in \text{ri dom } f$. In infinite dimensions it can be useless, since dom f could possibly have no interior points. But we shall show later that every function $f \in \Gamma(X)$ has a nonempty subdifferential on a dense subset of dom f (see Corollary 4.2.13).

From Propositions 3.1.3 and 3.2.12 we immediately get the following result providing an estimate from above of the norm of the elements in ∂f.

Proposition 3.2.16 *Let $f \in \Gamma(X)$ be Lipschitz with constant k in an open set $V \ni x$. Then*

$$\|x^*\| \le k, \forall x^* \in \partial f(x).$$

As a last remark we observe that the subdifferential keeps a fundamental property of the derivative of a convex function.

Proposition 3.2.17 *Let $f \in \Gamma(X)$. Then $0^* \in \partial f(\bar{x})$ if and only if $\bar{x}$ minimizes f on X.*

Proof. Obvious from the definition of subdifferential. $\square$

3.3 Gâteaux and Fréchet derivatives and the subdifferential

Definition 3.3.1 Let $f \colon X \to (-\infty, \infty]$ and $x \in \text{dom } f$. Then f is said to be *Gâteaux differentiable* at x if there exists $x^* \in X^*$ such that

$$f'(x;d) = \langle x^*, d \rangle, \; \forall d \in X.$$

And f is said to be *Fréchet differentiable* at x if there exists $x^* \in X^*$ such that

$$\lim_{d \to 0} \frac{f(x+d) - f(x) - \langle x^*, d \rangle}{\|d\|} = 0.$$

Gâteaux differentiability of f at x implies in particular that all the tangent lines to the graph of f at the point $(x, f(x))$, along all directions, lie in the same plane; Fréchet differentiability means that this plane is "tangent" to the graph at the point $(x, f(x))$.

Exercise 3.3.2 Show that if f is Gâteaux differentiable at x, the functional $x^* \in X^*$ given by the definition is unique. Show that Fréchet differentiability of f at x implies Gâteaux differentiability of f at x and that f is continuous at x. The opposite does not hold in general, as the example below shows.

Example 3.3.3 Let

$$f(x, y) = \begin{cases} 1 & \text{if } y \geq x^2 \text{or } y = 0, \\ 0 & \text{otherwise .} \end{cases}$$

Then all directional derivatives of f vanish at the origin, but f is not continuous at $(0, 0)$, so that it is not Fréchet differentiable at the origin.

However, for convex functions in finite dimensions, the notions of Fréchet and Gâteaux differentiability agree, as we shall see.

We shall usually denote by $\nabla f(x)$ the unique $x^* \in X^*$ in the definition of Gâteaux differentiability. If f is Fréchet differentiable at x, we shall preferably use the symbol $f'(x)$ to indicate its Fréchet derivative at x.

Now a first result about Gâteaux differentiability in the convex case. Remember that the limit defining the directional derivative exists for every direction d; thus, in order to have Gâteaux differentiability, we only need to show that the limit is finite in any direction, and that there are no "angles".

Proposition 3.3.4 *Let $f \in \Gamma(X)$. Then f is Gâteaux differentiable at $x \in X$ if and only if $d \mapsto f'(x; d)$ upper bounded in a neighborhood of the origin and*

$$\lim_{t \to 0} \frac{f(x + td) - f(x)}{t}, \quad \forall d \in X,$$

exists and is finite (as a two-sided limit).

Proof. The "only if" part is obvious. As far as the other one is concerned, observe that the equality between the right and left limits above means that $f'(x; -d) = -f'(x, d)$. Thus the function $d \mapsto f'(x; d)$, which is always sublinear, is in this case linear too. Upper boundedness next guarantees that $d \mapsto f'(x; d)$ is also continuous, and we conclude. $\qquad \square$

The next exercise shows that Fréchet and Gâteaux differentiability do not agree in general for convex functions.

Exercise 3.3.5 Let $X = l^1$ with the canonical norm and let $f(x) = \|x\|$. Then f is Gâteaux differentiable at a point $x = (x_1, x_2, \ldots,)$ if and only if $x_i \neq 0 \; \forall i$, and it is never Fréchet differentiable. $\nabla f(x) = x^* = (x_1^*, x_2^*, \ldots ,)$, where $x_n^* = \frac{x_n}{|x_n|} := \operatorname{sgn} x_n$.

Hint. If, for some i, $x_i = 0$, then the limit

$$\lim_{t \to 0} \frac{f(x + te_i) - f(x)}{t},$$

does not exist, since the right limit is different from the left one. If $x_i \neq 0$ $\forall i$, then for $\varepsilon > 0$, let N be such that $\sum_{i>N} |d_i| < \varepsilon$. For every small t,

$$\operatorname{sgn}(x_i + td_i) = \operatorname{sgn}(x_i), \quad \forall i \leq N.$$

Then

$$\left| \frac{\|x + td\| - \|x\|}{t} - \sum_{i \in \mathbb{N}} d_i \operatorname{sgn} x_i \right| < 2\varepsilon.$$

On the other hand, let x be such that $x_i \neq 0$ for all i and consider $d^n = (0, \ldots, -2x_n, \ldots)$. Then $d_n \to 0$, while

$$\left| \|x + d^n\| - \|x\| - \sum_{i \in \mathbb{N}} d_i^n \operatorname{sgn} x_i \right| = \|d^n\|,$$

showing that f is not Fréchet differentiable in x.

The concept of subdifferential extends the idea of derivative, in the sense explained in the following results.

Proposition 3.3.6 *Let $f \in \Gamma(X)$. If f is Gâteaux differentiable at x, then $\partial f(x) = \{\nabla f(x)\}$.*

Proof. By definition, $\forall d \in X$,

$$\lim_{t \to 0} \frac{f(x + td) - f(x)}{t} = \langle \nabla f(x), d \rangle.$$

As the function $0 < t \mapsto \frac{f(x+td)-f(x)}{t}$ is increasing,

$$\frac{f(x + td) - f(x)}{t} \geq \langle \nabla f(x), d \rangle,$$

whence

$$f(x + td) \geq f(x) + \langle \nabla f(x), td \rangle, \quad \forall td \in X,$$

showing that $\partial f(x) \ni \nabla f(x)$. Now, let $x^* \in \partial f(x)$. Then

$$f(x + td) \geq f(x) + \langle x^*, td \rangle, \quad \forall d \in X, \forall t > 0,$$

hence

$$\lim_{t \to 0^+} \frac{f(x + td) - f(x)}{t} := \langle \nabla f(x), d \rangle \geq \langle x^*, d \rangle, \quad \forall d \in X,$$

whence $x^* = \nabla f(x)$.   □

Proposition 3.3.7 *Let $f \in \Gamma(X)$. If f is continuous at x and if $\partial f(x) = \{x^*\}$, then f is Gâteaux differentiable at x and $\nabla f(x) = x^*$.*

Proof. First, observe that $d \mapsto f'(x;d)$ is everywhere continuous as $x \in$ int dom f. Next, let $X \ni d$ be a (norm one) fixed direction. Let us consider the linear functional, defined on span $\{d\}$,

$$l_d(h) = af'(x;d) \quad \text{if } h = ad.$$

Then $l_d(h) \leq f'(x;h)$ for all h in span $\{d\}$. The equality holds for $h = ad$ and $a > 0$, while $l_d(-d) = -f'(x;d) \leq f'(x;-d)$. By the Hahn–Banach theorem (see Theorem A.1.1), there is a linear functional $x_d^* \in X^*$ agreeing with l_d on span $\{d\}$, and such that $\langle x_d^*, h \rangle \leq f'(x;h) \; \forall h \in X$. Then $x_d^* \in \partial f(x)$, so that $x_d^* = x^*$. As by construction $\langle x^*, d \rangle = f'(x;d) \; \forall d \in X$, it follows that f is Gâteaux differentiable at x and $x^* = \nabla f(x)$. □

It may be worth noticing that in the previous result the assumption that f is continuous at x cannot be dropped. A set A (with empty interior) can have at a point x the normal cone reduced to the unique element zero (see Exercise A.1.8). Thus the indicator function of A is not Gâteaux differentiable at x, but $\partial I_A(x) = \{0\}$. Observe also that if dom f does have interior points, it is not possible that at a point x where f is not continuous, $\partial f(x)$ is a singleton (see Remark 3.2.9).

Recall that, denoting by $\{e_1, \ldots, e_n\}$ the canonical basis in $\mathbb{R}^n$, the *partial derivatives* of f at x are defined as follows:

$$\frac{\partial f}{\partial x_i}(x) = \lim_{t \to 0} \frac{f(x + te_i) - f(x)}{t},$$

whenever the limit exists and is finite. Then we have the following proposition.

Proposition 3.3.8 *Let $f \colon \mathbb{R}^n \to \mathbb{R}$ be convex. Then f is (Gâteaux) differentiable at $x \in \mathbb{R}^n$ if and only if the partial derivatives $\frac{\partial f}{\partial x_i}(x)$, $i = 1, \ldots, n$ exist.*

Proof. Suppose there exist the partial derivatives of f at x. As f is continuous, $\partial f(x) \neq \emptyset$. Let $x^* \in \partial f(x)$, and write $x_i^* = \langle x^*, e_i \rangle$. Then $\forall t \neq 0, f(x + te_i) - f(x) \geq tx_i^*$, hence

$$\frac{\partial f}{\partial x_i}(x) = \lim_{t \to 0^+} \frac{f(x + te_i) - f(x)}{t} \geq x_i^*,$$

$$\frac{\partial f}{\partial x_i}(x) = \lim_{t \to 0^-} \frac{f(x + te_i) - f(x)}{t} \leq x_i^*,$$

providing $x_i^* = \frac{\partial f}{\partial x_i}(x)$. Thus $\partial f(x)$ is a singleton, and we conclude with the help of Proposition 3.3.7. The opposite implication is an immediate consequence of Proposition 3.3.4. □

We shall see in Corollary 3.5.7 that Fréchet and Gâteaux differentiability actually agree for a convex function defined in a Euclidean space. The above proposition in turn shows that differentiability at a point is equivalent to the existence of the partial derivatives of f at the point.

3.4 The subdifferential of the sum

Let us consider the problem of minimizing a convex function f on a convex set C. This can be seen as the unconstrained problem of minimizing the function $f + I_C$. And $\bar{x} \in C$ is a solution of this problem if and only if $0 \in \partial(f + I_C)(\bar{x})$. Knowing this is not very useful unless $\partial(f + I_C) \subset \partial f + \partial I_C$. In such a case, we could claim the existence of a vector $x^* \in \partial f(\bar{x})$ such that $-x^*$ belongs to the normal cone of C at the point $\bar{x}$, a property that, at least when f is differentiable at $\bar{x}$, has a clear geometrical meaning. Unfortunately in general only the opposite relation holds true:

$$\partial(f + g) \supset \partial f + \partial g.$$

In the next exercise it can be seen that the desired relation need not be true.

Exercise 3.4.1 In $\mathbb{R}^2$ consider

$$A := \{(x, y) : y \geq x^2\},$$

$$B := \{(x, y) : y \leq 0\},$$

and their indicator functions I_A, I_B. Evaluate the subdifferential of I_A, I_B and of $I_A + I_B$ at the origin.

However, in some cases we can claim the desired result. Here is a first example:

Theorem 3.4.2 Let $f, g \in \Gamma(X)$ and let $\bar{x} \in \operatorname{int} \operatorname{dom} f \cap \operatorname{dom} g$. Then, for all $x \in X$

$$\partial(f + g)(x) = \partial f(x) + \partial g(x).$$

Proof. If $\partial(f + g)(x) = \emptyset$, there is nothing to prove. Otherwise, let $x^* \in \partial(f + g)(x)$. Then

$$f(y) + g(y) \geq f(x) + g(x) + \langle x^*, y - x \rangle, \ \forall y \in X. \tag{3.7}$$

Writing (3.7) in the form

$$f(y) - \langle x^*, y - x \rangle - f(x) \geq g(x) - g(y),$$

we see that

$$A := \{(y, a) : f(y) - \langle x^*, y - x \rangle - f(x) \leq a\},$$

$$B := \{(y, a) : g(x) - g(y) \geq a\}$$

are closed convex sets such that int $A \neq \emptyset$ and int $A \cap B = \emptyset$. From the Hahn–Banach theorem A.1.5, int A and B can be separated by a hyperplane that is not vertical, as is easy to see. Thus, there is an affine function $l(y) = \langle y^*, y \rangle + k$ such that

$$g(x) - g(y) \leq \langle y^*, y \rangle + k \leq f(y) - \langle x^*, y - x \rangle - f(x), \forall y \in X.$$

Setting $y = x$ we see that $k = \langle -y^*, x \rangle$, whence $\forall y \in X$,

$$g(y) \geq g(x) + \langle -y^*, y - x \rangle,$$

which gives $-y^* \in \partial g(x)$. Moreover, $\forall y \in X$,

$$f(y) \geq f(x) + \langle x^* + y^*, y - x \rangle,$$

so that $x^* + y^* \in \partial f(x)$. We thus have $x^* = -y^* + (x^* + y^*)$, with $-y^* \in \partial g(x)$ and $x^* + y^* \in \partial f(x)$.   □

Exercise 3.4.3 Let $f \colon X \to \mathbb{R}$ be convex and lower semicontinuous and let C be a closed convex set. Then $\bar{x} \in C$ is a solution of the problem of minimizing f over C if and only if there is $x^* \in \partial f(\bar{x})$ such that $-x^*$ is in the normal cone to C at $\bar{x}$.

In the chapter dedicated to duality, the previous result will be specified when the set C is characterized by means of inequality constraints; see Theorem 5.4.2.

3.5 The subdifferential multifunction

In this section we shall investigate some properties of the subdifferential of f, considered as a multivalued function (multifunction) from X to X^*.

Proposition 3.5.1 *Let $f \in \Gamma(X)$ and $x \in X$. Then $\partial f(x)$ is a (possibly empty) convex and weakly* closed subset of X^*. Moreover, if f is continuous at x, then ∂f is bounded on a neighborhood of x.*

Proof. Convexity follows directly from the definition. Now, let $x^* \notin \partial f(x)$. This means that there is $y \in X$ such that

$$f(y) - f(x) < \langle x^*, y - x \rangle.$$

By the definition of weak* topology, it follows that for each z^* in a suitable (weak*) neighborhood of x^*, the same inequality holds. This shows that $\partial f(x)$ is weakly* closed. Finally, if f is continuous at x, it is upper and lower bounded around x, and thus it is Lipschitz in a neighborhood of x (Corollary 2.2.19). From Proposition 3.2.16 we get local boundedness of ∂f.   □

As a consequence of this, the *multifunction* $x \mapsto \partial f(x)$ is convex, weakly* closed valued, possibly empty valued at some x and locally bounded around x if x is a continuity point of f. We investigate now some of its continuity properties, starting with a definition.

Definition 3.5.2 Let (X, τ), (Y, σ) be two topological spaces and let $F\colon X \to Y$ be a given multifunction. Then F is said to be $\tau - \sigma$ *upper semicontinuous* at $\bar{x} \in X$ if for each open set V in Y such that $V \supset F(\bar{x})$, there is an open set $I \subset X$ containing $\bar{x}$ such that, $\forall x \in I$,

$$F(x) \subset V.$$

F is said to be $\tau - \sigma$ *lower semicontinuous* at $\bar{x} \in X$ if for each open set V in Y such that $V \cap F(\bar{x}) \neq \emptyset$, there is an open set $I \subset X$ containing $\bar{x}$ such that, $\forall x \in I$,

$$F(x) \cap V \neq \emptyset.$$

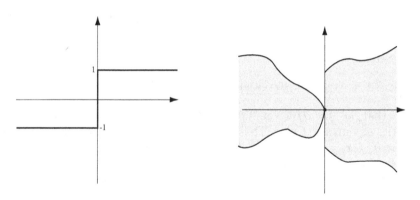

An upper semicontinuous multifunction not lower semicontinuous at 0.

A lower semicontinuous multifunction not upper semicontinuous at 0.

Figure 3.4.

Remark 3.5.3 The following facts are elementary to prove:

- If F is upper semicontinuous and if $F(x)$ is a singleton, then each *selection* of F (namely each function f such that $f(x) \in F(x), \forall x$) is continuous at x.
- Suppose $F(x)$ is a singleton for all x. Then if F is either upper semicontinuous or lower semicontinuous at a point, then it is continuous at that point, if it is considered as a *function*.

Exercise 3.5.4 Let X be a topological space and $f\colon X \to \mathbb{R}$ be a given function. Define the multifunction F on X as

$$F(x) = \{r \in \mathbb{R} : r \geq f(x)\},$$

i.e., the graph of F is the epigraph of f. Then F is upper semicontinuous at x if and only if f is lower semicontinuous at x.

The easy example of $f(x) = |x|$ shows that we cannot expect, in general, that ∂f be a lower semicontinuous multifunction. Instead, it enjoys upper semicontinuity properties, as we shall see in a moment.

Proposition 3.5.5 *Let $f \in \Gamma(X)$ be continuous and Gâteaux differentiable at x. Then the multifunction ∂f is norm-weak* upper semicontinuous at x.*

Proof. Let V be a weak* open set such that $V \supset \nabla f(x)$ and suppose there are a sequence $\{x_n\}$ converging to x and $x_n^* \in \partial f(x_n)$ such that $x_n^* \notin V$. As $\{x_n^*\}$ is bounded (see Proposition 3.5.1), it has a weak* limit x^* (it should be noticed that x^* is not necessarily limit of a subsequence). Now it is easy to show that $x^* \in \partial f(x) \subset V$, which is impossible.   □

Proposition 3.5.6 *Let $f \in \Gamma(X)$ be Fréchet differentiable at x. Then the multifunction ∂f is norm-norm upper semicontinuous at x.*

Proof. Setting

$$g(\cdot) = f(\cdot + x) - f(x) - \langle f'(x), \cdot - x \rangle,$$

we have that $\partial g(\cdot) = \partial f(\cdot + x) - f'(x)$. Clearly, ∂g enjoys the same continuity properties at zero as ∂f at x. Thus we can suppose, without loss of generality, that $x = 0$, $f(x) = 0$, $f'(x) = 0^*$. By way of contradiction, suppose there are $\varepsilon > 0$, $\{x_n\}$ converging to 0, $x_n^* \in \partial f(x_n)$ for all n, such that $\{x_n^*\}$ is bounded and $\|x_n^*\| > 3\varepsilon$. Then there are $d_n \in X$ such that $\|d_n\| = 1$ and

$$\langle x_n^*, d_n \rangle > 3\varepsilon.$$

By definition of Fréchet differentiability, there is $\delta > 0$ such that

$$|f(x)| \leq \varepsilon \|x\|,$$

for all x such that $\|x\| \leq \delta$. As $x_n^* \in \partial f(x_n)$, then

$$\langle x_n^*, x \rangle \leq f(x) - f(x_n) + \langle x_n^*, x_n \rangle, \quad \forall x \in X.$$

Set $y_n = \delta d_n$, with n so large that $|f(y_n)| < \varepsilon\delta$, $|\langle x_n^*, x_n \rangle| < \varepsilon\delta$. Then

$$3\varepsilon\delta < \langle x_n^*, y_n \rangle \leq f(y_n) - f(x_n) + \langle x_n^*, x_n \rangle \leq \varepsilon\delta + \varepsilon\delta + \varepsilon\delta,$$

a contradiction.   □

Corollary 3.5.7 *Let $f \colon \mathbb{R}^n \to \mathbb{R}$ be convex. Then Gâteaux and Fréchet differentiability agree at every point.*

Proof. From Propositions 3.5.5 and 3.5.6.   □

The next corollary shows a remarkable regularity property of the convex functions.

Corollary 3.5.8 *Let $f \in \Gamma(X)$ be Fréchet differentiable on an open convex set C. Then $f \in C^1(C)$.*

Proof. The *function* $f'(\cdot)$ is norm-norm continuous on C, being norm-norm upper semicontinuous as a multifunction.   □

Corollary 3.5.9 *Let $f \colon \mathbb{R}^n \to \mathbb{R}$ be convex and Gâteaux differentiable. Then $f \in C^1(\mathbb{R}^n)$.*

Proof. From Corollaries 3.5.7 and 3.5.8.   □

Proposition 3.5.10 *Let $f \in \Gamma(X)$ be continuous at $x \in X$. If there exists a selection h of ∂f norm-weak* continuous (norm-norm continuous) at x, then f is Gâteaux (Fréchet) differentiable at x.*

Proof. Let us start with Gâteaux differentiability. For every $y \in Y$,

$$\langle h(x), y - x \rangle \le f(y) - f(x), \quad \langle h(y), x - y \rangle \le f(x) - f(y),$$

from which

$$0 \le f(y) - f(x) - \langle h(x), y - x \rangle \le \langle h(y) - h(x), y - x \rangle. \tag{3.8}$$

Setting $y = x + tz$, for small $t > 0$, and dividing by t, we get

$$0 \le \frac{f(x + tz) - f(x)}{t} - \langle h(x), z \rangle \le \langle h(x + tz) - h(x), z \rangle.$$

Letting $t \to 0^+$, and using the fact that h is norm-weak* continuous,

$$0 \le f'(x; z) - \langle h(x), z \rangle \le 0.$$

From (3.8) we also deduce

$$0 \le f(y) - f(x) - \langle h(x), y - x \rangle \le \|h(x) - h(y)\| \|x - y\|,$$

whence f is Fréchet differentiable provided h is norm-norm continuous.   □

The next result extends to the subdifferential a well-known property of differentiable convex functions.

Definition 3.5.11 An operator $F \colon X \to X^*$ is said to be *monotone* if $\forall x, y \in X, \forall x^* \in F(x), \forall y^* \in F(y)$,

$$\langle x^* - y^*, x - y \rangle \ge 0.$$

Proposition 3.5.12 *Let $f \in \Gamma(X)$. Then ∂f is a monotone operator.*

Proof. From

$$\langle x^*, y - x \rangle \leq f(y) - f(x), \quad \langle y^*, x - y \rangle \leq f(x) - f(y),$$

we get the result by addition.   □

Proposition 3.5.12 can be refined in an interesting way.

Definition 3.5.13 A monotone operator $F \colon X \to X^*$ is said to be *maximal monotone* if $\forall y \in X$, $\forall y^* \notin F(y)$ there are $x \in X$, $x^* \in F(x)$ such that

$$\langle y^* - x^*, y - x \rangle < 0.$$

In other words, the graph of F is maximal in the class of the graph of monotone operators. We see now that the subdifferential is a maximal monotone operator.

Theorem 3.5.14 *Let $f \colon X \to \mathbb{R}$ be continuous and convex. Then ∂f is a maximal monotone operator.*

Proof. The geometric property of being maximal monotone does not change if we make a rotation and a translation of the graph of ∂f in $X \times X^*$. Thus we can suppose that $0 \notin \partial f(0)$ and we must find $x, x^* \in \partial f(x)$ such that $\langle x^*, x \rangle < 0$. As 0 is not a minimum point for f, there is $z \in X$ such that $f(0) > f(z)$. This implies that there exists $\bar{t} \in (0, 1]$ such that the directional derivative $f'(\bar{t}z; z) < 0$. Setting $x = \bar{t}z$, then $f'(x; x) < 0$. As $\partial f(x) \neq \emptyset$, if $x^* \in \partial f(x)$, then by Proposition 3.2.12 we get $\langle x^*, x \rangle < 0$.   □

The above result holds for every function f in $\Gamma(X)$, but the proof in the general case is much more delicate. The idea of the proof is the same, but the nontrivial point, unless f is real valued, is to find, referring to the above proof, z and $\bar{t}$ such that $f'(\bar{t}z; z) < 0$. One way to prove it relies on a variational principle, as we shall see later (see Proposition 4.2.14).

3.6 Twice differentiable functions

In the previous section we have considered the subdifferential multifunction ∂f, and its continuity properties, relating them to some regularity of the convex function f. In this section, we define an additional regularity requirement for a multifunction, when when applied to the subdifferential of f, provides "second order regularity" for the function f. Let us start with two definitions.

Definition 3.6.1 Let X be a Banach space and $f \in \Gamma(X)$. Suppose $\bar{x} \in$ int dom f. The subdifferential ∂f is said to be *Lipschitz stable* at $\bar{x}$ if $\partial f(\bar{x}) = \{\bar{p}\}$ and there are $\varepsilon > 0$, $K > 0$ such that

$$\|p - \bar{p}\| \leq K \|x - \bar{x}\|,$$

provided $\|x - \bar{x}\| < \varepsilon, p \in \partial f(x)$.

Definition 3.6.2 Let X be a Banach space and $f \in \Gamma(X)$. Suppose $\bar{x} \in$ int dom f. We say that ∂f is *Fréchet differentiable* at $\bar{x}$ if $\partial f(\bar{x}) = \{\bar{p}\}$ and there is a linear operator $T \colon X \to X^*$ such that

$$\lim_{x \to \bar{x}} \frac{\|p - \bar{p} - T(x - \bar{x})\|}{\|x - \bar{x}\|} = 0, \tag{3.9}$$

provided $p \in \partial f(x)$.

Definition 3.6.3 Let X be a Banach space and $f \in \Gamma(X)$. Suppose $\bar{x} \in$ int dom f. We say that f is *twice Fréchet differentiable* at $\bar{x}$ if $\partial f(\bar{x}) = \bar{p}$ and there is a quadratic form $Q(x) := \langle Ax, x \rangle$ ($A \colon X \to X^*$ linear bounded operator) such that

$$\lim_{x \to \bar{x}} \frac{f(x) - \langle \bar{p}, x - \bar{x} \rangle - (1/2)Q(x - \bar{x})}{\|x - \bar{x}\|^2} = 0. \tag{3.10}$$

The following lemma shows that if two convex functions are close on a given bounded set and one of them is convex and the other is regular, the subdifferential of the convex function can be controlled (in a smaller set) by the derivative of the regular one, another nice property of convex functions.

Lemma 3.6.4 *Let $f \colon X \to (-\infty, \infty]$ be convex. Let $\delta, a > 0$, let $g \colon B(0; a) \to \mathbb{R}$ be a Fréchet differentiable function and suppose $|f(x) - g(x)| \le \delta$ for $x \in B(0; a)$. Let $0 < r < R \le a$, let x be such that $\|x\| \le r$ and $x^* \in \partial f(x)$. Then*

$$d\big(x^*, \mathrm{co}\{g'(B(x; R - r))\}\big) \le \frac{2\delta}{R - r}.$$

If g is convex, we also have

$$d\big(x^*, \mathrm{co}\{\partial g(B(0; R))\}\big) \le \frac{2\delta}{R - r}.$$

Proof. Without loss of generality we can suppose $x^* = 0$. Let α be such that $\alpha < \|y^*\|$ for all $y^* \in \mathrm{co}\{g'(B(x; R - r))\}$. Then there exists d, with $\|d\| = 1$, such that $\langle -y^*, d \rangle > \alpha$ for all $y^* \in \mathrm{co}\{g'(B(x, R - r))\}$. We have

$$\delta \ge f(x + (R - r)d) - g(x + (R - r)d)$$
$$\ge f(x) - g(x) - (g(x + (R - r)d) - g(x)).$$

There is an $s \in (0, R - r)$ such that

$$\langle (R - r)g'(x + sd), d \rangle = g(x + (R - r)d) - g(x).$$

Thus

$$2\delta \ge (R - r)\langle -g'(x + sd), d \rangle \ge (R - r)\alpha.$$

Then

$$\alpha \leq \frac{2\delta}{R-r},$$

and this ends the proof of the first claim. About the second one, let d be such that $\|d\| = 1$ and $\langle -y^*, d \rangle > \alpha$ for all $y^* \in \mathrm{co}\{\partial g(B(0;R))\}$. Let $z^* \in \partial g(x + (R-r)d)$. Then

$$2\delta \geq g(x) - g(x + (R-r)d) \geq (R-r)\langle -z^*, d \rangle \geq (R-r)\alpha,$$

and we conclude as before.   □

Remark 3.6.5 The above result can be refined in a sharp way by using the Ekeland variational principle, as we shall see in Lemma 4.2.18.

We are ready for our first result, which appears to be very natural, since it states that the variation of the function minus its linear approximation is of quadratic order if and only if the variation of its subdifferential is of first order (thus extending in a natural way well-known properties of smooth functions).

Proposition 3.6.6 Let $\{\bar{p}\} = \partial f(\bar{x})$. Then the following two statements are equivalent:

(i) ∂f is Lipschitz stable at $\bar{x}$;
(ii)   There are $k > 0$ and a neighborhood $W \ni \bar{x}$ such that

$$|f(x) - f(\bar{x}) - \langle \bar{p}, x \rangle| \leq k(\|x - \bar{x}\|)^2,$$

for all $x \in W$.

Proof. First, let us observe that we can suppose, without loss of generality,

$$\bar{x} = 0, \quad f(\bar{x}) = 0, \quad \bar{p} = 0,$$

by possibly considering the function

$$\hat{f}(x) = f(x + \bar{x}) - f(\bar{x}) - \langle \bar{p}, x \rangle.$$

In this case observe that $h(x) \geq 0$, $\forall x \in X$. Let us prove that (i) implies (ii). Let $H, K > 0$ be such that, if $\|x\| \leq H$, $p \in \partial f(x)$, then

$$\|p\| \leq K\|x\|.$$

Since

$$0 = f(0) \geq f(x) + \langle p, -x \rangle,$$

we have

$$f(x) \leq \|p\|\|x\| \leq K\|x\|^2.$$

We now prove that (ii) implies (i). Suppose there are $a, K > 0$ such that

$$|f(x)| \le K\|x\|^2,$$

if $\|x\| \le a$. Now take x with $r := \|x\| \le (a/2)$. We have then

$$|f(x)| \le Kr^2.$$

We now apply Lemma 3.6.4 to f and to the zero function, with a, r as above, $R = 2r$ and $\delta = Kr^2$. We then get

$$\|p\| \le 2Kr = 2K\|x\|,$$

provided $\|x\| \le (a/2)$.   □

The following result connects Fréchet differentiability of ∂f with twice Fréchet differentiability of f. This result too is quite natural.

Proposition 3.6.7 *Let* $\bar{p} \in \partial f(\bar{x})$. *Then the following two statements are equivalent:*

(i)   *∂f is Fréchet differentiable at $\bar{x}$;*
(ii) *f is twice Fréchet differentiable at $\bar{x}$.*

Proof. As in the previous proposition, we can suppose

$$\bar{x} = 0, \quad f(\bar{x}) = 0, \quad \bar{p} = 0.$$

Let us show that (i) implies (ii). Assume there is an operator T as in (3.9), and let Q be the quadratic function associated to it: $Q(u) = \frac{1}{2}\langle Tu, u\rangle$. Setting $h(s) = f(sx)$ we have that

$$f(x)(-f(0) = 0) = h(1) - h(0) = \int_0^1 h'(s)\,ds = \int_0^1 f'(sx; x)\,ds.$$

Now, remembering that $f'(sx; x) = \sup_{p \in \partial f(sx)} \langle p, x \rangle$ (see Theorem 3.2.14), we then have

$$f(x) - \frac{1}{2}Q(x) = \int_0^1 \left[\sup_{p \in \partial f(sx)} \langle p, x \rangle - s\langle Tx, x \rangle \right] ds,$$

from which we get

$$\left| f(x) - \frac{1}{2}Q(x) \right| \le \int_0^1 \sup_{p \in \partial f(sx)} |\langle p - Tsx, x \rangle|\, ds;$$

from this, remembering (3.9), we easily get (3.10). The proof that (ii) implies (i) relies again on Lemma 3.6.4. There is a quadratic function Q of the form $Q(x) = \langle Tx, x \rangle$, such that there are $a, \varepsilon > 0$ with

$$\left| f(x) - \frac{1}{2}Q(x) \right| \le \varepsilon\|x\|^2,$$

if $\|x\| \leq a$. Now take x such that $r := \|x\| \leq \frac{a}{2}$. We have then

$$|f(x) - \frac{1}{2}Q(x)| \leq \varepsilon r^2.$$

We apply Lemma 3.6.4 to f and to the function $\frac{1}{2}Q$, with a, r as above, $R = r(1 + \sqrt{\varepsilon})$ and $\delta = \varepsilon r^2$. We then get

$$d(q, \mathrm{co}\{T(B(x, \sqrt{\varepsilon}r))\}) \leq \frac{2\varepsilon r^2}{\sqrt{\varepsilon}r},$$

provided $\|x\| \leq \frac{a}{2}$. But then

$$\|p - Tx\| \leq 2\sqrt{\varepsilon}\|x\| + \|T\|\sqrt{\varepsilon}\|x\|,$$

and from this we easily get (3.10). $\square$

3.7 The approximate subdifferential

There are both theoretical and practical reasons to define the concept of approximate subdifferential. On the one hand, the (exact) subdifferential does not exist at each point of dom f. On the other hand, it is also difficult to evaluate. To partly overcome these difficulties the notion of approximate subdifferential is introduced.

Definition 3.7.1 Let $\varepsilon \geq 0$ and $f \colon X \to (-\infty, \infty]$. Then $x^* \in X^*$ is said to be an ε-*subgradient* of f at x_0 if

$$f(x) \geq f(x_0) + \langle x^*, x - x_0 \rangle - \varepsilon.$$

The ε-*subdifferential* of f at x, denoted by $\partial_\varepsilon f(x)$, is the set of the ε-subgradients of f at x.

Clearly, the case $\varepsilon = 0$ recovers the definition of the (exact) subdifferential. Moreover,

$$\partial f(x) = \bigcap_{\varepsilon > 0} \partial_\varepsilon f(x).$$

Here is a first result.

Theorem 3.7.2 Let $f \in \Gamma(X)$, $x \in \text{dom } f$. Then $\emptyset \neq \partial_\varepsilon f(x)$ is a weak* closed and convex set, $\forall \varepsilon > 0$. Furthermore,

$$\partial_{\lambda\alpha + (1-\lambda)\beta} f(x) \supset \lambda \partial_\alpha f(x) + (1 - \lambda)\partial_\beta f(x),$$

for every $\alpha, \beta > 0$, for every $\lambda \in [0, 1]$.

Proof. To prove that $\partial_\varepsilon f(x) \neq \emptyset$, one exploits the usual separation argument of Lemma 2.2.16, by separating $(x, f(x) - \varepsilon)$ from epi f; proving the other claims is straightforward. $\square$

We provide two examples.

Example 3.7.3

$$f(x) = \begin{cases} -2\sqrt{x} & \text{if } x \geq 0, \\ \infty & \text{otherwise.} \end{cases}$$

It is not hard to see that for $\varepsilon > 0$, the ε-subdifferential of f at the origin is the half line $(-\infty, -\frac{1}{\varepsilon}]$, an unbounded set (not surprising, see Remark 3.2.9). On the other hand, the subdifferential of f at the origin is empty.

Example 3.7.4 Let $f(x) = |x|$. Then

$$\partial_\varepsilon f(x) = \begin{cases} [-1, -1 - \frac{\varepsilon}{x}] & \text{if } x < -\frac{\varepsilon}{2}, \\ [-1, 1] & \text{if } -\frac{\varepsilon}{2} \leq x \leq \frac{\varepsilon}{2}, \\ [1 - \frac{\varepsilon}{x}, 1] & \text{if } x > \frac{\varepsilon}{2}. \end{cases}$$

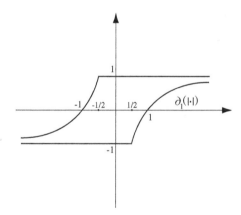

Figure 3.5. The approximate subdifferential $\partial_1(|\cdot|)(0)$.

The following result is easy and provides useful information.

Theorem 3.7.5 *Let* $f \in \Gamma(X)$. *Then* $0^* \in \partial_\varepsilon f(x_0)$ *if and only if*

$$\inf f \geq f(x_0) - \varepsilon.$$

Thus, whenever an algorithm is used to minimize a convex function, if we look for an ε-solution, it is enough that $0 \in \partial_\varepsilon f(x)$, a much weaker condition than $0 \in \partial f(x)$.

We now see an important connection between the ε-subdifferential and the directional derivatives (compare the result with Theorem 3.2.14).

Proposition 3.7.6 *Let $f \in \Gamma(X)$, $x \in \operatorname{dom} f$. Then, $\forall d \in X$,*

$$f'(x; d) = \lim_{\varepsilon \to 0^+} \sup\{\langle x^*, d \rangle : x^* \in \partial_\varepsilon f(x)\}.$$

Proof. Observe at first that, for monotonicity reasons, the limit in the above formula always exists. Now, let $\varepsilon > 0$ and $d \in X$; then, $\forall t > 0$, $\forall x^* \in \partial_\varepsilon f(x)$,

$$\frac{f(x + td) - f(x) + \varepsilon}{t} \geq \langle x^*, d \rangle.$$

Setting $t = \sqrt{\varepsilon}$, we get

$$\frac{f(x + \sqrt{\varepsilon} d) - f(x) + \varepsilon}{\sqrt{\varepsilon}} \geq \sup\{\langle x^*, d \rangle : x^* \in \partial_\varepsilon f(x)\}.$$

Taking the limit in the formula above,

$$f'(x; d) \geq \lim_{\varepsilon \to 0^+} \sup\{\langle x^*, d \rangle : x^* \in \partial_\varepsilon f(x)\},$$

which shows one inequality. To get the opposite one, it is useful to appeal again to a separation argument. Let $\alpha < f'(x; d)$ and observe that for $0 \leq t \leq 1$,

$$f(x + td) \geq f(x) + t\alpha.$$

Consider the line segment

$$S = \{(x, f(x) - \varepsilon) + t(d, \alpha) : 0 \leq t \leq 1\}.$$

S is a compact convex set disjoint from epi f. Thus there are $y^* \in X^*$, $r \in \mathbb{R}$ such that

$$\langle y^*, y \rangle + r f(y) > \langle y^*, x + td \rangle + r(f(x) - \varepsilon + t\alpha),$$

$\forall y \in \operatorname{dom} f$, $\forall t \in [0, 1]$. As usual, $r > 0$. Dividing by r and setting $x^* = -\frac{y^*}{r}$, we get

$$\langle x^*, d \rangle \geq \alpha - \varepsilon,$$

(with the choice of $y = x, t = 1$), and if $v \in X$ is such that $x + v \in \operatorname{dom} f$, setting $y = x + v$ and $t = 0$,

$$f(x + v) - f(x) + \varepsilon \geq \langle x^*, v \rangle,$$

which means $x^* \in \partial_\varepsilon f(x)$. The last two facts provide

$$\sup\{\langle x^*, d \rangle : x^* \in \partial_\varepsilon f(x)\} \geq \alpha - \varepsilon,$$

and this ends the proof.   □

We state, without proof, a result on the sum of approximate subdifferentials. To get an equality in the stated formula, one needs to add conditions as, for instance, int dom $f \cap$ int dom $g \neq \emptyset$.

Proposition 3.7.7 *Let $\varepsilon \geq 0$ and $x \in \operatorname{dom} f \cap \operatorname{dom} g$. Then*

$$\partial_\varepsilon (f + g)(x) \supset \cup \{\partial_\sigma f(x) + \partial_\delta g(x) : 0 \leq \sigma, 0 \leq \delta, \sigma + \delta \leq \varepsilon\}.$$

4

Minima and quasi minima

> *Rationality of thought imposes*
> *a limit on a person's concept*
> *of his relation to the cosmos.*
> (J. F. Nash, Autobiography)

Convexity plays a key role in minimization. First of all, a local minimum is automatically a global one. Secondly, for convex functions, the classical Fermat necessary condition for a local extremum becomes sufficient to characterize a global minimum.

In this chapter we deal with the problem of existence of a minimum point, and thus we quite naturally begin with stating and commenting on the Weierstrass existence theorem. We also show that in reflexive (infinite dimensional) Banach spaces convexity is a very important property for establishing existence of a global minimum under reasonable assumptions. There are however several situations, for example outside reflexivity, where to have a general existence theorem for a wide class of functions is practically impossible. Thus it is important to know that at least for "many" functions in a prescribed class, an existence theorem can be provided. A fundamental tool for getting this type of result is the Ekeland variational principle, probably one of the most famous results in modern nonlinear analysis. So, in this chapter we spend some time in analyzing this variational principle, and deriving some of its interesting consequences, mainly in the convex setting.

The problem we were alluding to of identifying classes of functions for which "most" of the problems have solutions will be discussed in detail in Chapter 11. The chapter ends with the description of some properties of the level sets of a convex function, and with a taste of the algorithms that can be used in order to find the minima of a convex function, in a finite dimensional setting.

4.1 The Weierstrass theorem

The next result is the fundamental Weierstrass theorem.

Theorem 4.1.1 *Let (X, τ) be a topological space, and assume $f: (X, \tau) \to (-\infty, \infty]$ is τ-lower semicontinuous. Suppose moreover there is $\bar{a} > \inf f$ such that $f^{\bar{a}}$ is τ-compact. Then f has absolute minima: $\operatorname{Min} f := \{\bar{x} : f(\bar{x}) \leq f(x), \forall x \in X\}$ is a nonempty set.*

Proof.

$$\operatorname{Min} f = \bigcap_{\bar{a} > a > \inf f} f^a.$$

Each f^a is nonempty and τ-closed (due to τ-lower semicontinuity of f); hence

$$\{f^a : \bar{a} > a > \inf f\}$$

is a family of nonempty, nested, τ-compact sets, and this entails nonemptiness of their intersection.   □

The previous theorem is surely a milestone in optimization. Thus, when we face an optimization problem, the challenge is to see if there is a topology τ on the set X in order to fulfill its assumptions. Observe that the two requested conditions, τ-lower semicontinuity of f, and having a τ-compact level set, go in opposite directions. Given a function f on X, in order to have f τ-lower semicontinuous we need many closed sets on X (i.e., the finer the topology τ with which we endow X, the better the situation), but to have a compact level set we need a topology rich in compact sets, which is the same as saying poor in open (and so, closed) sets. For instance, think of a continuous function (in the norm topology) defined on an infinite-dimensional Hilbert space. Clearly, each level set of f is a closed set. But also, no level set (at height greater than $\inf f$) is compact! To see this, observe that each f^a must contain a ball around a point x fulfilling $f(x) < a$. As is well known, compact sets in infinite-dimensional spaces do have empty interiors. Thus Weierstrass' theorem can *never* be applied in this setting, with the norm topology. Fortunately, we have other choices for the topology on the space. On the Banach space X, let us consider the weak topology. This is defined as the weakest topology making continuous all the elements of X^*, the continuous dual space of X. By the very definition, this topology is coarser than the norm topology, and strictly coarser in infinite dimensions, as it is not difficult to show. This implies that the weak topology will provide us more compact sets, but fewer closed sets. Thus, the following result is very useful.

Proposition 4.1.2 *Let X be a Banach space, and let $F \subset X$ be a norm closed and convex set. Then F is weakly closed.*

Proof. To prove the claim, we show that F^c, the complement of F, is weakly open. Remember that a subbasic family of open sets for the weak topology is given by

$$\{x \in X : \langle x^*, x \rangle < a,\ x^* \in X^*, a \in \mathbb{R}\}.$$

So, let $x \in F^c$. Being F closed and convex, we can strictly separate F from x (Theorem A.1.6): there are $x^* \in X^*$ and $a \in \mathbb{R}$ such that

$$F \subset \{x \in X : \langle x^*, x \rangle > a\} \text{ and } \langle x^*, x \rangle < a.$$

Thus the open set $\{x \in X : \langle x^*, x \rangle < a\}$ contains x and does not intersect F.
□

As a consequence of the previous results we can prove, for instance, the following theorem (some simple variant of it can be formulated as well):

Theorem 4.1.3 *Let X be a reflexive Banach space, let $f \in \Gamma(X)$. Suppose* $\lim_{\|x\| \to \infty} f(x) = \infty$. *Then the problem of minimizing f over X has solutions.*

Proof. As a consequence of the Banach–Alaoglu theorem, reflexivity guarantees that a weakly closed and bounded set is weakly compact.
□

Exercise 4.1.4 Let us take a nonempty closed convex set C in a Banach space X, and $x \in X$. The *projection* of x over C is the (possibly empty) set $p_C(x)$ of the points of C which are *nearest* to x:

$$p_C(x) = \{z \in C : \|z - x\| \le \|c - x\|, \forall c \in C\}.$$

Prove that $p_C(x) \ne \emptyset$, provided X is reflexive, and that it is a singleton if X is a Hilbert space. In this case, prove also that $y = P_C(x)$ if and only if $y \in C$ and

$$\langle x - y, c - y \rangle \le 0, \forall c \in C.$$

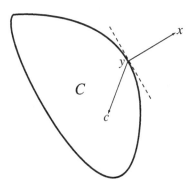

Figure 4.1. The projection y of x on the set C.

The concept of projection allows us to get a formula for the subdifferential of the distance function $d(\,\cdot\,,C)$, where C is a closed convex subset of a Hilbert space X.

Proposition 4.1.5 *Let X be a Hilbert space, C a nonempty closed convex subset of X, $x \in X$. Then*

$$\partial d(\,\cdot\,,C)(x) = \begin{cases} 0^* & \text{if } x \in \text{int } C, \\ N_C(x) \cap B^* & \text{if } x \in \partial C, \\ \dfrac{x - P_C(x)}{\|x - P_C(x)\|} & \text{if } x \notin C, \end{cases}$$

where, as usual, $N_C(x)$ is the normal cone at x to C and $P_C(x)$ is the projection of x over C.

Proof. To prove the claim, we appeal to the fact that

$$d(x,C) = (\|\,\cdot\,\| \,\nabla I_C)(x),$$

that the inf-convolution is exact at any point, and to Proposition 3.2.11, which provides a formula for the subdifferential of the inf-convolution at a point where it is exact. Let $x \in \text{int } C$. Setting $u = 0$, $v = x$, we have that $d(x,C) = \|u\| + I_C(v)$, $\partial\|u\| = B_{X^*}$, $\partial I_C(v) = \{0^*\}$, $\partial d(\,\cdot\,,C)(x) = \partial\|u\| \cap \partial I_C(v) = \{0^*\}$. Now, let us suppose x is in the boundary of C: $x \in \partial C$. Again take $u = 0$, $v = x$. This provides $\partial\|u\| = B_{X^*}$, $\partial I_C(v) = N_C(x)$, and thus $\partial d(\,\cdot\,,C)(x) = \partial\|u\| \cap \partial I_C(v) = B^* \cap N_C(x)$. Finally, let $x \notin C$. Then $d(x,C) = \|x - P_C(x)\| + I_C(p_C(x))$, $\partial\|x - P_C(x)\| = \frac{x - P_C(x)}{\|x - P_C(x)\|}$, $\partial I_C(P_C(x)) = N_C(P_C(x))$. But $\frac{x - P_C(x)}{\|x - P_C(x)\|} \in N_C(P_C(x))$, as it is seen in the Exercise 4.1.4, and this ends the proof. $\qquad\square$

Exercise 4.1.6 Let X be a reflexive Banach space and let $f: X \to (-\infty, \infty]$ be a lower semicontinuous, lower bounded function. Let $\varepsilon > 0$, $r > 0$ and $\bar{x} \in X$ be such that $f(\bar{x}) \leq \inf_X f + r\varepsilon$. Then, there exists $\hat{x} \in X$ enjoying the following properties:

(i) $\|\hat{x} - \bar{x}\| \leq r$;
(ii) $f(\hat{x}) \leq f(\bar{x})$;
(iii) $f(\hat{x}) \leq f(x) + \varepsilon\|\hat{x} - x\| \quad \forall x \in X$.

Hint. The function $g(x) = f(x) + \varepsilon\|\bar{x} - x\|$ has a minimum point $\hat{x}$. Check that $\hat{x}$ fulfills the required properties.

The following section is dedicated to extending the previous result to complete metric spaces.

4.2 The Ekeland variational principle

Due to the lack of a suitable topology to exploit the basic Weierstrass existence theorem, it is quite difficult, except for the reflexive case, to produce general existence results for minimum problems. So it is important to produce results guaranteeing existence at least in "many" cases. The word "many" of course can be given different meanings. The Ekeland variational principle, the fundamental result we describe in this section, allows us to produce a generic existence theorem. But its power goes far beyond this fact; its claim for the existence of a quasi minimum point with particular features has surprisingly many applications, not only in optimization, but also, for instance, in critical point and fixed point theory. Let us start by introducing a useful definition.

Definition 4.2.1 Let (X, d) be a metric space, let $f \colon X \to \mathbb{R}$ be lower semicontinuous. The *strong slope* of f at x, denoted by $|\nabla f|(x)$ is defined as

$$|\nabla f|(x) = \begin{cases} \limsup\limits_{y \to x} \frac{f(x) - f(y)}{d(x, y)} & \text{if } x \text{ is not a local minimum,} \\ 0 & \text{if } x \text{ is a local minimum.} \end{cases}$$

The next is an estimation from above of the strong slope.

Proposition 4.2.2 *Let X be a metric space, let $f \colon X \to \mathbb{R}$ be locally Lipschitz at $x \in X$, with Lipschitz constant L. Then $|\nabla f|(x) \leq L$.*

For a more regular function f we have:

Proposition 4.2.3 *Let X be a Banach space, let $f \colon X \to \mathbb{R}$ be Gâteaux differentiable at $x \in X$. Then $|\nabla f|(x) \geq \|\nabla f(x)\|_*$.*

Proof. Let $u \in X$ be such that $\|u\| = 1$ and $\langle \nabla f(x), -u \rangle \geq \|\nabla f(x)\|_* - \varepsilon$, for some small $\varepsilon > 0$. Then

$$\limsup\limits_{y \to x} \frac{f(x) - f(y)}{d(x, y)} \geq \lim\limits_{t \to 0} \frac{f(x) - f(x + tu)}{t} = \langle \nabla f(x), -u \rangle \geq \|\nabla f(x)\|_* - \varepsilon.$$

This allows us to complete the proof. $\square$

Clearly, every function f which is discontinuous at a point x but Gâteaux differentiable at the same point, provides an example when the inequality in the above proposition is strict. But with a bit more regularity we get

Proposition 4.2.4 *Let X be a Banach space, let $f \colon X \to \mathbb{R}$ be Fréchet differentiable at $x \in X$. Then $|\nabla f|(x) = \|f'(x)\|_*$.*

Proof. Write, for $y \neq x$,

$$f(y) = f(x) + \langle f'(x), y - x \rangle + \varepsilon_y \|y - x\|,$$

where $\varepsilon_y \to 0$ if $y \to x$. Then we get

$$\frac{f(x) - f(y)}{\|y - x\|} = \langle -f'(x), \frac{y - x}{\|y - x\|} \rangle + \varepsilon_y \le \|f'(x)\|_* + \varepsilon_y.$$

This shows that $|\nabla f|(x) \le \|f'(x)\|$ and, by means of Proposition 4.2.3, we can conclude. □

Propositions 4.2.3 and 4.2.4 explain the importance of the notion of strong slope (and also the notation used). In particular, for a Fréchet differentiable function, it generalizes the notion of norm of the derivative, to a purely metric setting. Beyond this, it has also interesting connections with nonsmooth differentials of nonconvex functions.

We can now introduce the variational principle.

Theorem 4.2.5 *Let (X, d) be a complete metric space and let $f \colon X \to (-\infty, \infty]$ be a lower semicontinuous, lower bounded function. Let $\varepsilon > 0$, $r > 0$ and $\bar{x} \in X$ be such that $f(\bar{x}) \le \inf_X f + r\varepsilon$. Then, there exists $\hat{x} \in X$ enjoying the following properties:*

(i) $d(\hat{x}, \bar{x}) \le r$;
(ii) $f(\hat{x}) \le f(\bar{x}) - \varepsilon d(\bar{x}, \hat{x})$;
(iii) $f(\hat{x}) < f(x) + \varepsilon d(\hat{x}, x) \ \forall x \ne \hat{x}$.

Proof. Let us define the following relation on $X \times X$:

$$x \preceq y \text{ if } f(x) \le f(y) - \varepsilon d(x, y).$$

It is routine to verify that $\preceq$ is reflexive, antisymmetric and transitive. Moreover, lower semicontinuity of f guarantees that $\forall x_0 \in X$, the set $A := \{x \in X : x \preceq x_0\}$ is a closed set. Let us now define

$$x_1 = \bar{x}, \quad S_1 = \{x \in X : x \preceq x_1\},$$
$$x_2 \in S_1 \text{ such that } f(x_2) \le \inf_{S_1} f + \frac{r\varepsilon}{4};$$

and recursively

$$S_n = \{x \in X : x \preceq x_n\},$$
$$x_{n+1} \in S_n \text{ such that } f(x_{n+1}) \le \inf_{S_n} f + \frac{r\varepsilon}{2(n+1)}.$$

For all $n \ge 1$, S_n is a nonempty closed set, and $S_n \supset S_{n+1}$. Let us now evaluate the size of the sets S_n. Let $x \in S_n$, for $n > 1$. Then $x \preceq x_n$ and $x \in S_{n-1}$, hence

$$f(x) \le f(x_n) - \varepsilon d(x, x_n),$$
$$f(x_n) \le f(x) + \frac{r\varepsilon}{2n},$$

giving

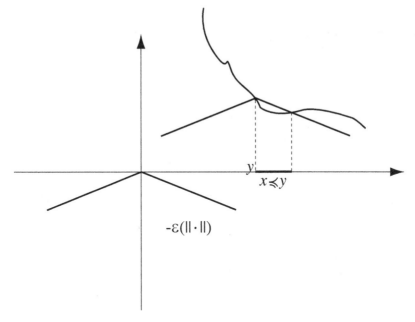

Figure 4.2. The $\preceq$ relation.

$$d(x, x_n) \leq \frac{r}{2n}.$$

In the same way it can be shown that if $x \in S_1$, then $d(x, x_1) = d(x, \bar{x}) \leq r$. Since X is a complete metric space and the sequence of the diameters of the sets S_n goes to zero, it follows that $\bigcap_{n \geq 1} S_n$ is a singleton (see Exercise 4.2.6). Let $\bigcap_{n \geq 1} S_n := \{\hat{x}\}$. Now, it is a pleasure to show that $\hat{x}$ has the required properties. The first and the second one immediately follow from the fact that $\hat{x} \in S_1$, while, to verify the third one, if we suppose the existence of $x \in X$ such that $f(\hat{x}) \geq f(x) + \varepsilon d(x, \hat{x})$, then $x \preceq \hat{x} \preceq x_n, \forall n$, implying $x \in \bigcap_{n \geq 1} S_n$ and so $x = \hat{x}$.   □

Exercise 4.2.6 Let (X, d) be a complete metric space, let $\{S_n\}$ be a sequence of nested closed sets such that diam $S_n \to 0$. Prove that $\bigcap S_n$ is a singleton.

Hint. Take $x_n \in S_n$ for all n. Then $\{x_n\}$ is a Cauchy sequence. Thus $\bigcap S_n$ is nonempty. Moreover, it cannot contain more than one point, as diam $S_n \to 0$.

The third condition of the Ekeland principle has many interesting, and sometimes rather surprising, consequences. At first, it shows that the *approximate* solution $\hat{x}$ of the problem of minimizing f is, at the same time, also the *unique exact* solution of a minimum problem, close to the original one, in a sense we shall specify in Chapter 11. Moreover, this approximate solution enjoys an important property with respect to the strong slope, as we now see.

Corollary 4.2.7 *Let X be a complete metric space. Let $f\colon X \to (-\infty, \infty]$ be lower semicontinuous and lower bounded. Let $\varepsilon, r > 0$ and $\bar{x} \in X$ be such that $f(\bar{x}) < \inf_X f + \varepsilon r$. Then there exists $\hat{x} \in X$ with the following properties:*

(i) $d(\hat{x}, \bar{x}) < r$;

(ii) $f(\hat{x}) \leq f(\bar{x})$;

(iii) $|\nabla f|(\hat{x}) < \varepsilon$.

Proof. It is enough to apply the principle, with suitable $0 < \varepsilon_0 < \varepsilon$, $0 < r_0 < r$. The last condition implies $|\nabla f|(\hat{x}) \leq \varepsilon_0$, as is easy to see.   □

From the previous results we deduce:

Corollary 4.2.8 *Let X be a Banach space, let $f\colon X \to \mathbb{R}$ be lower semicontinuous, lower bounded and Gâteaux differentiable. Given $\varepsilon, r > 0$ and $\bar{x} \in X$ such that $f(\bar{x}) < \inf_X f + \varepsilon r$, there exists $\hat{x} \in X$ with the following properties:*

(i) $d(\hat{x}, \bar{x}) < r$;

(ii) $f(\hat{x}) \leq f(\bar{x})$;

(iii) $\|\nabla f(\hat{x})\|_* < \varepsilon$.

Proof. From Proposition 4.2.3 and Corollary 4.2.7.   □

Corollary 4.2.9 *Let X be a Banach space, let $f\colon X \to \mathbb{R}$ be lower semicontinuous, lower bounded and Gâteaux differentiable. Then there exists a sequence $\{x_n\} \subset X$ such that*

(i) $f(x_n) \to \inf f$;

(ii) $\nabla f(x_n) \to 0^*$.

Sequences $\{x_n\}$ such that $\nabla f(x_n) \to 0^*$ are known in the literature as *Palais–Smale sequences*, and at level a if it happens that $f(x_n) \to a$. A function f is said to satisfy the Palais–Smale condition (at level a) if every Palais–Smale sequence with bounded values (at level a) has a limit point. This is a compactness assumption crucial in every abstract existence theorem in critical point theory. And the notion of strong slope is the starting point for a purely metric critical point theory. The above corollary claims the existence of Palais–Smale sequences at level $\inf f$.

The Ekeland principle has interesting consequences for convex functions too.

Theorem 4.2.10 *Let X be a Banach space, let $f \in \Gamma(X)$. Let $x \in \operatorname{dom} f$, $\varepsilon, r, \sigma > 0$, $x^* \in \partial_{\varepsilon r} f(x)$. Then there are $\hat{x} \in \operatorname{dom} f$ and $\hat{x}^* \in X^*$, such that*

(i) $\hat{x}^* \in \partial f(\hat{x})$;

(ii) $\|x - \hat{x}\| \leq \frac{r}{\sigma}$;

(iii) $\|\hat{x}^* - x^*\|_* \leq \varepsilon \sigma$;

(iv) $|f(x) - f(\hat{x})| \leq r(\varepsilon + \frac{\|x^*\|_*}{\sigma})$.

Proof. As $x^* \in \partial_{\varepsilon r} f(x)$, it holds, $\forall y \in X$,

$$f(y) \geq f(x) + \langle x^*, y - x \rangle - \varepsilon r.$$

Setting $g(y) = f(y) - \langle x^*, y \rangle$, we get

$$g(x) \leq \inf_X g + (\varepsilon \sigma) \frac{r}{\sigma}.$$

Applying the principle to the function g (and replacing r by $\frac{r}{\sigma}$, ε by $\sigma \varepsilon$), we have then the existence of an element $\hat{x} \in \operatorname{dom} f$ satisfying condition (ii). Let us find the right element in its subdifferential. Condition (iii) of the principle says that $\hat{x}$ minimizes the function $g(\cdot) + \varepsilon \sigma \| \cdot - \hat{x} \|$, so that

$$0^* \in \partial(g(\cdot) + \varepsilon \sigma \| \cdot - \hat{x} \|)(\hat{x}).$$

We can use the sum Theorem 3.4.2. We then get

$$0^* \in \partial g(\hat{x}) + \varepsilon \sigma B_{X^*} = \partial f(\hat{x}) - x^* + \varepsilon \sigma B_{X^*}.$$

This is equivalent to saying that there exists an element $\hat{x}^* \in \partial f(\hat{x})$ such that $\|\hat{x}^* - x^*\|_* \leq \varepsilon \sigma$. Finally, condition (iv) routinely follows from (ii), (iii) and from $x^* \in \partial_{\varepsilon r} f(x)$, $\hat{x}^* \in \partial f(\hat{x})$.   □

The introduction of a constant σ in the above result is not made with the intention of creating more entropy. For instance, the choice of $\sigma = \max\{\|x^*\|_*, 1\}$ allows controlling the variation of the function f, at the expense, of course, of controlling of the norm of $\hat{x}^*$. Thus the following useful result can be easily proved.

Corollary 4.2.11 *Let X be a Banach space, let $f \in \Gamma(X)$. Let $x \in \operatorname{dom} f$. Then there is a sequence $\{x_n\} \subset \operatorname{dom} \partial f$ such that*

$$x_n \to x \text{ and } f(x_n) \to f(x).$$

Proof. This follows from (ii) and (iv) of Theorem 4.2.10, with the above choice of σ, $\varepsilon = 1$, and $r = \frac{1}{n}$.   □

Corollary 4.2.12 *Let X be a Banach space, let $f \in \Gamma(X)$ be lower bounded, let $\varepsilon, r > 0$ and $\bar{x} \in \operatorname{dom} f$ be such that $f(\bar{x}) < \inf f + \varepsilon r$. Then there exist $\hat{x} \in \operatorname{dom} f$ and $\hat{x}^* \in \partial f(\hat{x})$, such that*

(i) $\|\bar{x} - \hat{x}\| < r$;

(ii) $\|\hat{x}^*\|_* < \varepsilon$.

Proof. We apply Theorem 4.2.10 to the point $x = \bar{x}$, and with $\sigma = 1$. Observe that $0^* \in \partial_{\varepsilon_0 r_0} f(\bar{x})$, with suitable $\varepsilon_0 < \varepsilon$ and $r_0 < r$.   □

Another very interesting consequence of the previous theorem is the following fact.

Corollary 4.2.13 *Let X be a Banach space and $f \in \Gamma(X)$. Then there exists a dense subset D of $\operatorname{dom} f$ such that $\partial f(x) \neq \emptyset$ for all $x \in D$.*

Proof. Fix any $r > 0$ and $x \in \operatorname{dom} f$. Find x^* in $\partial_{r/2} f(x)$. Apply Theorem 4.2.10 to x, x^*, with the choice of $\varepsilon = 1/2, \sigma = 1$. We get $\hat{x}$ such that $\partial f(\hat{x}) \neq \emptyset$ and such that $\|x - \hat{x}\| < r$, and this finishes the proof. □

The following proposition, beyond being interesting in itself, is useful in proving that the subdifferential of a function in $\Gamma(X)$ is a maximal monotone operator. Remember that in Theorem 3.5.14 we have already shown this result for a narrower class of functions. To prove it, we follow an idea of S. Simmons (see [Si]).

Proposition 4.2.14 *Let X be a Banach space, let $f \in \Gamma(X)$, and suppose $f(0) > \inf f$. Then there are $z \in \operatorname{dom} f$, $z^* \in \partial f(z)$ with the following properties:*

(i) $f(z) < f(0)$;
(ii) $\langle z^*, z \rangle < 0$.

Proof. Observe at first that (i) is an immediate consequence of (ii) and of the definition of subdifferential. So, let us establish the second property. Let $f(0) > a > \inf f$, and set

$$2k := \sup_{x \neq 0} \frac{a - f(x)}{\|x\|}.$$

It is obvious that $k > 0$. We shall prove later that $k < \infty$. By definition of k,

$$f(x) + 2k\|x\| \geq a, \ \forall x \in X.$$

Moreover, there exists $\bar{x}$ such that

$$k < \frac{a - f(\bar{x})}{\|\bar{x}\|},$$

providing

$$f(\bar{x}) + 2k\|\bar{x}\| < a + k\|\bar{x}\| \leq \inf\{f(x) + 2k\|x\| : x \in X\} + k\|\bar{x}\|.$$

We can then apply Corollary 4.2.12 with $\varepsilon = k$ ed $r = \|\bar{x}\|$. Hence there are $z \in \operatorname{dom} f$ and $w^* \in \partial(f(\cdot) + k\| \cdot \|)(z)$ such that

$$\|z - \bar{x}\| < \|\bar{x}\| \text{ and } \|w^*\| < k.$$

The first condition implies $z \neq 0$. By the sum Theorem 3.4.2 we also have

$$w^* = z^* + y^*,$$

with

$$z^* \in \partial f(z) \text{ and } y^* \in \partial(k\|\cdot\|)(z).$$

The last condition, by applying the definition of subdifferential, implies

$$0 \geq k\|z\| - \langle y^*, z \rangle,$$

whence

$$\langle y^*, z \rangle \geq k\|z\|.$$

We then get

$$\langle z^*, z \rangle = \langle w^*, z \rangle - \langle y^*, z \rangle < k\|z\| - k\|z\| \leq 0.$$

To conclude, we must verify that $k < \infty$. It is enough to consider the case when $f(x) < a$. Let $x^* \in X^*, \alpha \in \mathbb{R}$ be such that $f(y) \geq \langle x^*, y \rangle - \alpha, \forall y \in X$. The existence of such an affine function minorizing f relies on the fact that $f \in \Gamma(X)$ (Corollary 2.2.17). We then have

$$a - f(x) \leq |a| + |\alpha| + \|x^*\|_* \|x\|,$$

whence

$$\frac{a - f(x)}{\|x\|} \leq \frac{|a| + |\alpha|}{d(0, f^a)} + \|x^*\|_*,$$

and this ends the proof.   □

Exercise 4.2.15 Prove the following generalization of Theorem 4.4.1. Let $f \in \Gamma(X)$. Then ∂f is a maximal monotone operator.

Hint. Use the proof of Theorem 4.1.1 and the previous proposition.

To conclude this section, we want to get a result on the characterization of the epigraph of $f \in \Gamma(X)$, which improves upon Theorem 2.2.21. There, it was proved that the epigraph can be characterized as the intersection of the epigraphs of all the affine functions minorizing f. Here we prove that we can just consider *very particular* affine functions minorizing f, in order to have the same characterization.

To prove our result, we first must show the following lemma.

Lemma 4.2.16 *Let C be a closed convex set, and $x \notin C$. Then, for every $k > 0$, there exist $c \in C$, $c^* \in \partial I_C(c)$ such that*

$$\langle c^*, x - c \rangle \geq k.$$

Proof. Let $d = d(x, C)$, let $\alpha > k + d + 2$ and let $\bar{x} \in C$ be such that $\|\bar{x} - x\| < d(1 + \frac{1}{\alpha})$. Let

$$S = \{(tx + (1 - t)\bar{x}, t\alpha + (1 - t)(-1)) : 0 \leq t \leq 1\}.$$

Then $S \cap \text{epi}\, I_C = \emptyset$ and they can be strictly separated. Thus there exist $x^* \in X^*, r^* \in \mathbb{R}$ and $h \in \mathbb{R}$ such that $(x^*, r^*) \neq (0^*, 0)$ and

$$\langle (x^*, r^*), (c, r) \rangle \geq h > \langle (x^*, r^*), (u, \beta) \rangle,$$

for all $c \in C$, $r \geq 0$, $(u, \beta) \in S$. Taking any $c \in C$ and $r > 0$ big enough in the above inequalities shows that $r^* \geq 0$. And taking $c = \bar{x} = u$ shows that actually $r^* > 0$. Setting $y^* = -\frac{x^*}{r^*}$, and putting at first $(u, \beta) = (\bar{x}, -1)$ and then $(u, \beta) = (x, \alpha)$ in the above inequalities, we finally get

$$y^* \in \partial_1 I_C(\bar{x}) \text{ and } \langle y^*, x - c \rangle > \alpha, \quad \forall c \in C.$$

Thanks to Theorem 4.2.10 ($\varepsilon, r, \sigma = 1$), we have the existence of $c \in C$, $c^* \in \partial I_C(c)$ such that

$$\|c - \bar{x}\| \leq 1 \text{ and } \|c^* - y^*\|_* \leq 1.$$

Thus

$$\langle c^*, x - c \rangle = \langle c^* - y^*, x - c \rangle + \langle y^*, x - c \rangle > \alpha - (\|x - \bar{x}\| + \|\bar{x} - c\|)$$
$$\geq \alpha - (d(1 + \frac{1}{\alpha}) + 1) > k.$$

$\square$

Theorem 4.2.17 Let $f \in \Gamma(X)$. Then, for all $x \in X$,

$$f(x) = \sup\{f(y) + \langle y^*, x - y \rangle : (y, y^*) \in \partial f\}.$$

Proof. Observe at first that from the previous lemma the conclusion easily follows for the indicator function of a given closed convex set. Next, let us divide the proof into two parts. At first we prove the claim for $\bar{x} \in \text{dom } f$, then for $\bar{x}$ such that $f(\bar{x}) = \infty$, which looks a bit more complicated. Thus, given $\bar{x} \in \text{dom } f$ and $\eta > 0$, we need to find $(y, y^*) \in \partial f$ such that

$$f(y) + \langle y^*, \bar{x} - y \rangle \geq f(\bar{x}) - \eta.$$

Fix ε such that $2\varepsilon^2 < \eta$ and separate epi f from $(\bar{x}, f(\bar{x}) - \varepsilon^2)$. We then find $x^* \in \partial_{\varepsilon^2} f(\bar{x})$ (using the standard separation argument seen for the first time in Lemma 2.2.16). From Theorem 4.2.10 we have the existence of y, $y^* \in \partial f(y)$, such that

$$\|x^* - y^*\| \leq \varepsilon \text{ and } \|\bar{x} - y\| \leq \varepsilon.$$

Thus

$$f(y) + \langle y^*, \bar{x} - y \rangle \geq f(\bar{x}) + \langle x^* - y^*, y - \bar{x} \rangle - \varepsilon^2 \geq f(\bar{x}) - \eta.$$

This shows the first part of the claim. Suppose now $f(\bar{x}) = \infty$, and fix $k > 0$. We need to find $(y, y^*) \in \partial f$ such that

$$f(y) + \langle y^*, \bar{x} - y \rangle \geq k.$$

We shall apply Lemma 4.2.16 to $C = \text{epi}\, f$ and to $x = (\bar{x}, k)$. We then see that there exist $(x, r) \in \text{epi}\, f$, $(x^*, r^*) \in \partial I_{\text{epi}\, f}(x, r)$ such that

$$\langle (x^*, r^*), (\bar{x}, k) - (x, r) \rangle \geq 2. \tag{4.1}$$

Moreover, the condition $(x^*, r^*) \in \partial I_{\text{epi}\, f}(x, r)$ amounts to saying that

$$\langle (x^*, r^*), (y, \beta) - (x, r) \rangle \leq 0, \tag{4.2}$$

for all $(y, \beta) \in \text{epi}\, f$. From (4.2) it is easy to see that $r^* \leq 0$ and, with the choice of $(y, \beta) = (x, f(x))$, we see that $r = f(x)$. Suppose now $r^* < 0$. Then we can suppose, without loss of generality, that $r^* = -1$. Thus $(x^*, -1)$ supports epi f at $(x, f(x))$ and this means that $x^* \in \partial f(x)$. Moreover, from (4.1) we get

$$\langle x^*, \bar{x} - x \rangle + (-1)(k - f(x)) \geq 2,$$

i.e.,

$$f(x) + \langle x^*, \bar{x} - x \rangle \geq k + 2 > k,$$

so that we have shown the claim in the case $r^* < 0$. It remains to see the annoying case when $r^* = 0$. In such a case (4.1) and (4.2) become

$$\langle x^*, \bar{x} - x \rangle \geq 2, \langle x^*, y - x \rangle \leq 0, \forall y \in \text{dom}\, f. \tag{4.3}$$

Set $d = \|x - \bar{x}\|$ and $a = \frac{1}{\|x^*\|_*}$. Let $y^* \in \partial_a f(x)$, and observe that from (4.3) we have that for all $t > 0$, $z_t^* := y^* + tx^* \in \partial_a f(x)$. From Theorem 4.2.10 there exist y_t, $y_t^* \in \partial f(y_t)$ such that

$$\|x - y_t\| \leq a, \text{ and } \|z_t^* - y_t^*\|_* \leq 1.$$

As $\{y_t : t > 0\}$ is a bounded set, there exists b such that $f(y_t) \geq b$ for all $t > 0$. We then get

$$\begin{aligned}
f(y_t) + \langle y_t^*, \bar{x} - y_t \rangle &= f(y_t) + \langle y_t^* - z_t^*, \bar{x} - y_t \rangle + \langle z_t^*, \bar{x} - y_t \rangle \\
&\geq b - (d + a) - \|y^*\|(d + a) + t(\langle x^*, \bar{x} - x \rangle + \langle x^*, x - y_t \rangle) \\
&\geq b - (d + a) - \|y^*\|(d + a) + t.
\end{aligned}$$

Then we can choose t big enough to make the following inequality be true:

$$b - (1 + \|y^*\|)(d + a) + t \geq k,$$

and this ends the proof. □

We conclude by improving the result of the Lemma 3.6.4, once again with a beautiful argument following from the Ekeland variational principle.

Lemma 4.2.18 Let $f \colon X \to (-\infty, \infty]$ be convex. Let $\delta, a > 0$, $g \colon B(0, a) \to \mathbb{R}$ a Gâteaux function and suppose $|f(x) - g(x)| \leq \delta$ for $x \in B(0; a)$. Let

$0 < r < R \le a$, *let x be such that $\|x\| \le r$ and $x^* \in \partial f(x)$. Then both the following estimates hold:*

$$d(x^*, \nabla g(B(x; R - r))) \le \frac{2\delta}{R - r},$$

$$d(x^*, \nabla g(RB)) \le \frac{2\delta}{R - r}.$$

The same holds if g is convex and real valued, provided we replace ∇g with ∂g.

Proof. Without loss of generality we can suppose $f(x) = 0$ and $x^* = 0^*$. Then $g(x) < \delta$ and, if $\|u\| \le R$, $g(u) > -\delta$ (since f is nonnegative). It follows that

$$g(x) < \inf g + \frac{2\delta}{R - r}(R - r),$$

on the ball rB. To conclude, it is enough to use Corollary 4.2.8 (or Corollary 4.2.12 for the convex case).   □

4.3 Minimizing a convex function

In this section we want to analyze some properties of the level sets of a convex function, and to give a flavor of how one can proceed in looking for a minimum of a convex function defined on a Euclidean space. We do not go into the details of this topic; the interested reader is directed to excellent books treating this important problem in a systematic way, such as the one by Hiriart-Urruty–Lemaréchal [HUL]. We start by considering the level sets.

4.3.1 Level sets

We begin by establishing a result which actually could be derived by subsequent, more general statements, but which we prefer to present here, and to prove it with an elementary argument.

Proposition 4.3.1 *Let $f \colon \mathbb{R}^n \to (-\infty, \infty]$ be a convex, lower semicontinuous function. Suppose $\mathrm{Min}\, f$ is nonempty and compact. Then f^a is bounded for all $a > \inf f$ and $\forall \varepsilon > 0$ there exists $a > \inf f$ such that $f^a \subset B_\varepsilon[\mathrm{Min}\, f]$. Moreover, if $\{x_n\}$ is such that $f(x_n) \to \inf f$, then $\{x_n\}$ has a limit point which minimizes f. And if $\mathrm{Min}\, f$ is a singleton x, then $x_n \to x$.*

Proof. Let $r > 0$ be such that $\mathrm{Min}\, f \subset (r-1)B$ and, without loss of generality, suppose $0 \in \mathrm{Min}\, f$ and $f(0) = 0$. By contradiction, suppose there are $a > \inf f$ and $\{x_n\}$ such that $f(x_n) \le a$ and $\|x_n\| \to \infty$. It is an easy matter to verify that the sequence $\{\frac{rx_n}{\|x_n\|}\}$ is such that $f(\frac{rx_n}{\|x_n\|}) \to 0$, as a consequence of convexity of f. Then $\{\frac{rx_n}{\|x_n\|}\}$ has a subsequence converging to a point $\bar{x}$ of norm

r, and $\bar{x}$ minimizes f, by lower semicontinuity of f. But this is impossible. Now suppose there is $\varepsilon > 0$ such that for all n there is x_n such that $f(x_n) \leq \inf f + \frac{1}{n}$ and $d(x_n, \operatorname{Min} f) > \varepsilon$ for all n. Then $\{x_n\}$ is bounded, thus it has a cluster point which minimizes f, against the fact that $d(x_n, \operatorname{Min} f) > \varepsilon)$ for all n. To conclude, we must show that if $\operatorname{Min} f$ is a singleton, say x and $f(x_n) \to \inf f$, then $\{x_n\}$ converges to x. This is a purely topological argument. Suppose not; then there are $a > 0$ and a subsequence $\{y_n\}$ of $\{x_n\}$ such that $\|y_n - x\| \geq a$ for all n. As $\{y_n\}$ is bounded, it has a limit point which minimizes f, so that this limit point must be x, against the assumption $\|y_n - x\| \geq a$ for all n. $\square$

The first result we present shows that the level sets of a convex lower semicontinuous function "cannot be too different". Next, we inquire about the connections between the local shape of the boundary of a level set, at a point x, the descent directions at the point x, and the subdifferential of f at x. For the first result, recall the definition of recession cone given in Definition 1.1.15.

Proposition 4.3.2 *Let $f \in \Gamma(X)$ and suppose $f^a, f^b \neq \emptyset$. Then $0^+(f^a) = 0^+(f^b)$.*

Proof. Let $z \in f^a$, $x \in 0^+(f^a)$ and fix $y \in f^b$. We must show that $f(x+y) \leq b$. As $(1 - \frac{1}{n})y + \frac{1}{n}(z + nx) \to y + x$, we have

$$f(y + x) \leq \liminf f\left(\left(1 - \frac{1}{n}\right)y + \frac{1}{n}(z + nx)\right) \leq \liminf\left(\left(1 - \frac{1}{n}\right)b + \frac{1}{n}a\right) = b,$$

and this ends the proof. $\square$

Remark 4.3.3 Consider a separable Hilbert space with basis $\{e_n : n \in \mathbb{N}\}$, and the function

$$f(x) = \sum_{n=1}^{\infty} \frac{\langle x, e_n \rangle^2}{n^4}.$$

From the previous proposition (but it is easily seen directly, too), $0^+(f^a) = \{0\} \; \forall a > 0$, as $0^+(f^0) = \{0\}$. However f^a is unbounded for all $a > 0$, and this shows that Proposition 4.3.1 and Proposition 1.1.16 fail in infinite dimensions.

Proposition 4.3.4 *Let $f : X \to (-\infty, \infty]$ be convex and lower semicontinuous. Suppose there is $b > \inf f$ such that f^b is bounded. Then f^a is bounded for all $a > \inf f$.*

Proof. In the finite dimensional case the result is an immediate consequence of Proposition 4.3.2, since $0^+(f^a) = 0^+(f^b) = \{0\}$ and this is equivalent to saying that f^a is bounded (moreover, the condition $b > \inf f$ can be weakened to $f^b \neq \emptyset$). In the general case, let $a > b$, let r be such that $f^b \subset (r - 1)B$ and take a point $\bar{x}$ such that $f(\bar{x}) < b$. With the usual translation of the axes we can suppose, without loss of generality, $\bar{x} = 0$, $f(0) = 0$ and consequently

$b > 0$. This clearly does not affect boundedness of the level sets. Let y be such that $\|y\| = \frac{r(a+1)}{b}$. Then $z = \frac{b}{a+1} y$ has norm r. It follows that

$$b < f(z) \leq \frac{b}{a+1} f(y),$$

whence $f(y) \geq a + 1$. This shows that $f^a \subset \frac{r(a+1)}{b} B$.   □

The next proposition is quite simple.

Proposition 4.3.5 *Let $f \colon X \to (-\infty, \infty]$ be convex, lower semicontinuous. Let $b > \inf f$ be such that f^b is bounded. Then, for every $r > 0$ there exists $c > b$ such that $f^c \subset B_r[f^b]$.*

Proof. Without loss of generality, suppose $f(0) = 0$. Let $k > 0$ be such that $f^{b+1} \subset kB$, and $k > r(b+1)$. The choice of $c = b + \frac{rb}{k-r}$ works since, if $x \in f^c$, then $\frac{b}{c} x \in f^b$. Moreover,

$$\left\| x - \frac{b}{c} x \right\| \leq k \frac{c-b}{c} = r.$$

                □

Exercise 4.3.6 Let $f \colon X \to \mathbb{R}$ be convex and continuous, where X is a Euclidean space. Let C be a closed convex subset of X. Let $a \in \mathbb{R}$ be such that $f^a \neq \emptyset$ and suppose $0^+(C) \cap 0^+(f^a) = \{0\}$. Then $f(C)$ is closed.

Hint. Suppose $\{y_n\} \subset f(C)$ and $y_n \to y$. Let $c_n \in C$ be such that $y_n = f(c_n)$. Show that $\{c_n\}$ must be bounded.

Exercise 4.3.7 Let $f \in \Gamma(X)$, X a Banach space. Suppose $a > \inf f$. Then $f^a = \mathrm{cl}\{x : f(x) < a\}$.

Hint. Let x be such that $f(x) = a$ and let z be such that $f(z) < a$. Look at f on the segment $[x, z]$.

We now see that, given a point x, the directions y such that $f'(x; y) < 0$ are those for which the vector goes "into" the level set relative to x.

Proposition 4.3.8 *Let $f \colon X \to (-\infty, \infty]$ be convex and lower semicontinuous. Let x be a point where f is (finite and) continuous. Then*

$$\{y : f'(x; y) < 0\} = \{y : \exists \lambda > 0, z, \ f(z) < f(x) \ and \ y = \lambda(z - x)\}.$$

Proof. Let $A = \{y : f'(x; y) < 0\}$ and let $B = \{y : \exists \lambda > 0, z, f(z) < f(x) \text{ and } y = \lambda(z - x)\}$. Observe that both A and B are cones. Now, let $y \in B$. Then there are $\lambda > 0$ and z such that $y = \lambda(z - x)$ and $f(z) < f(x)$. Since A is a cone we can suppose, without loss of generality, $\lambda < 1$. We have that $f(\lambda z + (1 - \lambda)x) < f(x)$ for all λ. Thus $f(x + y) - f(x) < 0$, which implies $f'(x; y) < 0$ so that $y \in A$. Now, let $y \in A$. Then $f(x + ty) - f(x) < 0$ for small $t > 0$. The conclusion follows.   □

We now want to say something on the following topic. As is well known, if a function f is smooth, and one considers a point x where ∇f does not vanish, then $\nabla f(x)$ is perpendicular to the tangent plane to the level set at height $f(x)$. In the convex case, this means that the gradient is a vector in the normal cone at x to the level set at height $f(x)$. Moreover the direction of $\nabla f(x)$ is a descent direction. At least for small $t > 0$ we have $f(x - t\nabla f(x)) < f(x)$. But what happens in the nonsmooth case? The following example shows that things can be different.

Example 4.3.9 This is an example showing that in the nonsmooth case a direction opposite to one subgradient at a point of a given function is not necessarily a descent direction for the function itself, not even locally. Let

$$f(x, y) = 2|x| + |y|,$$

let $p = (0, 2)$, and let the direction v be $v = (1, 1)$. It is straightforward to see that $v \in \partial f(p)$ and that for no $t > 0$ does $p - tv$ belong to the level set relative to p.

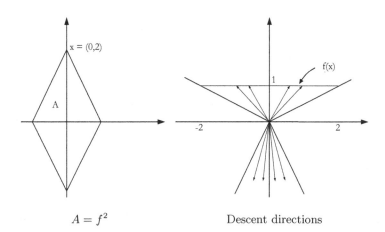

$$A = f^2 \qquad\qquad \text{Descent directions}$$

Figure 4.3.

Also in the nonsmooth case, however, it is true that, if $x^* \in \partial f(x)$, then x^* is in the normal cone at x to the level set at height $f(x)$, as is easy to see. But actually it is possible to provide much more precise information, and this is what we are going to do now.

The result of the next exercise will be used in the proposition following it.

Exercise 4.3.10 Let X be a Banach space, $x \in X$, $0^* \neq x^*$. Set $H = \{z : \langle x^*, z \rangle \geq \langle x^*, x \rangle\}$. Prove that

$$N_H(x) = \mathbb{R}_-\{x^*\}.$$

Hint. Let $z^* \in N_H(x)$. Then $\langle z^*, u \rangle = \langle z^*, x + u - x \rangle \leq 0$, for all u such that $\langle x^*, u \rangle = 0$. It follows that $\langle z^*, u \rangle = 0$, for all u such that $\langle x^*, u \rangle = 0$. Derive the conclusion.

Theorem 4.3.11 *Let X be a Banach space, let $f \colon X \to (-\infty, \infty]$ be convex and lower semicontinuous. Let x be a point where f is (finite and) continuous and suppose $f(x) = a > \inf f$. Then*

$$N_{f^a}(x) = \text{cone}\{\partial f(x)\}.$$

Proof. The fact that $N_{f^a}(x)$ contains the cone generated by the subdifferential of f on x is easy to see and is true also if $f(x) = \inf f$. To see the opposite inclusion, let $0^* \neq x^* \in N_{f^a}(x)$. Since $\langle x^*, z - x \rangle \leq 0$ for all $z \in f^a$, it follows that $\langle x^*, z - x \rangle < 0$ for all $z \in \text{int } f^a$. Otherwise, for some $z \in \text{int } f^a$ we would have $\langle x^*, z - x \rangle = 0$. This would imply that x^* has a local maximum at z, but in this case it would be $x^* = 0^*$. From this we have that $f(z) < f(x)$ implies $\langle x^*, z \rangle < \langle x^*, x \rangle$ and this in turn implies that if $\langle x^*, z \rangle \geq \langle x^*, x \rangle$, then $f(z) \geq f(x)$. In other words, f has a minimum on x over the set $H = \{z : \langle x^*, z \rangle \geq \langle x^*, x \rangle\}$. It follows, by using the sum theorem (since f is continuous at x) that

$$0^* \in \partial(f + I_H)(x) = \partial f(x) + N_H(x).$$

Now, as suggested by Exercise 4.3.10, $N_H(x) = \mathbb{R}_-\{x^*\}$. Thus there are $t \geq 0$ and $z^* \in \partial f(x)$ such that $x^* = tz^*$, and this ends the proof.   □

If X is finite dimensional, it is enough to assume that $\partial f(x) \neq \emptyset$, but in this case one must take the closure of the cone generated by $\partial f(x)$ (see [Ro, Theorem 23.7]).

4.3.2 Algorithms

Usually, even if we know that the set of the minima of a (convex) function is nonempty, it is not easy or even possible to directly find a minimum point (for instance by solving the problem $0^* \in \partial f(x)$.) For this reason, several algorithms were developed in order to build up sequences of points approximating a solution (in some sense). In this section we shall consider some of these procedures. We are then given a convex function $f \colon \mathbb{R}^n \to \mathbb{R}$ with a nonempty set of minimizers, and we try to construct sequences $\{x_k\}$ approximating $\text{Min } f$. The sequences $\{x_k\}$ will be built up in the following fashion:

$$x_0 \text{ arbitrary}, \quad x_{k+1} = x_k - \lambda_k d_k.$$

The vector d_k is assumed to be of norm one, so that λ_k is the *length* of the step at time k. Of course, both the choices of λ_k and d_k are crucial for good behavior of the algorithm. As far as λ_k is concerned, it is clear that it must not be too small, as in such a case the sequence $\{x_k\}$ could converge to something not minimizing the function. And if it converges to a solution, its convergence could be much too slow. On the other hand, it should not be too big, as in this case the algorithm need not converge. On the other side, $-d_k$ represents the *direction* along which we build up the element x_{k+1}, starting from x_k. Usually, it is a vector d_k such that $-d_k$ has the same direction as a vector $v_k \in \partial f(x_k)$. In the smooth case, this choice guarantees that the function decreases at each step, at least if λ is sufficiently small. In the nonsmooth case, we have seen in Example 4.3.9 that this does not always happen.

Theorem 4.3.12 *Let $\{\lambda_k\}$ be such that*

$$\lambda_k \mapsto 0, \tag{4.4}$$

$$\sum_{k=0}^{\infty} \lambda_k = \infty. \tag{4.5}$$

Let

$$v_k \in \partial f(x_k),$$

and let

$$d_k = \begin{cases} \frac{v_k}{\|v_k\|} & \text{if } v_k \neq 0, \\ 0 & \text{if } v_k = 0. \end{cases}$$

Moreover, suppose $\mathrm{Min}\, f$ *is a nonempty bounded set. Then*

$$\lim_{k \to +\infty} d(x_k, \mathrm{Min}\, f) = 0 \quad \text{and} \quad \lim_{k \to +\infty} f(x_k) = \inf f.$$

Proof. First, observe that if for some k it is $d_k = 0$, then we have reached a minimum point. In this case the sequence could possibly become constant, but it is not necessary to assume this. The result holds also in the case the algorithm does not stop. Simply observe that if $d_k = 0$, then $x_{k+1} = x_k$. Thus, we can assume, without loss of generality, that $d_k \neq 0$ for all k. Moreover, observe that the equality $\lim_{k \to +\infty} f(x_k) = \inf f$ is an easy consequence of the first part of the claim.

Now, suppose there are $a > 0$ and k such that

$$d(x_k, \mathrm{Min}\, f) \geq a > 0. \tag{4.6}$$

This implies, in view of Proposition 4.3.1, that there exists $c > 0$ such that $f(x_k) \geq \inf f + c$. Since, for all x,

$$f(x) \geq f(x_k) + \langle v_k, x - x_k \rangle,$$

we have that

$$\langle v_k, x - x_k \rangle \leq 0, \forall x \in f^{\inf f + c}.$$

Since f is continuous and $\mathrm{Min}\, f$ is compact, there exists $r > 0$ such that $B_r[\mathrm{Min}\, f] \subset f^{\inf f + c}$. Take $\bar{x} \in \mathrm{Min}\, f$ and consider the point $\bar{x} + rd_k \in B_r[\mathrm{Min}\, f]$. Then

$$\langle v_k, \bar{x} + r d_k - x_k \rangle \leq 0,$$

and also

$$\langle d_k, \bar{x} + r d_k - x_k \rangle \leq 0,$$

providing

$$\langle d_k, \bar{x} - x_k \rangle \leq -r.$$

Thus (4.6) implies

$$\begin{aligned}
\|\bar{x} - x_{k+1}\|^2 &= \|\bar{x} - x_k\|^2 + 2\lambda_k \langle d_k, \bar{x} - x_k \rangle + \lambda_k^2 \\
&\leq \|\bar{x} - x_k\|^2 - 2r\lambda_k + \lambda_k^2 \quad\quad\quad (4.7) \\
&\leq \|\bar{x} - x_k\|^2 - r\lambda_k,
\end{aligned}$$

eventually. From this we obtain in particular that, if (4.6) holds and k is large enough,

$$d(x_{k+1}, \operatorname{Min} f) \leq d(x_k, \operatorname{Min} f). \quad\quad\quad (4.8)$$

Now suppose, by contradiction, there is $a > 0$ such that, for all large k,

$$d(x_k, \operatorname{Min} f) \geq a > 0. \quad\quad\quad (4.9)$$

From (4.7) we then get

$$\|\bar{x} - x_{k+i}\|^2 \leq \|\bar{x} - x_k\|^2 - r \sum_{j=k}^{k+i-1} \lambda_j \to -\infty,$$

which is impossible. It follows that $\liminf d(x_k, \operatorname{Min} f) = 0$. Now, fix $a > 0$ and K such that $\lambda_k < a$ for $k \geq K$. There is $k > K$ such that $d(x_k, \operatorname{Min} f) < a$. This implies

$$d(x_{k+1}, \operatorname{Min} f) < 2a.$$

Now, two cases can occur:

(i) $d(x_{k+2}, \operatorname{Min} f) < a$;
(ii) $d(x_{k+2}, \operatorname{Min} f) \geq a$.

In the second case, from (4.8) we can conclude that

$$d(x_{k+2}, \operatorname{Min} f) \leq d(x_{k+1}, \operatorname{Min} f) < 2a.$$

Thus, in any case, we have that

$$d(x_{k+2}, \operatorname{Min} f) < 2a.$$

By induction, we conclude that $d(x_n, \operatorname{Min} f) \leq 2a$ for all large n, and this ends the proof.   □

With some changes in the above proof, it can be seen that the same result holds if we take $v_k \in \partial_{\varepsilon_k} f(x_k)$, for any sequence $\{\varepsilon_k\}$ converging to zero.

The above result can be refined if Min f has interior points.

Corollary 4.3.13 *With the assumptions of Theorem 4.3.12, if moreover* int Min $f \neq \emptyset$, *then* $v_k = 0$ *for some* k.

Proof. Suppose, by way of contradiction, $v_k \neq 0$ for all k. Let $\bar{x} \in$ int Min f. Then there is $r > 0$ such that $B[\bar{x}; r] \subset$ Min f. Let $\tilde{x}_k = \bar{x} + r d_k$. Then $\tilde{x}_k \in B[\bar{x}; r] \subset$ Min f, hence $f(\tilde{x}_k) = \inf f$. Moreover,

$$f(y) \geq f(x_k) + \langle v_k, y - x_k \rangle \quad \forall y \in \mathbb{R}^n,$$

providing

$$f(\tilde{x}_k) \geq f(x_k) + \langle v_k, \tilde{x}_k - x_k \rangle.$$

Moreover, $f(x_k) \geq \inf f = f(\tilde{x}_k)$, hence

$$\langle v_k, \tilde{x}_k - x_k \rangle \leq 0.$$

We repeat what we did in the first part of Theorem 4.3.12 to get that

$$\|x_{k+s} - \bar{x}\|^2 \leq \|x_k - \bar{x}\|^2 - r \sum_{i=k}^{k+s-1} \lambda_i \to -\infty,$$

which provides the desired contradiction.   □

The results above concern the case when f has a nonempty and bounded set of minimizers. The next result instead takes into account the case when the set of the minimizers of f in unbounded. As we shall see, we must put an extra condition on the size of the length steps λ_k. Thus, we shall suppose as before

$$\lambda_k \to 0,$$

$$\sum_{k=0}^{+\infty} \lambda_k = \infty,$$

$$v_k \in \partial f(x_k),$$

$$d_k = \begin{cases} 0 & \text{if } v_k = 0, \\ \dfrac{v_k}{\|v_k\|} & \text{if } v_k \neq 0. \end{cases}$$

Moreover, suppose

$$\sum_{k=0}^{\infty} \lambda_k^2 < \infty. \tag{4.10}$$

Then, the following result holds:

Theorem 4.3.14 *If* Min f *is nonempty, then the sequence* $\{x_k\}$ *converges to an element belonging to the set* Min f.

Proof. As in Theorem 4.3.12, we consider the case when $d_k \neq 0$ for all k. Let $\bar{x} \in$ Min f. Then

$$
\begin{aligned}
\|x_{k+1} - \bar{x}\|^2 &= \|x_k - \bar{x} - \lambda_k d_k\|^2 \\
&= \|x_k - \bar{x}\|^2 + 2\langle x_k - \bar{x}, -\lambda_k d_k \rangle + \lambda_k{}^2 \\
&\leq \|x_k - \bar{x}\|^2 + 2\frac{\lambda_k}{\|v_k\|}\langle v_k, \bar{x} - x_k \rangle + \lambda_k{}^2.
\end{aligned}
\tag{4.11}
$$

Moreover,

$$
f(y) - f(x_k) \geq \langle v_k, y - x_k \rangle \quad \forall y \in \mathbb{R}^n \quad \forall k,
$$

whence

$$
0 \geq \inf f - f(x_k) \geq \langle v_k, \bar{x} - x_k \rangle \quad \forall k.
\tag{4.12}
$$

From (4.11) we get

$$
\begin{aligned}
\|x_{k+1} - \bar{x}\|^2 &\leq \|x_k - \bar{x}\|^2 + \lambda_k{}^2 \\
&\leq \|x_0 - \bar{x}\|^2 + \sum_{i=0}^{k} \lambda_i{}^2.
\end{aligned}
\tag{4.13}
$$

From (4.13) and (4.10) we see that the sequence $\{x_k\}$ is bounded. This implies that the sequence $\{v_k\}$ is also bounded, as f is Lipschitz on a ball containing $\{x_k\}$. We see now that there is a subsequence $\{x_{k_j}\}$ such that

$$
a_{k_j} := \langle v_{k_s}, \bar{x} - x_{k_s} \rangle \to 0.
\tag{4.14}
$$

Otherwise, from (4.12) there would be $b > 0$ and $K \in \mathbb{R}$ such that

$$
a_k \leq -b \quad \forall k > K.
$$

From (4.11) we get

$$
\|x_{k+1} - \bar{x}\|^2 \leq \|x_0 - \bar{x}\|^2 + 2\sum_{i=0}^{k} \frac{\lambda_i}{\|v_i\|}\langle v_i, \bar{x} - x_i \rangle + \sum_{i=0}^{k} \lambda_i{}^2,
$$

implying

$$
\lim_{k \to \infty} \|x_{k+1} - \bar{x}\|^2 = -\infty,
$$

which is impossible. Thus, from (4.14) and (4.12) we get that

$$
f(x_{k_s j}) \to f(\bar{x}).
$$

As $\{x_{k_j}\}$ is bounded, it has a subsequence (still labeled by k_j) converging to some element x^*. Hence

$$\lim_{j \to \infty} f(x_{k_j}) = f(x^*)$$

implying $x^* \in \text{Min } f$. It remains to prove that the whole sequence $\{x_k\}$ converges to x^*. From the fact that $\bar{x}$ is arbitrary in (4.11) and (4.12), we can put x^* there instead of $\bar{x}$. Given $\varepsilon > 0$, there exists $K_1 \in \mathbb{R}$ such that, if $k_j > K_1$,

$$\left\| x_{k_j} - x^* \right\|^2 < \frac{\varepsilon}{2}, \quad \text{and} \quad \sum_{i=k_j}^{\infty} \lambda_i^2 < \frac{\varepsilon}{2}.$$

Then, from (4.11) and (4.12) we get that

$$\left\| x_{k_j+n} - x^* \right\|^2 \le \left\| x_{k_j} - x^* \right\|^2 + \sum_{i=k_j}^{k_j+n-1} \lambda_i^2 < \varepsilon \quad \forall n \ge 1.$$

This implies $x_k \to x^*$. □

5

The Fenchel conjugate

Ουκ έστ' εραστῆσ όστισ ουκ αεὶ φιλεῖ
(He is not a lover who does not love forever)
(Euripides, "The Trojan Women")

In the study of a (constrained) minimum problem it often happens that another problem, naturally related to the initial one, is useful to study. This is the so-called duality theory, and will be the subject of the next chapter.

In this one, we introduce a fundamental operation on convex functions that allows building up a general duality theory. Given an extended real valued function f defined on a Banach space X, its Fenchel conjugate f^* is a convex and lower semicontinuous function, defined on the dual space X^* of X. After defining it, we give several examples and study its first relevant properties. Then we observe that we can apply the Fenchel conjugation to f^* too, and this provides a new function, again defined on X, and minorizing everywhere the original function f. It coincides with f itself if and only if $f \in \Gamma(X)$, and is often called the convex, lower semicontinuous relaxation (or regularization) of f. Moreover, there are interesting connections between the subdifferentials of f and f^*; we shall see that the graphs of the two subdifferentials are the same. Given the importance of this operation, a relevant question is to evaluate the conjugate of the sum of two convex functions. We then provide a general result in this sense, known as the Attouch–Brézis theorem.

5.1 Generalities

As usual, we shall denote by X a Banach space, and by X^* its topological dual.

Definition 5.1.1 Let $f \colon X \to (-\infty, \infty]$ be an arbitrary function. The *Fenchel conjugate* of f is the function $f^* \colon X^* \to [-\infty, \infty]$ defined as

$$f^*(x^*) := \sup_{x \in X} \{\langle x^*, x \rangle - f(x)\}.$$

We have that

$$(x^*, \alpha) \in \text{epi} \, f^* \iff f(x) \geq \langle x^*, x \rangle - \alpha, \, \forall x \in X,$$

which means that the points of the epigraph of f^* parameterize the affine functions minorizing f. In other words, if the affine function $l(x) = \langle x^*, x \rangle - \alpha$ minorizes f, then the affine function $m(x) = \langle x^*, x \rangle - f^*(x^*)$ fulfills

$$l(x) \leq m(x) \leq f(x).$$

We also have that

$$\text{epi} \, f^* = \bigcap_{x \in X} \text{epi}\{\langle \cdot, x \rangle - f(x)\}.$$

Observe that even if f is completely arbitrary, its conjugate is a convex function, since $\text{epi}\{\langle \cdot, x \rangle - f(x)\}$ is clearly a convex set for every $x \in X$. Furthermore, as $\text{epi}\{\langle \cdot, x \rangle - f(x)\}$ is for all x, a closed set in $X^* \times \mathbb{R}$ endowed with the product topology inherited by the weak* topology on X^* and the natural topology on $\mathbb{R}$, it follows that for any arbitrary f, $\text{epi} \, f^* \subset X^* \times \mathbb{R}$ is a closed set in the above topology.

A geometrical way to visualize the definition of f^* can be captured by observing that

$$-f^*(x^*) = \sup\{\alpha : \alpha + \langle x^*, x \rangle \leq f(x), \forall x \in X\}.$$

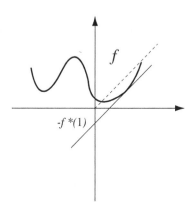

Figure 5.1.

For,

$$f^*(x^*) = \inf\{-\alpha : \alpha + \langle x^*, x \rangle \leq f(x), \forall x \in X\}$$
$$= -\sup\{\alpha : \alpha + \langle x^*, x \rangle \leq f(x), \forall x \in X\}.$$

Example 5.1.2 Here we see some examples of conjugates.

(a) The conjugate of an affine function: for $a \in X^*$, $b \in \mathbb{R}$, let $f(x) = \langle a, x \rangle + b$; then

$$f^*(x^*) = \begin{cases} -b & \text{if } x^* = a, \\ \infty & \text{otherwise.} \end{cases}$$

(b) $f(x) = \|x\|$, $f^*(x^*) = I_{B^*}(x^*)$.

(c) Let X be a Hilbert space and $f(x) = \frac{1}{2}\|x\|^2$, then $f^*(x^*) = \frac{1}{2}\|x^*\|_*^2$, as one can see by looking for the maximizing point in the definition of the conjugate.

(d) $f(x) = I_C(x)$, $f^*(x^*) = \sup_{x \in C} \langle x^*, x \rangle := \sigma_C(x^*)$; σ_C is a positively homogeneous function, called the *support* function of C. If C is the unit ball of the space X, then $f^*(x^*) = \|x^*\|_*$. If C is a cone, the support function of C is the indicator function of the cone C°, the polar cone of C, which is defined as $C^\circ = \{x^* \in X^* : \langle x^*, x \rangle \leq 0, \forall x \in C\}$. Observe that C° is a weak*-closed convex cone.

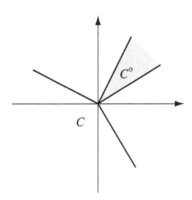

Figure 5.2. A cone C and its polar cone C°.

Exercise 5.1.3 Find f^*, for each f listed: (a) $f(x) = e^x$, (b) $f(x) = x^4$, (c) $f(x) = \sin x$, (d) $f(x) = \max\{0, x\}$, (e) $f(x) = -x^2$, (f) $f(x, y) = xy$,

(g) $f(x) = \begin{cases} e^x & \text{if } x \geq 0, \\ \infty & \text{otherwise,} \end{cases}$ (h) $f(x) = \begin{cases} x \ln x & \text{if } x \geq 0, \\ \infty & \text{otherwise,} \end{cases}$

(i) $f(x) = \begin{cases} 1 & \text{if } x \geq 0, \\ -1 & \text{otherwise;} \end{cases}$ (j) $f(x) = (x^2 - 1)^2$,

(k) $f(x) = \begin{cases} 0 & \text{if } |x| \leq 1, \\ (x^2 - 1)^2 & \text{otherwise.} \end{cases}$

The next proposition summarizes some elementary properties of f^*; we leave the easy proofs as an exercise.

Proposition 5.1.4 *We have:*

(i) $f^*(0) = -\inf f$;

(ii) $f \le g \Rightarrow f^* \ge g^*$;

(iii) $(\inf_{j \in J} f_j)^* = \sup_{j \in J} f_j^*$;

(iv) $(\sup_{j \in J} f_j)^* \le \inf_{j \in J} f_j^*$;

(v) $\forall r > 0, (rf)^*(x^*) = rf^*(\frac{x^*}{r})$;

(vi) $\forall r \in \mathbb{R}, (f + r)^*(x^*) = f^*(x^*) - r$;

(vii) $\forall \hat{x} \in X$, if $g(x) := f(x - \hat{x})$, then $g^*(x^*) = f^*(x^*) + \langle x^*, \hat{x} \rangle$.

Example 5.1.5 Let $f(x) = x$, $g(x) = -x$. Then $(\max\{f, g\})^*(x^*) = I_{[-1,1]}$, $\min\{f^*, g^*\}(x^*) = 0$ if $|x| = 1$, ∞ elsewhere. Thus the inequality in the fourth item above can be strict, which is almost obvious from the fact that in general $\inf_{j \in J} f_j^*$ need not be convex.

Example 5.1.6 Let $g \colon \mathbb{R} \to (-\infty, \infty]$ be an even function. Let $f \colon X \to \mathbb{R}$ be defined as $f(x) = g(\|x\|)$. Then

$$f^*(x^*) = g^*(\|x^*\|_*).$$

For,

$$f^*(x^*) = \sup_{x \in X}\{\langle x^*, x \rangle - g(\|x\|)\} = \sup_{t \ge 0} \sup_{\|x\| = t}\{\langle x^*, x \rangle - g(\|x\|)\}$$

$$= \sup_{t \ge 0}\{t\|x^*\|_* - g(t)\} = \sup_{t \in \mathbb{R}}\{t\|x^*\|_* - g(t)\} = g^*(\|x^*\|_*).$$

Exercise 5.1.7 Let X be a Banach space, $f(x) = \frac{1}{p}\|x\|^p$, with $p > 1$. Then $f^*(x^*) = \frac{1}{q}\|x^*\|_*^q$ $(\frac{1}{p} + \frac{1}{q} = 1)$.

The case $p = 2$ generalizes Example 5.1.2 (c).

Exercise 5.1.8 Let X be a Banach space, let $A \colon X \to X$ be a linear, bounded and invertible operator. Finally, let $f \in \Gamma(X)$ and $g(x) = f(Ax)$. Evaluate g^*.

Hint. $g^*(x^*) = f^*((A^{-1})^*)(x^*)$.

Exercise 5.1.9 Evaluate f^* when f is

$$f(x) = \begin{cases} -\sqrt{x} & \text{if } x \ge 0, \\ \infty & \text{otherwise,} \end{cases} \qquad f(x,y) = \begin{cases} -2\sqrt{xy} & \text{if } x \ge 0, y \ge 0, \\ \infty & \text{otherwise.} \end{cases}$$

Exercise 5.1.10 Let X be a Banach space. Suppose $\lim_{\|x\| \to \infty} \frac{f(x)}{\|x\|} = \infty$. Prove that $\operatorname{dom} f^* = X^*$ and that the supremum in the definition of the conjugate of f is attained if X is reflexive.

Exercise 5.1.11 Let X be a Banach space and let $f \in \Gamma(X)$. Then the following are equivalent:

(i) $\lim_{\|x\| \to \infty} f(x) = \infty$;

(ii)   there are $c_1 > 0, c_2$ such that $f(x) \geq c_1\|x\| - c_2$;
(iii) $0 \in \text{int dom } f^*$.

Find an analogous formulation for the function $f(x) - \langle x^*, x \rangle$, where $x^* \in X^*$.

Hint. Suppose $f(0) = 0$, and let r be such that $f(x) \geq 1$ if $\|x\| \geq r$. Then, for x such that $\|x\| > r$, we have that $f(x) \geq \frac{\|x\|}{r}$. Moreover, there exists $\hat{c} < 0$ such that $f(x) \geq \hat{c}$ if $\|x\| \leq r$. Then $f(x) \geq \frac{\|x\|}{r} + \hat{c} - 1$ for all x. This shows that (i) implies (ii).

Exercise 5.1.12 Let $f \in \Gamma(X)$. Then $\lim_{\|x^*\|_* \to \infty} \frac{f^*(x^*)}{\|x^*\|_*} = \infty$ if and only if f is upper bounded on all the balls. In particular this happens in finite dimensions, if and only if f is real valued. On the contrary, in infinite dimensions there are continuous real valued convex functions which are not bounded on the unit ball.

Hint. Observe that the condition $\lim_{\|x^*\|_* \to \infty} \frac{f^*(x^*)}{\|x^*\|_*} = \infty$ is equivalent to having that for each $k > 0$, there is c_k such that $f^*(x^*) \geq k\|x^*\|_* - c_k$. On the other hand, f is upper bounded on kB if and only if there exists c_k such that $f(x) \leq I_{kB}(x) + c_k$.

5.2 The bijection between $\Gamma(X)$ and $\Gamma^*(X^*)$

Starting from a given arbitrary function f, we have built its conjugate f^*. Of course, we can apply the same conjugate operation to f^*, too. In this way, we shall have a new function f^{**}, defined on X^{**}. But we are not interested in it. We shall instead focus our attention to its *restriction* to X, and we shall denote it by f^{**}. Thus

$$f^{**} \colon X \to [-\infty, \infty]; f^{**}(x) = \sup_{x^* \in X^*} \{\langle x^*, x \rangle - f^*(x^*)\}.$$

In this section, we study the connections between f and f^{**}.

Proposition 5.2.1 *We have $f^{**} \leq f$.*

Proof. $\forall x \in X, \forall x^* \in X^*$,

$$\langle x^*, x \rangle - f^*(x^*) \leq f(x).$$

Taking the supremum over $x^* \in X^*$ in both sides provides the result.   □

Definition 5.2.2 We define the *convex, lower semicontinuous regularization* of $f \colon X \to (-\infty, \infty]$ to be the function $\hat{f}$ such that

$$\text{epi } \hat{f} = \text{cl co epi } f.$$

The definition is consistent because the convex hull of an epigraph is still an epigraph. Clearly, $\hat{f}$ is the largest convex (the closure of a convex set is convex) and lower semicontinuous function minorizing f: if $g \leq f$ and g is convex and lower semicontinuous, then $g \leq \hat{f}$. For, epi g is a closed convex set containing epi f, hence it contains cl co epi f.

Remark 5.2.3 If f is convex, then $\hat{f} = \bar{f}$. If $f \in \Gamma(X)$, then $f = \hat{f}$. This easily follows from

$$\text{epi } f = \text{cl co epi } f.$$

Observe that we always have $\hat{f} \geq f^{**}$, as $f^{**} \leq f$ and f^{**} is convex and lower semicontinuous.

The next theorem provides a condition to ensure that $\hat{f}$ and f^{**} coincide. Exercise 5.2.5 shows that such a condition is not redundant.

Theorem 5.2.4 *Let $f: X \to (-\infty, \infty]$ be such that there are $x^* \in X^*, \alpha \in \mathbb{R}$ with $f(x) \geq \langle x^*, x\rangle + \alpha, \forall x \in X$. Then $\hat{f} = f^{**}$.*

Proof. The claim is obviously true if f is not proper, as in such a case, both f^{**} and $\hat{f}$ are constantly ∞. Then we have that $\forall x \in X$,

$$\hat{f}(x) \geq f^{**}(x) \geq \langle x^*, x\rangle + \alpha.$$

The last inequality follows from the fact that $f \geq g \implies f^{**} \geq g^{**}$ and that the biconjugate of an affine function coincides with the affine function itself. Thus $f^{**}(x) > -\infty$ for all x. Let us suppose now, for the sake of contradiction, that there is $x_0 \in X$ such that $f^{**}(x_0) < \hat{f}(x_0)$. It is then possible to separate $(x_0, f^{**}(x_0))$ and epi $\hat{f}$. If $\hat{f}(x_0) < \infty$, we then get the existence of $y^* \in X^*$ such that

$$\langle y^*, x_0\rangle + f^{**}(x_0) < \langle y^*, x\rangle + \hat{f}(x) \leq \langle y^*, x\rangle + f(x), \forall x \in X.$$

(To be sure of this, take a look at the proof of Theorem 2.2.21). This implies

$$f^{**}(x_0) < \langle -y^*, x_0\rangle - \sup_{x \in X}\{\langle -y^*, x\rangle - f(x)\} = \langle -y^*, x_0\rangle - f^*(-y^*),$$

which is impossible. We then have to understand what is going on when $\hat{f}(x_0) = \infty$. In the case that the separating hyperplane is not vertical, one concludes as before. In the other case, we have the existence of $y^* \in X^*, c \in \mathbb{R}$ such that

(i) $\langle y^*, x\rangle - c < 0 \, \forall x \in \text{dom } f$;
(ii) $\langle y^*, x_0\rangle - c > 0$.

Then

$$f(x) \geq \langle x^*, x\rangle + \alpha + t(\langle y^*, x\rangle - c), \forall x \in X, t > 0,$$

and this in turn implies, by conjugating twice, that

$$f^{**}(x) \geq \langle x^*, x\rangle + \alpha + t(\langle y^*, x\rangle - c), \forall x \in X, t > 0.$$

But then

$$f^{**}(x_0) \geq \langle x^*, x_0 \rangle + \alpha + t(\langle y^*, x_0 \rangle - c), \ \forall t > 0,$$

which implies $f^{**}(x_0) = \infty$.   □

Exercise 5.2.5 Let

$$f(x) = \begin{cases} -x^2 & \text{if } x \leq 0, \\ \infty & \text{otherwise.} \end{cases}$$

Find f^{**} and $\hat{f}$.

Proposition 5.2.6 *Let* $f: X \to [-\infty, \infty]$ *be a convex function and suppose* $f(x_0) \in \mathbb{R}$. *Then* f *is lower semicontinuous at* x_0 *if and only if* $f(x_0) = f^{**}(x_0)$.

Proof. We always have $f^{**}(x_0) \leq f(x_0)$ (Proposition 5.2.1). Now, suppose f is lower semicontinuous at x_0. Let us see first that $\bar{f}$ cannot assume value $-\infty$ at any point. On the contrary, suppose there is z such that $\bar{f}(z) = -\infty$. Then $\bar{f}$ is never real valued, and so $\bar{f}(x_0) = -\infty$, against the fact that f is lower semicontinuous and real valued at x_0. It follows that $\bar{f}$ has an affine minorizing function; thus

$$\bar{f} = \hat{\bar{f}} = (\bar{f})^{**} \leq f^{**}.$$

As $\bar{f}(x_0) = f(x_0)$, we finally have $f(x_0) = f^{**}(x_0)$. Suppose now $f(x_0) = f^{**}(x_0)$. Then

$$\liminf f(x) \geq \liminf f^{**}(x) \geq f^{**}(x_0) = f(x_0),$$

and this shows that f is lower semicontinuous at x_0.   □

The function

$$f(x) = \begin{cases} -\infty & \text{if } x = 0, \\ \infty & \text{otherwise,} \end{cases}$$

shows that the assumption $f(x_0) \in \mathbb{R}$ is *not* redundant in the above proposition. A more sophisticated example is the following one. Consider an infinite dimensional Banach space X, take $x^* \in X^*$ and a linear discontinuous functional l on X. Define

$$f(x) = \begin{cases} l(x) & \text{if } \langle x^*, x \rangle \geq 1, \\ \infty & \text{otherwise.} \end{cases}$$

Then f is continuous at zero, and it can be shown that $f^{**}(x) = -\infty$ for all x. Observe that f is lower semicontinuous at no point of its effective domain. This is the case because it can be shown that if there is at least a point of the effective domain of f where f is lower semicontinuous, then $f(x) = f^{**}(x)$

for all x such that f is lower semicontinuous (not necessarily real valued) at x ([Si2, Theorem 3.4]).

The next proposition shows that iterated application of the conjugation operation *does not* provide new functions.

Proposition 5.2.7 *Let $f: X \to (-\infty, \infty]$. Then $f^* = f^{***}$.*

Proof. As $f^{**} \leq f$, one has $f^* \leq f^{***}$. On the other hand, by definition of f^{***}, we have $f^{***}(x^*) = \sup_x\{\langle x^*, x \rangle - f^{**}(x)\}$, while, for all $x \in X$, $f^*(x^*) \geq \langle x^*, x \rangle - f^{**}(x)$, and this allows to conclude. □

Denote by $\Gamma^*(X^*)$ the functions of $\Gamma(X^*)$ which are conjugate of some function of $\Gamma(X)$. Then, from the previous results we get:

Theorem 5.2.8 *The operator $*$ is a bijection between $\Gamma(X)$ and $\Gamma^*(X^*)$.*

Proof. If $f \in \Gamma(X)$, f^* cannot be $-\infty$ at any point. Moreover, f^* cannot be identically ∞ as there is an affine function $l(\cdot)$ of the form $l(x) = \langle x^*, x \rangle - r$ minorizing f (Corollary 2.2.17), whence $f^*(x^*) \leq r$. These facts imply that $*$ actually acts between $\Gamma(X)$ and $\Gamma^*(X^*)$. To conclude, it is enough to observe that if $f \in \Gamma(X)$, then $f = f^{**}$ (Proposition 5.2.4). □

Remark 5.2.9 If X is not reflexive, then $\Gamma^*(X^*)$ is a proper subset of $\Gamma(X^*)$. It is enough to consider a linear functional on X^* which is the image of no element of X via the canonical embedding of X into X^{**}; it belongs to $\Gamma(X^*)$, but it is not the conjugate of any function $f \in \Gamma(X)$.

5.3 The subdifferentials of f and f^*

Let us see, by a simple calculus in a special setting, how it is possible to evaluate the conjugate f^* of a function f, and the connection between the derivative of f and that of f^*. Let $f: \mathbb{R}^n \to (-\infty, \infty]$ be a convex function. Since $f^*(x^*) = \sup_{x \in X}\{\langle x^*, x \rangle - f(x)\}$, we start by supposing that f is superlinear ($\lim_{\|x\| \to \infty} \frac{f(x)}{\|x\|} = \infty$) and thus we have that the supremum in the definition of the conjugate is attained, for every x^*. To find a maximum point, like every student we assume that the derivative of f is zero at the maximum point, called $\bar{x}$. We get $x^* - \nabla f(\bar{x}) = 0$. We suppose also that ∇f has an inverse. Then $\bar{x} = (\nabla f)^{-1}(x^*)$. By substitution we get

$$f^*(x^*) = \langle x^*, (\nabla f)^{-1}(x^*) \rangle - f((\nabla f)^{-1}(x^*)).$$

We try now to determine $\nabla f^*(x^*)$. We get

$$\nabla f^*(x^*) = (\nabla f)^{-1}(x^*) + \langle J_{(\nabla f)^{-1}}(x^*), x^* \rangle - \langle J_{(\nabla f)^{-1}}(x^*), \nabla f(\nabla f)^{-1})(x^*) \rangle$$
$$= (\nabla f)^{-1}(x^*),$$

where $J_{(\nabla f)^{-1}}$ denotes the jacobian matrix of the function $(\nabla f)^{-1}$. Then we have the interesting fact that the derivative of f is the inverse of the derivative of f^*. This fact can be fully generalized to subdifferentials, as we shall see in a moment.

Proposition 5.3.1 *Let $f\colon X \to (-\infty, \infty]$. Then $x^* \in \partial f(x)$ if and only if $f(x) + f^*(x^*) = \langle x^*, x \rangle$.*

Proof. We already know that

$$f(x) + f^*(x^*) \geq \langle x^*, x \rangle, \ \forall x \in X, x^* \in X^*.$$

If $x^* \in \partial f(x)$, then

$$f(y) - \langle x^*, y \rangle \geq f(x) - \langle x^*, x \rangle, \ \forall y \in X,$$

whence, $\forall y \in X$,

$$\langle x^*, y \rangle - f(y) + f(x) \leq \langle x^*, x \rangle.$$

Taking the supremum over all y in the left side provides one implication. As to the other one, if $f(x) + f^*(x^*) = \langle x^*, x \rangle$, then from the definition of f^*, we have that

$$f(x) + \langle x^*, y \rangle - f(y) \leq \langle x^*, x \rangle, \ \forall y \in X,$$

which shows that $x^* \in \partial f(x)$. $\qquad\square$

Proposition 5.3.1 has some interesting consequences. At first,

Proposition 5.3.2 *Let $f\colon X \to (-\infty, \infty]$. If $\partial f(x) \neq \emptyset$, then $f(x) = f^{**}(x)$. If $f(x) = f^{**}(x)$, then $\partial f(x) = \partial f^{**}(x)$.*

Proof. $\forall x \in X, \forall x^* \in X^*$, we have

$$f^*(x^*) + f^{**}(x) \geq \langle x, x^* \rangle.$$

If $x^* \in \partial f(x)$, by Proposition 5.3.1 we get

$$f^*(x^*) + f(x) = \langle x^*, x \rangle.$$

It follows that $f^{**}(x) \geq f(x)$, and this shows the first part of the claim. Suppose now $f(x) = f^{**}(x)$. Then, using the equality $f^* = (f^{**})^*$,

$$x^* \in \partial f(x) \iff \langle x^*, x \rangle = f(x) + f^*(x^*) = f^{**}(x) + f^{***}(x^*)$$
$$\iff x^* \in \partial f^{**}(x).$$

$\qquad\square$

Another interesting consequence is the announced connection between the subdifferentials of f and f^*.

Corollary 5.3.3 *Let* $f\colon X \to (-\infty, \infty]$. *Then*

$$x^* \in \partial f(x) \Longrightarrow x \in \partial f^*(x^*).$$

If $f(x) = f^{**}(x)$, *then*

$$x^* \in \partial f(x) \text{ if and only if } x \in \partial f^*(x^*).$$

Proof. $x^* \in \partial f(x) \Longleftrightarrow \langle x^*, x \rangle = f(x) + f^*(x^*)$. Thus $x^* \in \partial f(x)$ implies $f^{**}(x) + f^*(x^*) \leq \langle x^*, x \rangle$, and this is equivalent to saying that $x \in \partial f^*(x^*)$. If $f(x) = f^{**}(x)$,

$$x^* \in \partial f(x) \Longleftrightarrow \langle x^*, x \rangle = f(x) + f^*(x^*) = f^{**}(x) + f^*(x^*)$$
$$\Longleftrightarrow x \in \partial f^*(x^*).$$

$\square$

Thus, for a function $f \in \Gamma(X)$, it holds that $x^* \in \partial f(x)$ if and only if $x \in \partial f^*(x^*)$.

The above conclusion suggests how to draw the graph of the conjugate of a given function $f\colon \mathbb{R} \to \mathbb{R}$. We can construct the graph of its subdifferential, we "invert" it and we "integrate", remembering that, for instance, $f^*(0) = -\inf f$. See Figures 5.3–5.5 below.

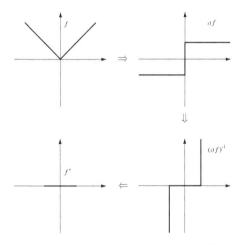

Figure 5.3. From the function to its conjugate through the subdifferentials.

A similar relation holds for approximate subdifferentials. For the following generalization of Proposition 5.3.1 holds:

Proposition 5.3.4 *Let* $f \in \Gamma(X)$. *Then* $x^* \in \partial_\varepsilon f(x)$ *if and only if* $f^*(x^*) + f(x) \leq \langle x^*, x \rangle + \varepsilon$. *Hence,* $x^* \in \partial_\varepsilon f(x)$ *if and only if* $x \in \partial_\varepsilon f^*(x^*)$.

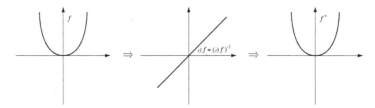

Figure 5.4. Another example.

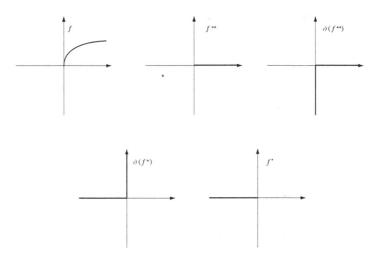

Figure 5.5. ... and yet another one.

Proof. $x^* \in \partial_\varepsilon f(x)$ if and only if

$$f(x) + \langle x^*, y \rangle - f(y) \le \langle x^*, x \rangle + \varepsilon, \ \forall y \in X,$$

if and only if $f(x) + f^*(x^*) \le \langle x^*, x \rangle + \varepsilon$. The second claim follows from $f = f^{**}$.   □

The previous proposition allows us to show that only in exceptional cases can the approximate subdifferential be a singleton (a nonempty, small set indeed).

Proposition 5.3.5 *Let* $f \in \Gamma(X)$ *and suppose there are* $x \in \text{dom } f$, $x^* \in X^*$ *and* $\bar\varepsilon > 0$ *such that* $\partial_{\bar\varepsilon} f(x) = \{x^*\}$. *Then* f *is an affine function.*

Proof. As a first step one verifies that $\partial_\varepsilon f(x) = \{x^*\}$ for all $\varepsilon > 0$. This is obvious if $\varepsilon < \bar\varepsilon$, because $\partial_\varepsilon f(x) \ne \emptyset$, and due to monotonicity. Furthermore, the convexity property described in Theorem 3.7.2 implies that $\partial_\varepsilon f(x)$ is a singleton also for $\varepsilon > \bar\varepsilon$. For, take $\sigma < \bar\varepsilon$ and suppose $\partial_\varepsilon f(x) \ni y^* \ne x^*$, for some $\varepsilon > \bar\varepsilon$. An easy but tedious calculation shows that being $\partial_\sigma f(x) \ni x^*$,

$\partial_{\bar{\varepsilon}}f(x) \ni \frac{\varepsilon-\bar{\varepsilon}}{\varepsilon-\sigma}x^* + \frac{\bar{\varepsilon}-\sigma}{\varepsilon-\sigma}y^* \neq x^*$, a contradiction. It follows, by Proposition 5.3.4, that if $y^* \neq x^*$,

$$f^*(y^*) > \langle y^*, x \rangle - f(x) + \varepsilon, \ \forall \varepsilon > 0,$$

and this implies $\operatorname{dom} f^* = \{x^*\}$. We conclude that f must be an affine function. □

5.4 The conjugate of the sum

Proposition 5.4.1 *Let $f, g \in \Gamma(X)$. Then*

$$(f\nabla g)^* = f^* + g^*.$$

Proof.

$$(f\nabla g)^*(x^*) = \sup_{x \in X}\left\{\langle x^*, x \rangle - \inf_{x_1+x_2=x}\{f(x_1) + g(x_2)\}\right\}$$

$$= \sup_{\substack{x_1 \in X \\ x_2 \in X}} \left\{\langle x^*, x_1 \rangle + \langle x^*, x_2 \rangle - f(x_1) - g(x_2)\right\} = f^*(x^*) + g^*(x^*).$$

□

Proposition 5.4.1 offers a good idea for evaluating $(f + g)^*$. By applying the above formula to f^*, g^* and conjugating, we get that

$$(f^*\nabla g^*)^{**} = (f^{**} + g^{**})^* = (f + g)^*.$$

So that if $f^*\nabla g^* \in \Gamma(X^*)$, then

$$(f + g)^* = f^*\nabla g^*.$$

Unfortunately we know that the inf-convolution operation between functions in $\Gamma(X)$ does not always produce a function belonging to $\Gamma(X)$; besides the case when at some point it is valued $-\infty$, it is not always lower semicontinuous. The next important theorem, due to Attouch–Brézis (see [AB]), provides a sufficient condition to get the result.

Theorem 5.4.2 *Let X be a Banach space and X^* its dual space. Let $f, g \in \Gamma(X)$. Moreover, let*

$$F := \mathbb{R}^+(\operatorname{dom} f - \operatorname{dom} g)$$

be a closed vector subspace of X. Then

$$(f + g)^* = f^*\nabla g^*,$$

and the inf-convolution is exact.

Proof. From the previous remark, it is enough to show that the inf-convolution is lower semicontinuous; in proving this we shall also see that it is exact (whence, in particular, it never assumes the value $-\infty$). We start by proving the claim in the particular case when $F = X$. From Exercise 2.2.4 it is enough to show that the level sets $(f^*\nabla g^*)^a$ are weak* closed for all $a \in \mathbb{R}$. On the other hand,

$$(f^*\nabla g^*)^a = \bigcap_{\varepsilon>0} C_\varepsilon := \{y^* + z^* : f^*(y^*) + g^*(z^*) \le a + \varepsilon\}.$$

It is then enough to show that the sets C_ε are weak* closed. Fixing $r > 0$, let us consider

$$K_{\varepsilon r} := \{(y^*, z^*) : f^*(y^*) + g^*(z^*) \le a + \varepsilon \text{ and } \|y^* + z^*\|_* \le r\}.$$

Then $K_{\varepsilon r}$ is a closed set in the weak* topology. Setting $T(y^*, z^*) = y^* + z^*$, we have that

$$C_\varepsilon \cap rB_{X^*} = T(K_{\varepsilon r}).$$

Since T is continuous from $X^* \times X^*$ to X^* (with the weak* topologies), if we show that $K_{\varepsilon r}$ is bounded (hence weak* compact), then $C_\varepsilon \cap rB^*$ is a weak* compact set, for all $r > 0$. The Banach–Dieudonné–Krein–Smulian theorem then guarantees that C_ε is weak* closed (See Theorem A.2.1 in Appendix B). Let us then show that $K_{\varepsilon r}$ is bounded. To do this, we use the uniform boundedness theorem. Thus, it is enough to show that $\forall y, z \in X$, there is a constant $C = C(y, z)$ such that

$$|\langle(y^*, z^*), (y, z)\rangle| = |\langle y^*, y\rangle + \langle z^*, z\rangle| \le C, \forall(y^*, z^*) \in K_{\varepsilon r}.$$

By assumption there is $t \ge 0$ such that $y - z = t(u - v)$ with $u \in \operatorname{dom} f$ and $v \in \operatorname{dom} g$. Then

$$
\begin{aligned}
|\langle y^*, y\rangle + \langle z^*, z\rangle| &= |t\langle y^*, u\rangle + t\langle z^*, v\rangle + \langle y^* + z^*, z - tv\rangle| \\
&\le |t(f(u) + f^*(y^*) + g(v) + g^*(z^*))| + r\|z - tv\| \\
&\le |t(a + \varepsilon + f(u) + g(v))| + r\|z - tv\| = C(y, z).
\end{aligned}
$$

The claim is proved in the case when $F = X$. Let us now turn to the general case. Suppose $u \in \operatorname{dom} f - \operatorname{dom} g$. Then $-u \in F$ and so there are $t \ge 0$ and $v \in \operatorname{dom} f - \operatorname{dom} g$ such that $-u = tv$. It follows that

$$0 = \frac{1}{1+t}u + \frac{t}{1+t}v \in \operatorname{dom} f - \operatorname{dom} g.$$

Hence $\operatorname{dom} f \cap \operatorname{dom} g \ne \emptyset$ and after a suitable translation, we can suppose that $0 \in \operatorname{dom} f \cap \operatorname{dom} g$, whence $\operatorname{dom} f \subset F$, $\operatorname{dom} g \subset F$. Let $i: F \to X$ be the canonical injection of F in X and let $i^*: X^* \to F^*$ be its adjoint operator: $\langle i^*(x^*), d\rangle = \langle x^*, i(d)\rangle$. Let us consider the functions

$$\tilde{f}: F \to (-\infty, \infty], \ \tilde{f} := f \circ i, \quad \tilde{g}: F \to (-\infty, \infty], \ \tilde{g} := g \circ i.$$

We can apply the first step of the proof to them. We have

$$(\tilde{f} + \tilde{g})^*(z^*) = (\tilde{f}^* \nabla \tilde{g}^*)(z^*),$$

for all $z^* \in F^*$. It is now easy to verify that if $x^* \in X^*$,

$$f^*(x^*) = \tilde{f}^*(i^*(x^*)), \quad g^*(x^*) = \tilde{g}^*(i^*(x^*)),$$
$$(f + g)^*(x^*) = (\tilde{f} + \tilde{g})^*(i^*(x^*)), \quad (f^* \nabla g^*)(x^*) = (\tilde{f}^* \nabla \tilde{g}^*)(i^*(x^*)),$$

(in the last one we use that i^* is onto).

For instance, we have

$$\tilde{f}^*(i^*(x^*)) = \sup_{z \in F}\{\langle i^*(x^*), z \rangle - \tilde{f}(z)\} = \sup_{z \in F}\{\langle x^*, i(z) \rangle - f(i(z))\}$$
$$= \sup_{x \in X}\{\langle x^*, x \rangle - f(x)\},$$

where the last inequality holds as dom $f \subset F$. The others follow in the same way. Finally, the exactness at a point $x^* \in$ dom $f^* \nabla g^*$ follows from the compactness, previously shown, of $K_{\varepsilon, \|x^*\|_*}$, with $a = (f^* \nabla g^*)(x^*)$ and $\varepsilon > 0$ arbitrary. This allows us to conclude. $\qquad\square$

Besides its intrinsic interest, the previous theorem yields the following sum rule for the subdifferentials which generalizes Theorem 3.4.2.

Theorem 5.4.3 *Let $f, g \in \Gamma(X)$. Moreover, let*

$$F := \mathbb{R}_+(\text{dom } f - \text{dom } g)$$

be a closed vector space. Then

$$\partial(f + g) = \partial f + \partial g.$$

Proof. Let $x^* \in \partial(f + g)(x)$. We must find $y^* \in \partial f(x)$ and $z^* \in \partial g(x)$ such that $y^* + z^* = x^*$. By the previous result there are y^*, z^* such that $y^* + z^* = x^*$ and fulfilling $f^*(y^*) + g^*(z^*) = (f + g)^*(x^*)$. As $x^* \in \partial(f + g)(x)$ we have (Proposition 5.3.1)

$$\langle y^*, x \rangle + \langle z^*, x \rangle = \langle x^*, x \rangle = (f + g)(x) + (f + g)^*(x^*)$$
$$= f(x) + f^*(y^*) + g(x) + g^*(z^*).$$

This implies (why?)

$$\langle y^*, x \rangle = f(x) + f^*(y^*) \text{ and } \langle z^*, x \rangle = g(x) + g^*(z^*),$$

and we conclude. $\qquad\square$

The previous generalization is useful, for instance, in the following situation: suppose we have a Banach space Y, a (proper) closed subspace X and two continuous functions $f, g \in \Gamma(X)$ fulfilling the condition $\operatorname{int} \operatorname{dom} f \cap \operatorname{dom} g \neq \emptyset$. It can be useful sometimes to consider the natural extensions $\tilde{f}, \tilde{g} \in \Gamma(Y)$ of f and g (by defining them ∞ outside X). In such a case the previous theorem can be applied, while Theorem 3.4.2 obviously cannot.

Exercise 5.4.4 Let

$$f(x, y) = \begin{cases} -\sqrt{xy} & \text{if } x \leq 0, y \leq 0, \\ \infty & \text{otherwise,} \end{cases}$$

$$g(x, y) = \begin{cases} -\sqrt{-xy} & \text{if } x \geq 0, y \leq 0, \\ \infty & \text{otherwise.} \end{cases}$$

Find $(f + g)^*$ and $f^* \nabla g^*$.

Exercise 5.4.5 Given a nonempty closed convex set K,

$$d^*(\cdot, K) = \sigma_K + I_{B^*}.$$

Hint. Remember that $d(\cdot, K) = (\| \ \| \nabla I_K)(\cdot)$ and apply Proposition 5.4.1.

Exercise 5.4.6 Let X be a reflexive Banach space. Let $f, g \in \Gamma(X)$. Let

$$\lim_{\|x\| \to \infty} \frac{f(x)}{\|x\|} = \infty.$$

Then $(f \nabla g) \in \Gamma(X)$.

Hint. Try to apply the Attouch–Brézis theorem to f^*, g^*.

5.5 Sandwiching an affine function between a convex and a concave function

In this section we deal with the following problem: suppose we are given a Banach space X and two convex, lower semicontinuous extended real valued functions f and g such that $f(x) \geq -g(x) \ \forall x \in X$. The question is: when is it possible to find an affine function m with the property that

$$f(x) \geq m(x) \geq -g(x),$$

for all $x \in X$? It is clear that the problem can be restated in an equivalent, more geometric, way: suppose we can separate the sets $\operatorname{epi} f$ and $\operatorname{hyp}(-g)$ with a nonvertical hyperplane. With a standard argument this provides the affine function we are looking for. And, clearly, the condition $f \geq -g$ gives some hope to be able to make such a separation.

In order to study the problem, let us first observe the following simple fact.

Proposition 5.5.1 *Let $y^* \in X^*$. Then $y^* \in \{p : f^*(p) + g^*(-p) \leq 0\}$ if and only if there exists $a \in \mathbb{R}$ such that*

$$f(x) \geq \langle y^*, x \rangle + a \geq -g(x),$$

for all $x \in X$.

Proof. Suppose $f^*(y^*) + g^*(-y^*) \leq 0$. Then, for all $x \in X$,

$$\langle y^*, x \rangle - f(x) + g^*(-y^*) \leq 0,$$

i.e.,

$$f(x) \geq \langle y^*, x \rangle + a,$$

with $a = g^*(-y^*)$. Moreover

$$a = g^*(-y^*) \geq \langle -y^*, x \rangle - g(x),$$

for all $x \in X$, implying $\langle y^*, x \rangle + a \geq -g(x)$, for all $x \in X$. Conversely, if $f(x) \geq \langle y^*, x \rangle + a$ and $\langle y^*, x \rangle + a \geq -g(x)$ for all x, then

$$-a \geq f^*(y^*), \quad a \geq \langle -y^*, x \rangle - g(x),$$

for all x, implying $f^*(y^*) + g^*(-y^*) \leq 0$. $\qquad\square$

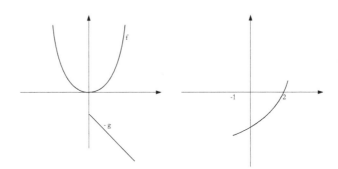

$$f(x) = \tfrac{1}{2}x^2, g(x) = \begin{cases} x+3 & \text{if } x \geq 0, \\ \infty & \text{otherwise.} \end{cases} \qquad\qquad f^*(\,\cdot\,) + g^*(-\,\cdot\,).$$

Figure 5.6.

It follows in particular that the set of the "slopes" of the affine functions sandwiched between f and $-g$ is a weak* closed and convex set, as it is the zero level set of the function $h(\cdot) = f^*(\cdot) + g^*(-\,\cdot)$. Now, observe that $\inf_x (f+g)(x) \geq 0$ if and only if $(f+g)^*(0^*) \leq 0$. Thus, *if*

$$(f + g)^*(0^*) = (f^* \nabla g^*)(0^*)$$

and the epi-sum is exact, then $\inf_x (f + g)(x) \geq 0$ is equivalent to saying that there exists $y^* \in X^*$ such that

$$(f^* \nabla g^*)(0^*) = f^*(y^*) + g^*(-y^*) \leq 0.$$

Thus a sufficient condition to have an affine function sandwiched between f and $-g$ is that the assumption of the Attouch–Brezis theorem be satisfied.

Now we specialize to the case when X is a Euclidean space. In this case the condition $f \geq -g$ implies that

$$\mathrm{ri}\,\mathrm{epi}\,f \cap \mathrm{ri}\,\mathrm{hyp}(-g) = \emptyset.$$

Then we can apply Theorem A.1.13 to separate the sets epi f and hyp$(-g)$. However, this does not solve the problem, as it can happen that the separating hyperplane is vertical. So, let us now see a sufficient condition in order to assure that the separating hyperplane is not vertical, which amounts to saying that the affine function we are looking for is finally singled out.

Proposition 5.5.2 *Suppose*

$$\mathrm{ri}\,\mathrm{dom}\,f \cap \mathrm{ri}\,\mathrm{dom}(-g) \neq \emptyset.$$

Then there exists y^* such that $f^*(y^*) + g^*(-y^*) \leq 0$.

Proof. Let us use the Attouch–Brezis theorem, as suggested at the beginning of the section. Thus, we must show that

$$F := \mathbb{R}_+(\mathrm{dom}\,f - \mathrm{dom}\,g)$$

is a subspace. As is suggested in the next exercise, it is enough to show that if $x \in F$, then $-x \in F$. We can suppose, without loss of generality, that $0 \in \mathrm{ri}\,\mathrm{dom}\,f \cap \mathrm{ri}\,\mathrm{dom}\,g$. As $x \in F$, there are $l > 0$, $u \in \mathrm{dom}\,f$ and $v \in \mathrm{dom}\,g$ such that $x = l(u - v)$. As $0 \in \mathrm{ri}\,\mathrm{dom}\,f \cap \mathrm{ri}\,\mathrm{dom}\,g$, there is $c > 0$ small enough such that $-cu \in \mathrm{dom}\,f$, $-cv \in \mathrm{dom}\,g$. Thus $-cu - (-cv) \in \mathrm{dom}\,f - \mathrm{dom}\,g$. Then

$$\frac{l}{c}(-cu - (-cv)) = -x \in F.$$

$\square$

Exercise 5.5.3 Let A be a convex set containing zero. Then $\bigcup_{\lambda > 0} \lambda A$ is a convex cone. Moreover, if $x \in \bigcup_{\lambda > 0} \lambda A$ implies $-x \in \bigcup_{\lambda > 0} \lambda A$, then $\bigcup_{\lambda > 0} \lambda A$ is a subspace.

Hint. Call $F = \bigcup_{\lambda > 0} \lambda A$. It has to be shown that $x, y \in F$ implies $x + y \in F$. There are positive l_1, l_2 and $u, v \in A$ such that $x = l_1 u$, $y = l_2 v$. Then $x/l_1 \in A$, $y/l_2 \in A$ and $\frac{1}{l_1 + l_2}(x + y)$ is a convex combination of x/l_1 and y/l_2.

We now give some pretty examples showing that the affine function separating epi f and hyp$(-g)$ need not exist, unless some extra condition is imposed.

Example 5.5.4

$$f(x) = \begin{cases} -\sqrt{x} & \text{if } x \geq 0, \\ \infty & \text{otherwise}, \end{cases}$$

$$g(x) = \begin{cases} 0 & \text{if } x = 0, \\ \infty & \text{otherwise}. \end{cases}$$

Here $\inf(f + g) = 0$, and $\mathrm{ri}(\mathrm{dom}\, f) \cap \mathrm{ri}(\mathrm{dom}\, g) = \emptyset$.

Example 5.5.5

$$f(u, v) = \begin{cases} -1 & \text{if } uv \geq 1, u \geq 0, \\ \infty & \text{otherwise}, \end{cases}$$

$$g(u, v) = \begin{cases} 0 & \text{if } u \geq 0, v = 0, \\ \infty & \text{otherwise}. \end{cases}$$

Here we have $\mathrm{dom}\, f \cap \mathrm{dom}\, g = \emptyset$.

Example 5.5.6

$$f(u, v) = \begin{cases} u & \text{if } v = -1, \\ \infty & \text{otherwise}, \end{cases}$$

$$g(u, v) = \begin{cases} 0 & \text{if } v = 0, \\ \infty & \text{otherwise}. \end{cases}$$

The Example 5.5.4 can induce the idea that the separator must be vertical as the two effective domains do intersect at a point. So, it could be argued that, if the two domain are far apart, the property could hold. But in Example 5.5.6 the distance between $\mathrm{dom}\, f$ and $\mathrm{dom}\, g$ is 1.

In the last two examples the domains of f and g do not intersect, while in the first example a crucial role is played by the fact that $\inf(f + g) = 0$. In the following example $\inf(f + g) > 0$, and yet there is no affine separator. Observe that such example could not be provided in one dimension (see Remark 2.2.15).

Example 5.5.7

$$f(u, v) = \begin{cases} 1 - 2\sqrt{uv} & \text{if } u, v \geq 0, \\ \infty & \text{otherwise}, \end{cases}$$

$$g(u, v) = \begin{cases} 1 - 2\sqrt{-uv} & \text{if } u \leq 0, v \geq 0, \\ \infty & \text{otherwise}. \end{cases}$$

A straightforward calculation shows

$$f^*(u^*, v^*) = \begin{cases} -1 & \text{if } u^* \leq 0, u^* v^* \geq 1, \\ \infty & \text{otherwise,} \end{cases}$$

$$g^*(u^*, v^*) = \begin{cases} -1 & \text{if } u^* \geq 0, u^* v^* \leq -1, \\ \infty & \text{otherwise.} \end{cases}$$

Our finite dimensional argument actually holds, without any changes in the proof, provided we assume that at least one of the sets epi f, hyp$(-g)$ has an interior point. In particular, the assumption in Proposition 5.5.2 becomes, in infinite dimensions, int dom $f \cap$ dom $g \neq \emptyset$. To conclude, let me mention that this section is inspired by my work with Lewis [LeL], where we studied the more general problem of giving sufficient conditions under which the slope of the affine function between f and $-g$ is in the range (or in the closure of the range) of the Clarke subdifferential of a locally Lipschitz function h such that $f \geq h \geq -g$.

6

Duality

Vergine Madre, Figlia del tuo figlio
Umile ed alta più che creatura
Termine fisso d'eterno consiglio
Tu sei colei che l'umana natura
Nobilitasti sì che il suo fattore
non disdegnò di farsi sua fattura
(D. Alighieri, "La Commedia", Canto XXXIII)

As we anticipated in the previous chapter, this one is dedicated to introducing a general scheme for a duality theory. This means that we associate to a given problem another one, with the idea that the second can provide useful information on the original one. For instance, we shall see that the value of the dual problem always provides a lower bound for the value of the original one. This can sometimes be useful when it is difficult to find the infimum of the initial problem. But what makes this approach even more interesting is the fact that quite often the dual problem also has a concrete interpretation in view of the initial one.

Just to provide an example, we shall see that finding an optimal strategy for a player in a zero sum finite, two person game (zero sum means that what one gains is what the other one pays) can be reduced to a linear programming problem, and that its dual is exactly the problem the other player must solve to find an optimal strategy for himself. Thus, after introducing our general duality scheme, we specialize to the convex case and single out an interesting class of problems, the regular ones, providing a list of their properties.

Next, we prove that a problem of a special form is regular, and from this we see that it is possible, via duality theory, to get without much effort the Euler equation for a problem in calculus of variations. We also consider the case of convex programming, and introduce the Lagrangean, Lagrange multipliers and so on. The program will be completed in Chapter 7, which focuses on linear programming.

6.1 The setting

Throughout this chapter, two Banach spaces X, P are given, together with their dual spaces X^*, P^*, respectively. We shall also consider the natural duality between the product spaces $X \times P$ and $X^* \times P^*$: $\langle (x^*, p^*), (x, p) \rangle :=$ $\langle x^*, x \rangle + \langle p^*, p \rangle$.

We are interested in formulating a *dual* problem to the minimum problem:

$$(\mathcal{P}) \qquad\qquad \inf_{x \in X} f(x), \qquad\qquad (6.1)$$

where $f : X \to [-\infty, \infty]$ is a given function. As usual, we say that $(\mathcal{P})$ has solutions if there is $\bar{x} \in X$ such that $-\infty < f(\bar{x}) \leq f(x)\ \forall x \in X$. Suppose also we are given a function $F : X \times P \to [-\infty, \infty]$ with the property that $F(x, 0_P) = f(x)$. Such a function F gives rise to a family (parameterized by $p \in P$) of problems:

$$(\mathcal{P}_p) \qquad\qquad \inf_{x \in X} F(x, p). \qquad\qquad (6.2)$$

We shall denote by $\inf(p)$ the value $\inf_{x \in X} F(x, p)$ and by $\mathrm{Min}(p)$ the (possibly empty) set of the solutions of $(\mathcal{P}_p)$.

Thus $\inf(0_P)$ and $\mathrm{Min}(0_P)$ are the basic objects of our initial problem.

The family $(\mathcal{P}_p)$ allows us to define dual problems.

Consider $F^* : X^* \times P^* \to [-\infty, \infty]$ and define the *dual problem* $(\mathcal{P}^*)$ to $(\mathcal{P})$ in the following way:

$$(\mathcal{P}^*) \qquad\qquad \sup_{p^* \in P^*} \{ -F^*(0_{X^*}, p^*) \}. \qquad\qquad (6.3)$$

Denote by $\sup(0_{X^*})$ and $\mathrm{Max}(0_{X^*})$ its value and the set of its solutions. The dual problem represents a maximum problem for a concave function, which is naturally equivalent to a minimum problem for a convex function (when expressing it as a maximum problem in a book where we almost always speak of minima, it was not an easy choice where to put pluses and minuses).

The problem $(\mathcal{P}^*)$ too is naturally embedded in a family (parameterized by $x^* \in X^*$) of dual problems $(\mathcal{P}^*_{x^*})$:

$$\sup_{p^* \in P^*} \{ -F^*(x^*, p^*) \}. \qquad\qquad (6.4)$$

This allows dualizing the problem $(\mathcal{P}^*)$, to finally get the bidual problem:

$$(\mathcal{P}^{**}) \qquad\qquad \inf_{x \in X} F^{**}(x, 0_P). \qquad\qquad (6.5)$$

Thus, if $F \in \Gamma(X \times P)$ the bidual problem is exactly the initial problem (otherwise one can speak about the relaxation of the initial problem). It is clear, but worth emphasizing, that the form of the dual problem of a problem $(\mathcal{P})$ is strongly affected by the choice of the parameterized family $(\mathcal{P}_p)$ one defines. To change this family means having a different dual problem. Observe that for some p, it can happen that $F(x, p) = \infty$ for all x. In this case, of course, the value of the minimum problem is ∞. This typically happens in constrained problems, when the constraint set is empty. An analogous situation can clearly occur for the dual problem.

For the reader's convenience we state in a proposition some previous results which will be frequently used below.

Proposition 6.1.1 *Let* $f \colon X \to [-\infty, \infty]$ *be a convex function and suppose* $f(x) \in \mathbb{R}$. *Then*

(i) $f(x) = f^{**}(x)$ *if and only if* f *is lower semicontinuous at* x *(Exercise 5.2.6).*

(ii) $f(x) = f^{**}(x)$ *implies* $\partial f(x) = \partial f^{**}(x)$ *(Proposition 5.3.2).*

(iii) $\partial f(x) \neq \emptyset$ *implies* f *lower semicontinuous at* x; *this in particular implies* $f(x) = f^{**}(x)$ *and so* $\partial f(x) = \partial f^{**}(x)$.

(iv)   *However it can happen that* $f(x) > f^{**}(x)$ *at a certain point* x, *and* $\partial f^{**}(x) \neq \emptyset$.

6.2 Fundamentals

First of all, let us make the following easy, yet crucial, remark. From the very definition of conjugate function, we have that

$$F(x, 0_P) + F^*(0_{X^*}, p^*) \geq \langle (0_{X^*}, p^*), (x, 0_P) \rangle = 0,$$

for all $x \in X$, $p^* \in P^*$. This immediately implies the following:

Proposition 6.2.1 *We have*

$$\inf(0_P) \geq \sup(0_{X^*}).$$

Thus the value of the dual problem provides a lower bound to the value of the initial one. The difference $\inf(0_P) - \sup(0_{X^*})$, always nonnegative, is called *the duality gap.* The interesting case is when the two values agree. In such a situation, one says that *there is no duality gap.*

Proposition 6.2.2 *The following are equivalent:*

(i) $\bar{x}$ *solves the initial problem, and* $\overline{p^*}$ *solves the dual problem;*

(ii) $(\bar{x}, \overline{p^*})$ *minimizes* $(x, p^*) \mapsto F(x, 0_P) + F^*(0_{X^*}, p^*)$;

We can summarize the previous remarks by means of the following:

Proposition 6.2.3 *The following are equivalent:*

(i) $F(\bar{x}, 0_P) + F^*(0_{X^*}, \overline{p^*}) = 0$;

(ii) $\bar{x}$ *solves the initial problem,* $\overline{p^*}$ *solves the dual problem, and there is no duality gap;*

(iii) $(0_{X^*}, \overline{p^*}) \in \partial F(\bar{x}, 0_P)$.

Proof. The equivalence between (i) and (iii) follows from Proposition 5.3.1.

$\square$

Let us see now some more refined relations between two problems in duality.

Proposition 6.2.4 $\inf^*(p^*) = F^*(0_{X^*}, p^*)$.

Proof.

$$\inf{}^*(p^*) = \sup_{p \in P}\{\langle p^*, p\rangle - \inf(p)\} = \sup_{p \in P}\{\langle p^*, p\rangle - \inf_{x \in X} F(x, p)\}$$

$$= \sup_{p \in P}\sup_{x \in X}\{\langle p^*, p\rangle - F(x, p)\} = F^*(0_{X^*}, p^*).$$

$\square$

Thus the Fenchel conjugate of the value function is, with a change of sign, the function to be maximized in the associated dual problem. This observation yields the following:

Corollary 6.2.5 $\sup(0_{X^*}) = \inf{}^{**}(0_P)$.

Proof. By Proposition 6.2.4,

$$\sup(0_{X^*}) = \sup_{p^* \in P^*}\{-F^*(0_{X^*}, p^*)\} = \sup_{p^* \in P^*}\{\langle 0_P, p^*\rangle - \inf{}^*(p^*)\} = \inf{}^{**}(0_P).$$

$\square$

From the previous result, we can once more get the known relation

Corollary 6.2.6 $\sup(0_{X^*}) \leq \inf(0_P)$.

Here is a second, general result.

Proposition 6.2.7 *Suppose* $\inf{}^{**}(0_P) \in \mathbb{R}$. *Then* $\mathrm{Max}(0_{X^*}) = \partial \inf{}^{**}(0_P)$.

Proof. Let $p^* \in \mathrm{Max}(0_{X^*})$. From Proposition 6.2.4

$$-\inf{}^*(p^*) = -F^*(0_{X^*}, p^*) = \sup_{q^* \in P^*} -F^*(0_{X^*}, q^*)$$

$$= \sup_{q^* \in P^*}\{\langle q^*, 0_P\rangle - \inf{}^*(q^*)\} = \inf{}^{**}(0_P),$$

giving

$$\inf{}^{**}(0_P) + \inf{}^*(p^*) = \langle 0_P, p^*\rangle,$$

whence $p^* \in \partial \inf{}^{**}(0_P)$, and conversely.

$\square$

Thus the solution set of the dual problem is connected to the subdifferential of the biconjugate of the value (inf) function of the initial problem. It is then quite interesting to know when the function inf coincides with its biconjugate $\inf^{**}$ (at least at the point 0_P), an equality that also entails $\partial \inf^{**}(0_P) = \partial \inf(0_P)$. This clearly suggests paying particular attention to the convex case, and this is what we shall do in the next section.

6.3 The convex case

The results of the previous section hold for general problems. Now we specialize to the convex case. To start with, we enrich the information contained in the Proposition 6.2.3.

Theorem 6.3.1 *Let $F \in \Gamma(X \times P)$. Then the following are equivalent:*

(i) $F(\bar{x}, 0_P) + F^*(0_{X^*}, \overline{p^*}) = 0$.
(ii) $\bar{x}$ *solves the initial problem,* $\bar{p}^*$ *solves the dual problem, and there is no duality gap.*
(iii) $(\bar{x}, 0_P) \in \partial F^*(0_{X^*}, \bar{p}^*)$.
(iv) $(0_{X^*}, \bar{p}^*) \in \partial F(\bar{x}, 0_P)$.

Proof. Since $F \in \Gamma(X \times P)$, Corollary 5.3.3 entails that (i) is equivalent to (iii) and (iv).   □

We have seen in Corollary 6.2.5 that there is no duality gap if and only if the value function coincides at 0_P with its biconjugate. This surely happens if the value function is convex and lower semicontinuous at 0_P. Thus we now turn our attention to cases when the value function fulfills these conditions. We start by investigating convexity.

Proposition 6.3.2 *Let $F \in \mathcal{F}(X \times P)$. Then $\inf \colon P \to [-\infty, \infty]$ is a convex function.*

Proof. Let $p_1, p_2 \in P$, $\lambda \in (0, 1)$ and $\inf(p_1), \inf(p_2) \in [-\infty, \infty]$. If

$$\max\{\inf(p_1), \inf(p_2)\} = \infty,$$

there is nothing to prove. Suppose then that $\max\{\inf(p_1), \inf(p_2)\} < \infty$ and let $a > \inf(p_1)$ and $b > \inf(p_2)$. Then there are $x_1, x_2 \in X$ such that $F(x_1, p_1) \leq a$ and $F(x_2, p_2) \leq b$. It follows that

$$\inf(\lambda p_1 + (1 - \lambda)p_2) \leq F(\lambda x_1 + (1 - \lambda)x_2), \lambda p_1 + (1 - \lambda)p_2)$$
$$\leq \lambda F(x_1, p_1) + (1 - \lambda)F(x_2, p_2) \leq \lambda a + (1 - \lambda)b.$$

We conclude, since $a > \inf(p_1)$ and $b > \inf(p_2)$, arbitrary.   □

Remark 6.3.3 The proof above relies on the fact that

$$\text{epi}_s \inf = \text{proj}_{P \times \mathbb{R}} \text{epi}_s F,$$

and that the projection of a convex set is convex as well. Thus, we have seen that convexity of F (in both variables!) guarantees convexity of the value function $\inf(\cdot)$. On the other hand, to have $\inf(\cdot) \in \Gamma(P)$ it is not enough to assume that $F \in \Gamma(X \times P)$. To begin with, easy examples show that the function inf can assume the value $-\infty$. Moreover, lower semicontinuity of the value inf does *not* follow, in general, from the same property of F, as the next example shows.

Example 6.3.4 Let X be a separable Hilbert space, with basis $\{e_n : n \in \mathbb{N}\}$, and let $P = \mathbb{R}$. Let $x^* = \sum_{n=1}^{\infty} \frac{1}{n} e_n$, $f_0(x) = \max\{-1, \langle x^*, x \rangle\}$, and define $g(x) = \sum_{n=1}^{\infty} \frac{(x, e_n)^2}{n^4}$. Finally, let

$$F(x, p) = \begin{cases} f_0(x) & \text{if } g(x) \leq p, \\ \infty & \text{otherwise.} \end{cases}$$

It is not difficult to verify that $\inf(p) = \infty$ if $p < 0$, $\inf(p) = -1$ if $p > 0$, while $\inf(0) = 0$.

The next proposition summarizes some previous claims:

Proposition 6.3.5 *Let $F \in \Gamma(X \times P)$ and suppose $\inf(0_P) \in \mathbb{R}$. Then the following are equivalent:*

(i)   inf *is lower semicontinuous at 0_P;*
(ii) $\inf(0_P) = \sup(0_{X^*})$.

Proof. Let us start by showing that (i) implies (ii). From Corollary 6.2.5 it is enough to verify that $\inf(0_P) = \inf^{**}(0_P)$. But inf is convex (Proposition 6.3.2) and lower semicontinuous at 0_P by assumption. Thus $\inf(0_P) = \inf^{**}(0_P)$ (see Proposition 6.1.1). Conversely, if $\inf(0_P) = \sup(0_{X^*}) = \inf^{**}(0_P)$, then inf is lower semicontinuous at 0_P, since it coincides, at that point, with its biconjugate (see Proposition 6.1.1). □

The condition $\inf(0_P) \in \mathbb{R}$, needed only to show that (i) implies (ii), is aimed at avoiding degenerate situations, like the following one. It is possible to have a family of constrained problems with no feasible points, for every p around 0_P. This means that $\inf(\cdot)$ is valued ∞ around 0_P, and continuous at 0_P. Analogously the same can happen for the dual problem. Thus in this case (i) is true, while (ii) is not, and there is a duality gap. Notwithstanding these are pathological situations, one cannot ignore them, as the value functions are usually extended-real valued.

Thus, the fact that there is no duality gap can be expressed in an equivalent form, by saying that the function inf is lower semicontinuous at 0_P. Let us now summarize the results established concerning the value functions.

The value of the dual problem always provides a lower bound to the value of the initial problem. Moreover, in the convex case, there is no duality gap (i.e., the two values coincide) if and only if the value function inf, related to the initial problem, is lower semicontinuous at the point 0_P.

In the following section we shall pay attention to the solution multifunction of the problems, and we shall single out a class of well behaved problems.

6.4 Regular problems

We assume throughout this section that $F \in \Gamma(X \times P)$.

Definition 6.4.1 We say that the problem $(\mathcal{P})$ is *regular* if $\inf(0_P) = \sup(0_{X^*}) \in \mathbb{R}$ and if the dual problem $(\mathcal{P}^*)$ has solutions.

Thus a problem $(\mathcal{P})$ in a given duality scheme is regular whenever there is no duality gap, and the associated dual problem has solutions. Let us now see a characterization of regularity.

Proposition 6.4.2 *The following conditions are equivalent:*

(i) $(\mathcal{P})$ *is regular.*
(ii) $\partial \inf(0_P) \neq \emptyset$.

Proof. If $(\mathcal{P})$ is regular, then $\inf(0_P) = \inf^{**}(0_P)$ and $\emptyset \neq \text{Max}(0_{X^*}) = \partial \inf^{**}(0_P) = \partial \inf(0_P)$. Conversely, $\partial \inf(0_P) \neq \emptyset$ implies $\mathbb{R} \ni \inf(0_P)$ and the value function $\inf(\cdot)$ is lower semicontinuous at 0_P; moreover, $\emptyset \neq \partial \inf(0_P) = \partial \inf^{**}(0_P) = \text{Max}(0_{X^*})$, thus the dual problem has solutions and the problem $(\mathcal{P})$ is regular. $\square$

Thus an equivalent way to define regularity is to say that the value function has a nonempty subdifferential at zero. We now give a condition providing regularity.

Proposition 6.4.3 *If* $\inf(0_P) \in \mathbb{R}$ *and if*

> there exists $x_0 \in X$ such that $p \mapsto F(x_0, p)$ is finite and continuous at 0_P,   (6.6)

then the problem $(\mathcal{P})$ is regular.

Proof. $\inf(0_P) \in \mathbb{R}$ by assumption. As the value function $\inf(\cdot)$ is a convex function, it is enough to show that it is continuous at 0_P in order to have also that $\partial \inf(0_P) \neq \emptyset$, and this, in view of Proposition 6.4.2, will conclude the proof. Now, from (6.6), the function $p \mapsto F(x_0, p)$ is continuous at 0_P, hence there are a neighborhood $I(0_P)$ and $m \in \mathbb{R}$ such that $F(x_0, p) \leq m \, \forall p \in I(0_P)$. Then $\inf(p) \leq m, \forall p \in I(0)$, whence the convex function inf is upper bounded in a neighborhood of 0_P and thus continuous at 0_P (see Lemma 2.1.1). $\square$

We now study an interesting problem, and we prove that under suitable assumptions, it is regular. Suppose we are given two Banach spaces X, Y, a linear bounded operator $L: X \to Y$, a function $H: X \times Y \to (-\infty, \infty]$ and suppose we have to minimize

$$f(x) = H(x, Lx).$$

The parameter space P will be any closed subspace of Y containing

$$\{Lx : H(x, Lx) < \infty\}.$$

The function F is defined as $F(x, p) = H(x, Lx + p)$. Let us start by finding the dual problem:

$$
\begin{aligned}
&F^*(x^*, p^*)\\
&= \sup_{\substack{x \in X \\ p \in P}} \{\langle x^*, x\rangle + \langle p^*, p\rangle - H(x, Lx + p)\}\\
&= \sup_{\substack{x \in X \\ p \in P}} \{\langle x^* - L^*p^*, x\rangle + \langle L^*p^*, x\rangle + \langle p^*, p\rangle - H(x, Lx + p)\}\\
&= \sup_{\substack{x \in X \\ p \in P}} \{\langle x^* - L^*p^*, x\rangle + \langle p^*, p + Lx\rangle - H(x, Lx + p)\}\\
&\qquad\qquad\qquad\qquad\qquad\qquad\qquad\qquad \text{(setting } Lx + p = y)\\
&= \sup_{\substack{x \in X \\ p \in P}} \{\langle x^* - L^*p^*, x\rangle + \langle p^*, y\rangle - H(x, y)\} = H^*(x^* - L^*p^*, p^*).
\end{aligned}
$$

Thus the dual problem consists in maximizing

$$-H^*(-L^*p^*, p^*).$$

Suppose both problems have solutions and that there is no duality gap. If $\bar{x}$ is a solution of the initial problem and $\bar{p}^*$ of the dual problem, then

$$(-L^*\bar{p}^*, \bar{p}^*) \in \partial H(\bar{x}, L\bar{x}).$$

For, from Theorem 6.3.1, we have that

$$F(\bar{x}, 0_P) + F^*(0_{X^*}, \bar{p}^*) = 0,$$

and here this becomes

$$H(\bar{x}, L\bar{x}) + H^*(-L^*\bar{p}^*, \bar{p}^*) = 0.$$

On the other hand, we have

$$0 = \langle -L^*\bar{p}^*, \bar{x}\rangle + \langle \bar{p}^*, L\bar{x}\rangle = \langle(-L^*\bar{p}^*, \bar{p}^*), (\bar{x}, L\bar{x})\rangle.$$

Thus

$$H^*(-L^*\bar{p}^*, \bar{p}^*) + H(\bar{x}, L\bar{x}) = \langle(-L^*\bar{p}^*, \bar{p}^*), (\bar{x}, L\bar{x})\rangle,$$

and this is equivalent to the condition provided above.

We specialize now to a particular case to show regularity. We suppose H has the form

$$H(x, y) = \begin{cases} h(x, y) & \text{if } x \in C, y \in D, \\ \infty & \text{otherwise,} \end{cases}$$

where $C \subset X$, $D \subset Y$ are two nonempty closed convex sets, and h is a continuous (real valued) function on $C \times D$. Then our problem becomes

$(\mathcal{P}_h)$ $$\inf_{x \in C: Lx \in D} h(x, Lx)$$

We make the basic assumption that

$$P := \mathbb{R}_+(D - LC)$$

is a closed subspace of Y. In such a case $F(x, p) = \infty$ if $p \notin P$, thus the parameter space will be the subspace P (*nomen omen*). The dual problem is to maximize $-H^*(-L^*p^*, p^*)$ on the space P^*. In order to have it nontrivial, we assume the existence of $p_0^* \in P^*$ such that $(-L^*p_0^*, p_0^*) \in \text{dom } H^*$.

Under this assumption, we shall prove that the given problem is regular. The proof uses, once again, a smart separation argument. The geometrical idea is the following. Let K be the projection, on the space $X^* \times \mathbb{R}$, of the epigraph of F^*:

$$K = \text{proj}_{X^* \times \mathbb{R}} \text{epi } F^* = \{(x^*, r) : \text{ there is } p^* \in P^* \text{ with } r \geq F^*(x^*, p^*)\}.$$

Thus $(0_{X^*}, - \inf(0_P)) \in K$ if and only if there exists $\bar{p}^* \in P^*$ such that

$$- \inf(0_P) \geq F^*(0_{X^*}, \bar{p}^*),$$

or, equivalently

$$\inf(0_P) \leq \sup(0_{X^*}).$$

Since the opposite inequality is always true, this shows that there is no duality gap. Moreover, the element $\bar{p}^*$ found above must be optimal for the dual problem, and this shows that the dual problem has solutions. No duality gap and the existence of solutions for the dual problem is exactly what we mean by a regular problem. Summarizing, regularity is equivalent to saying that there is $\bar{p}^*$ such that $(0_{X^*}, \bar{p}^*, - \inf(0_P)) \in \text{epi } F^*$, i.e.,

$$(0_{X^*}, - \inf(0_P)) \in K.$$

To prove this, we start with the following lemma:

Lemma 6.4.4 *The convex set K above is a weak* closed subset of $X^* \times \mathbb{R}$.*

Proof. To prove that K is closed in the weak* topology, we use the Banach–Dieudonné–Krein–Smulian theorem (see Theorem A.2.1), claiming that it is enough to show that $K \cap kB_{X^* \times \mathbb{R}}$ is a weak* closed set, for all $k > 0$. So, let $\{(x_n^*, r_n)\}$, $n \in I$, where I is a directed set, be a net inside $K \cap rB_{X^* \times \mathbb{R}}$ and converging to (x^*, r). Let p_n^* be such that $r_n \geq F^*(x_n^*, p_n^*)$. Let $z \in P$. Then there are $\lambda \in \mathbb{R}, c \in C, d \in D$ such that $z = \lambda(d - Lc)$. Hence

$$
\begin{aligned}
\frac{1}{\lambda} \langle p_n^*, z \rangle &= \langle p_n^*, d - Lc \rangle + \langle x_n^*, c \rangle - \langle x_n^*, c \rangle \\
&\leq F^*(x_n^*, p_n^*) + F(c, d - Lc) - \langle x_n^*, c \rangle \\
&\leq r_n + h(c, d) + k\|c\| \leq h(c, d) + k(\|c\| + 1),
\end{aligned}
$$

showing that $\{\langle p_n^*, z \rangle\}$ is a bounded set (by a constant depending on z). By the uniform boundedness theorem it follows that $\{p_n^*\}$ is a bounded net, whence it has a limit point p^*, which is the element we are looking for. $\qquad\square$

We are able to prove the required result.

Lemma 6.4.5 $(0_{X^*}, -\inf(0_P)) \in K$.

Proof. For the sake of contradiction, suppose instead $(0_{X^*}, -\inf(0_P)) \notin K$. We then find $(x, t) \in X \times \mathbb{R}, c \in \mathbb{R}$ such that

$$
-t \inf(0_P) < c < tr + \langle x, x^* \rangle,
$$

$\forall (x^*, r)$ for which there is $p^* \in P^*$ such that $r \geq F^*(x^*, p^*)$. It follows, as usual, that $t \geq 0$. If $t = 0$, then $0 < \langle x^*, x \rangle$ for all x^* such that there is p^* with $(x^*, p^*) \in \operatorname{dom} F^*$. This implies that for no p^*, $(0_{X^*}, p^*) \in \operatorname{dom} F^*$, contradicting the fact that there is p^* such that $(-L^* p^*, p^*)$ belongs to $\operatorname{dom} H^*$. Dividing by $t > 0$ in the formula above and setting $-\bar{x} = \frac{x}{t}$, we easily arrive at the desired contradiction:

$$
\begin{aligned}
-\inf(0_P) < \frac{c}{t} &\leq \inf_{x^*, p^*} \{F^*(x^*, p^*) + \langle x^*, -\bar{x} \rangle\} \\
&\leq -\sup_{x^*, p^*} \{\langle x^*, \bar{x} \rangle + \langle p^*, 0_P \rangle - F^*(x^*, p^*)\} \\
&= -F(\bar{x}, 0_P) = -H(\bar{x}, L\bar{x}) \leq -\inf(0_P).
\end{aligned}
$$

$\qquad\square$

We summarize the result in the next theorem.

Theorem 6.4.6 *With the notations and the setting above, suppose*

$$
P := \mathbb{R}_+(D - LC)
$$

is a closed subspace of Y. Moreover, suppose there exists $p_0^* \in P^*$ such that $(-L^* p_0^*, p_0^*) \in \operatorname{dom} H^*$. Then the problem $(\mathcal{P}_h)$ is regular.

6.5 The Lagrangean

In the previous sections we considered a duality theory based on conjugating the function $F(\cdot\,,\cdot)$ *with respect to both variables.* Another interesting approach is provided by using the Fenchel conjugate with respect to the parameter variable only.

Definition 6.5.1 We call the *Lagrangean* of the problem $(\mathcal{P})$ the function

$$L\colon X \times P^* \to [-\infty, \infty]$$

defined as

$$-L(x, p^*) := \sup_{p \in P}\{\langle p^*, p\rangle - F(x, p)\}.$$

For each fixed $x \in X$ the function $p^* \mapsto -L(x, p^*)$ is then the Fenchel conjugate of the function $p \mapsto F(x, p)$. Thus it is convex and lower semicontinuous, no matter what the function F is.

Proposition 6.5.2 $\forall x \in X$,

$$p^* \mapsto L(x, p^*)$$

is concave and upper semicontinuous. If F is convex, then $\forall p^* \in P^*$

$$x \mapsto L(x, p^*)$$

is convex.

Proof. The second claim follows from Proposition 6.3.2. □

We shall now express the problems $(\mathcal{P})$ and $(\mathcal{P}^*)$ in terms of the Lagrangean L:

$$
\begin{aligned}
F^*(x^*, p^*) &= \sup_{x \in X, p \in P}\{\langle x^*, x\rangle + \langle p^*, p\rangle - F(x, p)\}\\
&= \sup_{x \in X}\{\langle x^*, x\rangle + \sup_{p \in P}\{\langle p^*, p\rangle - F(x, p)\}\\
&= \sup_{x \in X}\{\langle x^*, x\rangle - L(x, p^*)\},
\end{aligned}
$$

from which we get the formula

$$-F^*(0_{X^*}, p^*) = \inf_{x \in X} L(x, p^*). \tag{6.7}$$

Thus the dual problem $(\mathcal{P}^*)$ can be written, exploiting the Lagrangean, as

$$\sup_{p^* \in P^*} \inf_{x \in X} L(x, p^*).$$

Analogously, if $F \in \Gamma(X \times P)$, the function $p \mapsto F(x,p)$ coincides for each fixed $x \in X$ with its biconjugate. Hence, $\forall x \in X$,

$$F(x,p) = \sup_{p^* \in P^*} \{\langle p^*, p \rangle + L(x,p^*)\},$$

implying

$$F(x,0_P) = \sup_{p^* \in P^*} L(x,p^*). \tag{6.8}$$

It follows that the initial problem $(\mathcal{P})$ can be written as

$$\inf_{x \in X} \sup_{p^* \in P^*} L(x,p^*).$$

Thus the problems $(\mathcal{P})$ and $(\mathcal{P}^*)$ are written in terms of *minmax* and *maxmin* problems for the Lagrangean.

Definition 6.5.3 $(\bar{x}, \bar{p}^*) \in X \times P^*$ is said to be *a saddle point* for L if $\forall x \in X, \forall p^* \in P^*$,

$$L(\bar{x},p^*) \le L(\bar{x},\bar{p}^*) \le L(x,\bar{p}^*).$$

It is easy to verify that if F is proper, then $L(\bar{x},\bar{p}^*) \in \mathbb{R}$. Observe that the definition of saddle point is not symmetric in the two variables, (as for instance happens in critical point theory). Here there is a minimum problem with respect to the first variable (for a fixed value of the second one), and conversely a maximum problem with respect to the second variable (for a fixed value of the first one).

Proposition 6.5.4 *Let $F \in \Gamma(X \times P)$. The following are equivalent:*

- $(\bar{x}, \bar{p}^*)$ *is a saddle point for L;*
- $\bar{x}$ *is a solution for $(\mathcal{P})$, $\bar{p}^*$ is a solution for $(\mathcal{P}^*)$ and $\inf(0_P) = \sup(0_{X^*})$.*

Proof. Let $(\bar{x}, \bar{p}^*)$ be a saddle point for L. From (6.7) we get

$$L(\bar{x},\bar{p}^*) = \inf_x L(x,\bar{p}^*) = -F^*(0_{X^*}, \bar{p}^*),$$

while from (6.8),

$$L(\bar{x},\bar{p}^*) = \sup_{p^*} L(\bar{x},p^*) = F(\bar{x},0_P).$$

Hence

$$F(\bar{x},0_P) + F^*(0_{X^*}, \bar{p}^*) = 0,$$

and we conclude by appealing to Theorem 6.3.1. For the opposite implication, it is enough to observe that

$$F(\bar{x},0_P) = \sup_{p^* \in P^*} L(\bar{x},p^*) \ge L(\bar{x},\bar{p}^*) \ge \inf_{x \in X} L(x,\bar{p}^*) = -F^*(0_{X^*},\bar{p}^*).$$

From $\inf(0_P) = \sup(0_{X^*})$, i.e., $F(\bar{x},0_P) + F^*(0_{X^*}, \bar{p}^*) = 0$, in the inequalities above, the equality signs must hold everywhere, and so $(\bar{x}, \bar{p}^*)$ is a saddle point for L. $\qquad\square$

Proposition 6.5.5 *Let $F \in \Gamma(X \times P)$ and let the problem $(\mathcal{P})$ be regular. Then $\bar{x}$ is a solution of $(\mathcal{P})$ if and only if there exists $\bar{p}^* \in P^*$ such that $(\bar{x}, \bar{p}^*)$ is a saddle point for L.*

Proof. It is enough to observe that if $(\mathcal{P})$ is regular with solution $\bar{x}$, then there is at least a solution $\bar{p}^*$ of the dual problem $(\mathcal{P}^*)$, and there is no duality gap (i.e., $\inf(0_P) = \sup(0_{X^*})$), whence $(\bar{x}, \bar{p}^*)$ is a saddle point for L, as we saw in the previous Proposition. $\qquad\square$

6.6 Examples of dual problems

In this section we start to see some interesting examples of the use of duality theory. More examples are contained in the next chapter.

6.6.1 Convex programming

Let $C \subset X$ be a nonempty, closed convex set in the reflexive Banach space X, and suppose we are given a convex, lower semicontinuous function $k \colon C \to \mathbb{R}$ and another function $g \colon X \to \mathbb{R}^m$ which is continuous and with convex components. Let us consider the problem

$$\inf_{\substack{x \in C \\ g(x) \leq 0}} k(x) = \inf_{x \in X} f(x), \qquad (6.9)$$

where

$$f(x) := \begin{cases} k(x) & \text{if } x \in C \text{ and } g(x) \leq 0, \\ \infty & \text{otherwise.} \end{cases}$$

The condition $g(x) \leq 0$ must be read coordinatewise. Let the parameter space be $P = \mathbb{R}^m$; the parameterized family of problems we shall consider is defined by

$$F(x, p) := \begin{cases} k(x) & \text{if } x \in C \text{ and } g(x) \leq p, \\ \infty & \text{otherwise.} \end{cases}$$

Observe that $F(x, p) = \bar{k}(x) + I_W(x, p)$, where

$$\bar{k}(x) := \begin{cases} k(x) & \text{if } x \in C, \\ \infty & \text{otherwise,} \end{cases}$$

and

$$W := \{(z, q) \in X \times P : g(z) - q \leq 0\}.$$

W is a convex set and $F \in \Gamma(X \times \mathbb{R}^m)$. Let us write the associated dual problem:

$$F^*(0_{X^*}, p^*) = \sup_{\substack{x \in X \\ p \in \mathbb{R}^m}} \{\langle p^*, p \rangle - F(x, p)\} = \sup_{\substack{x \in C \\ p \in \mathbb{R}^m \\ g(x) \leq p}} \{\langle p^*, p \rangle - k(x)\}$$

and setting $p = g(x) + q$,

$$F^*(0_{X^*}, p^*) = \sup_{x \in C} \sup_{\mathbb{R}^m \ni q \geq 0} \{\langle p^*, g(x) \rangle + \langle p^*, q \rangle - k(x)\}$$

$$= \begin{cases} \sup_{x \in C} \{\langle p^*, g(x) \rangle - k(x)\} & \text{if } p^* \leq 0, \\ \infty & \text{otherwise.} \end{cases}$$

As a result (with a little abuse of notation),

$$-F^*(0_{X^*}, \lambda) = \begin{cases} \inf_{x \in C} \{\langle \lambda, g(x) \rangle + k(x)\} & \text{if } \lambda \geq 0, \\ -\infty & \text{otherwise.} \end{cases}$$

It follows that the dual problem $(\mathcal{P}^*)$ becomes

$$\sup_{\mathbb{R}^m \ni \lambda \geq 0} \inf_{x \in C} \{k(x) + \langle \lambda, g(x) \rangle\}. \tag{6.10}$$

We now generalize Exercise 3.4.3 by means of the following:

Theorem 6.6.1 *Suppose*

$$(CQ) \qquad\qquad \exists x_0 \in C \text{ such that } g_i(x_0) < 0, \forall i = 1, \ldots, n,$$

and that

$$\lim_{\substack{x \in C \\ \|x\| \to \infty}} k(x) = \infty.$$

(We assume that this condition is automatically fulfilled if C is a bounded set.) Then the problem $(\mathcal{P})$ has solutions, is regular, and $\forall \bar{x} \in \mathrm{Min}(0_P), \forall \bar{\lambda} \in \mathrm{Max}(0_{X^*})$, one has

$$\langle \bar{\lambda}, g(\bar{x}) \rangle = 0.$$

Finally the Lagrangean of $(\mathcal{P})$ is

$$L(x, \lambda) = \begin{cases} \infty & \text{if } x \notin C, \\ k(x) + \langle \lambda, g(x) \rangle & \text{if } x \in C \text{ and } \lambda \geq 0, \\ -\infty & \text{otherwise.} \end{cases}$$

Proof. Let us start by showing that $(\mathcal{P})$ is regular. We use Proposition 6.4.3. The point $x_0 \in C$ of condition (CQ) guarantees that the function $p \mapsto F(x_0, p)$ is (finite and) continuous in a neighborhood of $p = 0$. The coercivity condition on the objective function provides existence of a solution for $(\mathcal{P})$. Then there are solutions both for the problem and for its dual, and it remains to verify that if $\bar{x} \in \mathrm{Min}(0_P)$ and $\bar{\lambda} \in \mathrm{Max}(0_{X^*})$, then $\langle \bar{\lambda}, g(\bar{x}) \rangle = 0$. The inequality

$$\langle \bar{\lambda}, g(\bar{x}) \rangle \leq 0$$

follows from $g(\bar{x}) \leq 0$ and $\bar{\lambda} \geq 0$. The opposite follows from

$$\inf(0_P) = k(\bar{x}) = \sup(0_{X^*}) = -F^*(0_{X^*}, \bar{\lambda})$$
$$= \inf_{x \in C} \{ \langle \bar{\lambda}, g(x) \rangle + k(x) \} \leq \langle \bar{\lambda}, g(\bar{x}) \rangle + k(\bar{x}).$$

To find the Lagrangean,

$$-L(x, p^*) = \sup_{p \in \mathbb{R}^m} \{ \langle p^*, p \rangle - F(x, p) \} = \sup_{\substack{p \in \mathbb{R}^m \\ g(x) \leq p}} \{ \langle p^*, p \rangle - \bar{k}(x) \},$$

providing $L(x, \lambda) = \infty$ if $x \notin C$. Moreover, if $x \in C$, setting $p = g(x) + q$,

$$-L(x, \lambda) = \sup_{\mathbb{R}^m \ni q \geq 0} \{ \langle -\lambda, g(x) \rangle - \langle \lambda, q \rangle - k(x) \},$$

from which we conclude.   □

A solution $\bar{\lambda}$ of the dual problem is called, in this setting, a *Lagrange multiplier* for the initial problem. We remind the reader that the set of the Lagrange multipliers of a regular mathematical programming problem is the subdifferential of the value function at the origin (see Proposition 6.2.7).

The extremality condition $\langle \bar{\lambda}, g(\bar{x}) \rangle = 0$ provides the so-called Kuhn–Tucker conditions. As $\bar{\lambda} \geq 0$ and $g(\bar{x}) \leq 0$, the condition is then equivalent to $\bar{\lambda}_i = 0$ if $g_i(\bar{x}) < 0$. The multipliers connected with the inactive constraints must necessarily vanish.

Exercise 6.6.2 Write the extremality condition $(0_{X^*}, \bar{p}^*) \in \partial F(\bar{x}, 0_P)$ for the convex programming problem. In particular, try to understand the geometrical meaning of the condition in the simplified case when $C = \mathbb{R}^n$, there is only one constraint function g and k, g are differentiable.

Hint. Remember (or prove) that for a closed convex set A, $\partial I_A(x) = 0$ if $x \in \text{int } A$, and $\partial I_A(x)$ is the normal cone at the point x to the set A when x is in the boundary of A. Use the fact that $F(x, \bar{p}^*) = k(x) + I_W(x, \bar{p}^*)$ and that the set W is the level set, at height zero, of the function $h(x, p) = g(x) - p$. Then apply the sum rule and conclude that the multiplier $\lambda = -\bar{p}^*$ must be zero if $g(\bar{x}) < 0$, while $\nabla k(\bar{x}) = -\bar{\lambda} \nabla g(\bar{x})$, meaning that, (if $\bar{\lambda} \neq 0$), the two level surfaces $k(x) = k(\bar{x})$ and $g(x) = 0$ must be tangent at the point $\bar{x}$; moreover, the two gradients must have opposite directions.

Example 6.6.3 In Example 6.3.4 we considered the following: let X be a separable Hilbert space with basis $\{ e_n : n \in \mathbb{N} \}$, and let $P = \mathbb{R}$. Let $x^* = \sum_{i=1}^{\infty} \frac{1}{n} e_n$, $f_0(x) = \max\{ -1, \langle x^*, x \rangle \}$, and define $g(x) = \sum_{n=1}^{\infty} \frac{(x, e_n)^2}{n^4}$. Finally, let

$$F(x,p) = \begin{cases} f_0(x) & \text{if } g(x) \leq p, \\ \infty & \text{otherwise.} \end{cases}$$

Thus this is a convex programming problem (in infinite dimensions). We have already seen that

$$\inf(p) = \begin{cases} \infty & \text{if } p < 0 \\ 0 & \text{if } p = 0, \\ -1 & \text{otherwise.} \end{cases}$$

Clearly, the value function $\inf(\cdot)$ is not lower semicontinuous at $p = 0$, and there must be a duality gap. From the previous calculation, we can get that

$$F(x,\lambda) = \begin{cases} -1 & \text{if } \lambda \geq 0, \\ \infty & \text{otherwise.} \end{cases}$$

Thus the solution set for the dual problem is $[0, \infty)$. As expected, with a due change of sign (as we set $\lambda = -p^*$), this set is the subdifferential of $\inf^{**}$ at the origin. Thus, this is an example of a problem having a solution as well as its dual problem, but not regular, as there is a duality gap.

6.6.2 An example in the calculus of variations

We now want to provide another interesting example based on the duality scheme we developed to get Theorem 6.4.6. We shall make only heuristic calculations, without bothering too much about the precise assumptions which make them formally correct. Consider the interval $[0, 1]$ and the set of the functions $x(\cdot)$ which are absolutely continuous on $[0, 1]$. This means that they are differentiable almost everywhere, with a derivative $g \in L^1([0, 1])$, in such a way that $x(t) = x(0) + \int_0^t g(s) \, ds$. Such a function g is unique (in $L^1([0, 1])$) and it is usually denoted by x' (the derivative of x). Let X be the space of the absolutely continuous functions on $[0, 1]$, vanishing at the endpoints and with derivative in $L^2([0, 1])$. This can be made a Hilbert space, with inner product $\langle x, u \rangle = \int_0^1 x'(s)u'(s) \, ds$, generating the norm $\|x\|^2 = \int_0^1 (x'(s))^2 \, ds = \|x'\|^2_{L^2([0,1])}$. This norm is equivalent to the norm defined as $|x|^2 = \|x\|^2_{L^2([0,1])} + \|x'\|^2_{L^2([0,1])}$.

Let us now consider the following problem of the calculus of variations. Given the function $h\colon [0, 1] \times \mathbb{R} \times \mathbb{R} \to \mathbb{R}$, let $(\mathcal{P})$ be the problem of minimizing

$$\int_0^1 h(t, x(t), x'(t)) \, dt,$$

over the space X. Setting $Y = L^2([0, 1])$,

$$H(x, y) = \int_0^1 h(t, x(t), y(t)) \, dt$$

and $L\colon X \to Y$, $Lx = x'$, we want to translate in this example what means the optimality condition obtained before:

$$(-L^*\bar{p}^*, \bar{p}^*) \in \partial H(\bar{x}, L\bar{x}).$$

To begin with, we shall suppose that h is a continuous function, with continuous partial derivatives with respect to the second and third variable, and convex in the pair formed by the second and third variable for each fixed value of the first one. Suppose also that it is possible to differentiate under the integral sign (usually growth conditions on h are requested to make it possible). So that H becomes differentiable and we have

$$\langle \nabla H(x, y), (u, v) \rangle = \int_0^1 D_2 h(t, x(t), y(t)) u(t)\, dt + \int_0^1 D_3 h(t, x(t), y(t)) v(t)\, dt,$$

for each direction $u \in X$, $v \in Y$, having used the symbol $D_j h(t, x(t), y(t))$ to indicate the partial derivative, with respect to the j-th component of h evaluated at $(t, x(t), y(t))$.

So that the condition

$$(-L^*\bar{p}^*, \bar{p}^*) \in \partial H(\bar{x}, L\bar{x})$$

here becomes

$$\langle -L^*\bar{p}^*, u \rangle + \langle \bar{p}^*, v \rangle = \int_0^1 D_2 h(t, \bar{x}(t), \bar{x}'(t)) u(t)\, dt$$

$$+ \int_0^1 D_3 h(t, \bar{x}(t), \bar{x}'(t)) v(t)\, dt.$$

This must be true for $\forall u \in X, v \in Y$, and so

$$\langle -L^*\bar{p}^*, u \rangle_X = \int_0^1 D_2 h(t, \bar{x}(t), \bar{x}'(t)) u(t)\, dt$$

$$\langle \bar{p}^*, v \rangle_Y = \int_0^1 D_3 h(t, \bar{x}(t), \bar{x}'(t)) v(t)\, dt,$$

(we can get this by considering in the product space $X \times Y$ the directions $(u, 0)$ and $(0, v)$). Since the second equality holds for all $v \in L^2([0, 1])$, we come up to

$$\bar{p}^* = D_3 h(\cdot, \bar{x}(\cdot), \bar{x}'(\cdot)),$$

(equality in $L^2([0, 1])$), while the first one can be rewritten as

$$\int_0^1 D_2 h(t, \bar{x}(t), \bar{x}'(t)) u(t)\, dt = \langle -L^*\bar{p}^*, u \rangle = \langle -\bar{p}^*, Lu \rangle$$

$$= -\int_0^1 D_3 h(t, \bar{x}(t), \bar{x}'(t)) u'(t)\, dt.$$

As a result,

$$\int_0^1 D_2h(t,\bar{x}(t),\bar{x}'(t))u(t)\,dt = -\int_0^1 D_3h(t,\bar{x}(t),\bar{x}'(t))u'(t)\,dt, \forall u \in X.$$

This means that $t \mapsto D_3h(t,\bar{x}(t),\bar{x}'(t))$ is absolutely continuous; moreover, by integrating by parts, and appealing to a density lemma, we get

$$D_2h(t,\bar{x}(t),\bar{x}'(t)) = (D_3h(t,\bar{x}(t),\bar{x}'(t)))',$$

which is nothing other than the celebrated *Euler equation* for the calculus of variations problem $(\mathcal{P})$.

7

Linear programming and game theory

> *To blame others for one's misfortune*
> *is a sign of human ignorance,*
> *to blame oneself*
> *the beginning of understanding,*
> *not to blame anyone true wisdom.*
> (Epitteto)

In this chapter we shall consider the classical linear programming problem. An elegant way to derive the important duality result for linear programming, is to appeal to game theory. Since this mathematical theory is important and very beautiful, we introduce some concepts related to it. In particular, we prove von Neumann's theorem on the existence of mixed strategies for finite zero-sum games and we use it to prove the main duality result in linear programming. We also take a look at some cooperative theory, always in connection with some linear programming problem.

Of course, linear programming can be seen as a particular case of convex programming. However the results we prove here cannot be covered by those obtained in the general convex case.

7.1 Linear programming I

Let us now introduce the linear programming problems.

Suppose we have an $m \times n$ matrix A and vectors b, c belonging to $\mathbb{R}^m$ and $\mathbb{R}^n$, respectively. Then the problem $(\mathcal{P})$ is the following one:

$$(\mathcal{P}) \qquad \begin{array}{ll} \text{minimize} & \langle c, x \rangle \\ \text{such that} & x \in C, Ax \geq b \end{array} \qquad (7.1)$$

We shall analyze two cases: when $C = X$ and when $C = \{x \in \mathbb{R}^n : x \geq 0\}$. We start by exploiting the results already obtained in the study of the mathematical programming problem. Thus, setting $C = \{x \in \mathbb{R}^n : x \geq 0\}$, $k(x) = \langle c, x \rangle$ and $g(x) = b - Ax$, from (6.10) we get that the dual problem becomes

$$\sup_{\mathbb{R}^m \ni \lambda \geq 0} \inf_{x \in C} \{\langle c, x \rangle + \langle \lambda, b - Ax \rangle\} = \sup_{\mathbb{R}^m \ni \lambda \geq 0} \{\langle \lambda, b \rangle + \inf_{x \geq 0} \{\langle c - A^T \lambda, x \rangle\}\},$$

which can be equivalently stated as

$$
\begin{aligned}
\text{maximize} \quad & \langle \lambda, b \rangle, \\
\text{such that} \quad & \lambda \geq 0, \ A^T \lambda \leq c.
\end{aligned}
\tag{7.2}
$$

We have shown the following

Theorem 7.1.1 *Let A be an $m \times n$ matrix and let b, c be vectors belonging to $\mathbb{R}^m$ and $\mathbb{R}^n$, respectively. The following two linear programming problems are in duality :*

$$
\begin{aligned}
\text{minimize} \quad & \langle c, x \rangle \\
\text{such that} \quad & x \geq 0, \ Ax \geq b,
\end{aligned}
\tag{7.3}
$$

$$
\begin{aligned}
\text{maximize} \quad & \langle \lambda, b \rangle \\
\text{such that} \quad & \lambda \geq 0, \ A^T \lambda \leq c.
\end{aligned}
\tag{7.4}
$$

In exactly the same way we get

Theorem 7.1.2 *Let A be an $m \times n$ matrix and let b, c be vectors belonging to $\mathbb{R}^m$ and $\mathbb{R}^n$, respectively. The following two linear programming problems are in duality:*

$$
\begin{aligned}
\text{minimize} \quad & \langle c, x \rangle \\
\text{such that} \quad & Ax \geq b,
\end{aligned}
\tag{7.5}
$$

$$
\begin{aligned}
\text{maximize} \quad & \langle \lambda, b \rangle \\
\text{such that} \quad & \lambda \geq 0, A^T \lambda = c.
\end{aligned}
\tag{7.6}
$$

We shall now focus on problems of the type described by Theorem 7.1.1; later on we shall see some applications related to problems of the other type.

Example 7.1.3 This is a version of the (so called) diet problem. We must prepare a diet in order to minimize costs, with the following constraint set. Some experts say that the diet must contain a minimal amount b_j of nutrient n_j, $1 \leq j \leq m$. A nutrient could be some vitamin or protein. We have the choice of n foods. Each unit of food f_i contains the amount a_{ji} of nutrient n_j. The cost of a unit of f_i is c_i. We must choose the quantity $x_i \geq 0$ of food f_i. Thus the problem can be written in the following form:

$$
\begin{aligned}
\text{minimize} \quad & \langle c, x \rangle \\
\text{such that} \quad & x \geq 0, \ Ax \geq b.
\end{aligned}
$$

Now, let us change the scenario. Suppose a pharmaceutical firm decides to produce n_j pills (units) of nutrients for a diet, and its scope is to arrange things in order to maximize profits. Then it must decide the price λ_j of the pill n_j in order to maximize the earnings obtained by selling the amount of nutrients necessary for the diet, i.e., $\lambda_1 b_1 + \cdots + \lambda_m b_m$. The obvious constraint is that buying the pills for the diet costs no more than buying the food necessary for the diet itself. Each unit of food f_i provides the amount a_{1i} of nutrient $n_1, \ldots, a_{mi}$ of nutrient n_m, and so the condition to be imposed is

$$a_{1i}\lambda_1 + \cdots + a_{mi}\lambda_m \leq c_i, i \leq 1 \leq n.$$

Moreover, needless to say, $\lambda_j \geq 0$. Writing the above problem in the usual form, we come to the following:

$$\text{maximize} \quad \langle \lambda, b \rangle$$
$$\text{such that} \quad \lambda \geq 0, \ A^T \lambda \leq c.$$

As a result, we see that the two problems are in duality.

7.2 Zero sum games

Let us now speak about two player, finite, zero sum games. An $n \times m$ matrix P represents one game of this type in the following sense. Player one chooses a row i, player two a column j, and p_{ij} is the amount the second player pays to the first one. The first, fundamental, issue is to establish when a pair $(\bar{i}, \bar{j})$, i.e,. the choice of a row by the first player and of a column by the second one, can be considered as a solution for the game. To investigate this point, let us first observe two simple facts. It is clear that if the first player selects the first row and in some way the second one knows it, then she will react by choosing the column providing the value $\min_j a_{1j}$. So that the first player will be able to guarantee himself (at least) the quantity $\max_i \min_j a_{ij}$. This is called the *conservative value* of the first player. In the same way, and taking into account a change of sign, the conservative value of the second player will be $\min_j \max_i a_{ij}$. Now, let us observe the following.

Proposition 7.2.1 *Let X, Y be any sets and let $f \colon X \times Y \to \mathbb{R}$ be an arbitrary function. Then*

$$\sup_x \inf_y f(x, y) \leq \inf_y \sup_x f(x, y).$$

Proof. Observe that for all x, y,

$$\inf_y f(x, y) \leq f(x, y) \leq \sup_x f(x, y).$$

Thus

$$\inf_y f(x, y) \leq \sup_x f(x, y).$$

Since the left-hand side of the above inequality does not depend on x and the right-hand side on y, the thesis easily follows.   □

It is interesting to observe that the inequality provided by the above proposition is absolutely natural in view of the interpretation we can give to it in the context of game theory. Whatever the first player can guarantee himself against any possible choice of the second one (the conservative value of the first player) cannot be *more* than the maximum amount the second player agrees to pay no matter what the first one does (the conservative value of the second player).

The next theorem, though very simple, tells us interesting things.

Theorem 7.2.2 *Under the assumptions of Proposition 7.2.1, the following are equivalent:*

(i)   *The pair $(\bar{x}, \bar{y})$ fulfills*

$$f(x, \bar{y}) \leq f(\bar{x}, \bar{y}) \leq f(\bar{x}, y) \quad \forall x \in X, \forall y \in Y.$$

(ii)   *The following conditions are satisfied:*
(a) $\inf_y \sup_x f(x, y) = \sup_x \inf_y f(x, y)$;
(b) $\inf_y f(\bar{x}, y) = \sup_x \inf_y f(x, y)$;
(c) $\sup_x f(x, \bar{y}) = \inf_y \sup_x f(x, y)$.

Proof. Let us begin by seeing that (i) implies (ii). From (i) we get

$$\inf_y \sup_x f(x, y) \leq \sup_x f(x, \bar{y}) \leq f(\bar{x}, \bar{y}) \leq \inf_y f(\bar{x}, y) \leq \sup_x \inf_y f(x, y).$$

From Proposition 7.2.1 we can conclude that in the line above all inequalities are equalities, and thus (ii) holds. Conversely, suppose (ii) holds. Then

$$\inf_y \sup_x f(x, y) = \sup_x f(x, \bar{y}) \geq f(\bar{x}, \bar{y}) \qquad \text{by(c)}$$

$$\geq \inf_y f(\bar{x}, y) = \sup_x \inf_y f(x, y) \qquad \text{by(b)}.$$

So that, because of (a), we have all equalities and the proof is complete.   □

The above theorem looks a little ugly, at least as far as condition (ii) is concerned, but is quite interesting from the point of view of its consequences. First of all a (saddle) point $(\bar{x}, \bar{y})$ as in condition (i) can be seen as a good solution of the game; once it is proposed as an outcome, no player will object to it. Player two, once she is told that $\bar{x}$ will be the strategy used by the first one, agrees to play $\bar{y}$, *because it is her best choice*. The same is true for the first one. Thus a saddle point is a stable outcome of the game. But there is much more. Condition (ii) says that the players must solve two *independent* problems in order to find their optimal strategies. So they do not need to know what the opponent will do. Condition (ii) tells us one more interesting thing. If (x, y) and (z, w) are two saddle points, then (x, w) and (z, y) are also saddle points and f takes the same value at the saddle points, the so called rectangular property of the saddle points. This means that the two players *do not need* to coordinate their strategies. It must be remarked that games which are not zero sum do not usually enjoy these properties, and this creates several problems in their analysis and implementation. For instance, a game can have two stable outcomes (Nash equilibria), but with different values, so that the two players are *not* indifferent as to which one will be used as outcome of the game. Furthermore the rectangular property does not hold, so that lack of information/coordination can produce unstable situations (see also Appendix C: More game theory).

Coming back to a zero sum game described by a matrix, it is then clear that the pair $(\bar{i}, \bar{j})$ is a solution for the game if for all i, j,

$$p_{i\bar{j}} \leq p_{\bar{i}\bar{j}} \leq p_{\bar{i}j}.$$

In more general situations (for instance, when the available strategies form an infinite set, and so the existence of max/min is not always guaranteed), when the two conservative values agree, we say that the game has value. Let us stop for a moment to consider an example.

Exercise 7.2.3 Consider the game described by the following matrix P:

$$\begin{pmatrix} 4 & 3 & 1 \\ 7 & 5 & 8 \\ 8 & 2 & 0 \end{pmatrix}.$$

Clearly, 5 is the maximum amount the second player agrees to pay because she will pay in any case no more than that by playing the second column (against a possible loss of 8 by playing the two other columns). On the other hand, player one is able to guarantee himself at least 5 (rather than 1 or 0), just playing the second row. As a result, 5 is clearly the outcome of the game.

Here is a second interesting example:

Example 7.2.4 The game is described by the matrix

$$\begin{pmatrix} 0 & 1 & -1 \\ -1 & 0 & 1 \\ 1 & -1 & 0 \end{pmatrix}$$

and it is the familiar "scissors, paper, stone" game, with payment 1 to the winner. Here, it is not clear how to play it rationally. The matrix does not have a saddle point, and it is obvious that any pair of choices is not stable (one of the players, if not both, could argue). Nevertheless, this game should not be played totally randomly. For instance, when playing several times with the same player, if I lead her to think that I do not like playing stone (this means that I play stone with probability zero) she will react by playing only scissors, guaranteeing herself at least the draw. Thus, instead the players should choose rows and columns with probabilities suggested by some optimum rule. This is formalized by doing the following: suppose the first player has n possible moves (the rows of the matrix P), and the second one m (the columns). The first one will then choose a vector $x = (x_1, \ldots, x_n)$ in the n-simplex, his new strategy space. Similarly, the m-simplex is the strategy space for the second one. These enlarged strategy spaces are called the spaces of *mixed strategies* for the players. The new payment function (what the second one pays to the first) is then the expected value:

$$f(x, \lambda) = \sum_{i=1,\ldots,n, j=1,\ldots,m} x_i \lambda_j p_{ij}.$$

A solution of the game is then a saddle point for f, i.e., a pair $(\bar{x}, \bar{\lambda})$ verifying

$$f(x, \bar{\lambda}) \leq f(\bar{x}, \bar{\lambda}) \leq f(\bar{x}, \lambda),$$

for all x, λ in the suitable simplexes. Remember that the existence of a saddle point in particular guarantees that the conservative value of the first player $\max_x \min_\lambda f(x, \lambda)$ agrees with the conservative value of the second player, $\min_\lambda \max_x f(x, \lambda)$.

We now prove that these games always have an equilibrium. This is a celebrated result due to von Neumann, and one of the first basic results in game theory. Let us denote by S_m, S_n the m-th and the n-simplexes, respectively.

Theorem 7.2.5 *A two player, finite, zero sum game as described before always has equilibrium in mixed strategies.*

Proof. First, we can suppose that all the entries p_{ij} of the matrix P are positive. If this is not the case, we can add to all of them the same large quantity in order to make all the entries positive. This does not change the nature of the game. (If you think the second player does not like this, you can convince her to play just by giving her the large amount of money you are adding to the entries. It is intended that the same amount will be given back to you by the first player at the end of the game. By the way, how does this change the payoff function of the game?) Now, consider the vectors $p_1, \ldots, p_m$ of $\mathbb{R}^n$, where p_j denotes the j-th column of the matrix P. These vectors lie in the positive cone of $\mathbb{R}^n$. Call C the convex hull of these vectors, and set

$$Q_t := \{x \in \mathbb{R}^n : x_i \leq t \text{ for } i = 1, \ldots, n\}.$$

Now set

$$v = \sup\{t \geq 0 : Q_t \cap C = \emptyset\}.$$

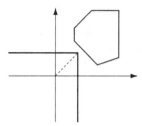

Figure 7.1. The sets C and Q_t.

It is easy to see that Q_v and C can be (weakly) separated by a hyperplane. There are coefficients $\bar{x}_1, \ldots, \bar{x}_n$, not all zero, and $b \in \mathbb{R}$ such that

$$\sum_{i=1}^{n} \bar{x}_i u_i \leq b \leq \sum_{i=1}^{n} \bar{x}_i w_i,$$

for all $u = (u_1, \ldots, u_n) \in Q_v$, $w = (w_1, \ldots, w_n) \in C$. It is straightforward to observe the following facts:

(i)   All $\bar{x}_i$ must be nonnegative and, since they cannot be all zero, we can assume $\sum \bar{x}_i = 1$. For, supposing that some $\bar{x}_i$ is negative implies $\sup\{\sum_{i=1}^{n} \bar{x}_i u_i : u \in Q_v\} = \infty$, which is impossible.

(ii) $b = v$. Obviously $b \geq v$. Suppose $b > v$, and take $a > 0$ so small that $b > v + a$. Then $\sup\{\sum_{i=1}^{n} \bar{x}_i u_i : u \in Q_{v+a}\} < b$, and this implies $Q_{v+a} \cap C = \emptyset$, contrary to the definition of v.

(iii) $Q_v \cap C \neq \emptyset$. On the contrary, suppose $Q_v \cap C = \emptyset$; this is equivalent to saying that $\max_i x_i > v$, for all $x \in C$. As $x \mapsto \max_i x_i$ is a continuous function, it assumes a minimum, say $a > v$, on the compact set C. But then $Q_l \cap C = \emptyset$, for all $l \leq a$, and this contradicts the definition of v.

Now let us consider the inequality

$$v \leq \sum_{i=1}^{n} \bar{x}_i w_i,$$

for $w = (w_1, \ldots, w_n) \in C$. As $w \in C$, then $w = \sum_{j=1}^{m} \lambda_j p_j$, for some $S_m \ni \lambda = (\lambda_1, \ldots, \lambda_m)$. Thus

$$f(\bar{x}, \lambda) = \sum_{i,j} \bar{x}_i \lambda_j p_{ij} \geq v, \qquad (7.7)$$

for all $\lambda \in S_m$. Now, let $\bar{w} \in Q_v \cap C$ (see (iii) above). As $\bar{w} \in C$, then $\bar{w} = \sum_{j=1}^{m} \bar{\lambda}_j p_j$, for some $S_m \ni \bar{\lambda} = (\bar{\lambda}_1, \ldots, \bar{\lambda}_m)$. Since $\bar{w} \in Q_v$, then $\bar{w}_i \leq v$ for all i. Thus, for all $x \in S_n$, we get

$$f(x, \bar{\lambda}) = \sum_{ij} x_i \bar{\lambda}_j p_{ij} \leq v. \qquad (7.8)$$

The inequality in (7.7) says that the first player can guarantee himself at least v, by playing $\bar{x}$. On the other hand, the inequality in (7.8) says that the second player can guarantee paying at most v, by playing $\bar{\lambda}$. Thus $(\bar{x}, \bar{\lambda})$ is a saddle point of the game and $v = f(\bar{x}, \bar{\lambda})$ is the value of the game.   □

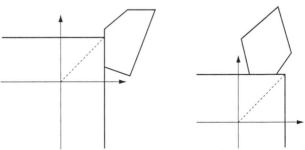

The first row is optimal for the first player.   The second row is optimal for the first player.

Figure 7.2.

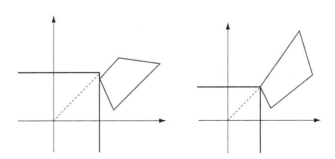

A saddle point in pure strategies.   What happens here?

Figure 7.3.

Observe that the above proof suggests a way to solve the game. The optimal strategy for the first player is given by the (normalized) coefficients characterizing the separating hyperplane; an optimal strategy for the second one can be obtained by considering a point lying in C and Q_v at the same time. As the point lies in C, it is a convex combination of the columns of the matrix. The coefficients of this convex combination then provide an optimal strategy for the second player. This remark is most useful when one of the players has only two available strategies. If both of them have at least three strategies, the calculations are not simple. Thus, some different trick must be invented. Here linear programming techniques play a role. Let us see how. The first player must choose $S_n \ni z = (z_1, \ldots, z_n)$ in such a way that

$$z_1 p_{1j} + \cdots + z_n p_{nj} \geq v, \quad 1 \leq j \leq m,$$

where v must be as large as possible. This is because the amount

$$z_1 p_{1j} + \cdots + z_n p_{nj}$$

is what player one will get if player two chooses column j. Thus the constraint set we impose means he will gain at least v, no matter which column will be played by the opponent and thus, no matter which *probability distribution* she will choose on the columns, being the payment function (of the second player to the first one) f, at a fixed x, an affine function of the variable λ. An affine function always assumes its maximum at a vertex of the simplex (maybe not only at some vertex, but this is irrelevant). And obviously, player one is interested in maximizing v. The second player instead has to find $S_m \ni \rho = (\rho_1, \ldots, \rho_m)$ such that

$$\rho_1 p_{i1} + \cdots + \rho_m p_{im} \leq u, \quad 1 \leq i \leq n,$$

where u must be as small as possible.

It turns out that the two problems are in duality, as we now see.

First, we suppose again, without loss of generality, that all coefficients of the matrix are positive. Then, it is enough to make a change of variable by setting $x_i = \frac{z_i}{v}$. Condition $\sum_{i=1}^m z_i = 1$ becomes $\sum_{i=1}^m x_i = \frac{1}{v}$. Then maximizing v is equivalent to minimizing $\sum_{i=1}^m x_i$.

Thus, denoting by 1_j the vector in $\mathbb{R}^j$ whose all coordinates are 1, we can write the first player problem in the following way:

$$\begin{aligned} \text{minimize} \quad & \langle 1_n, x \rangle \\ \text{such that} \quad & x \geq 0, \ P^T x \geq 1_m. \end{aligned} \tag{7.9}$$

In the same way, we see that the second player faces the following problem:

$$\begin{aligned} \text{maximize} \quad & \langle 1_m, \lambda \rangle \\ \text{such that} \quad & \lambda \geq 0, \ P\lambda \leq 1_n. \end{aligned} \tag{7.10}$$

We have thus two linear programming problems in duality with the choice of $c = 1_n, b = 1_m, A = P^T$.

We thus have seen that finding optimal mixed strategies is equivalent to solving a pair of linear programming problems in duality. In the next section instead we see how it is possible to derive a duality result for linear programming from the von Neumann theorem on game theory.

Exercise 7.2.6 A square matrix P is called skew symmetric if $p_{ij} = -p_{ji}$ for all i, j. Clearly, a skew symmetric matrix represents a *fair* game, in the sense that both players have the same opportunities. What player one can get, for instance, from row i is what the second one can get from column i. Prove that the value of the associated game must be zero and that the optimal strategies of the players are the same. Prove also that if it is known that all rows (columns) must be played with positive probability, then $x = (x_1, \ldots, x_n)$ is optimal if and only if it solves the system

$$\begin{cases} \langle x, p_j \rangle = 0, & \forall j \\ x_i > 0, & \sum x_i = 1, \end{cases}$$

where p_j denotes, as usual, the column j of the matrix P.

Exercise 7.2.7 Pierluigi and Carla play the following game: they both have a sheet of paper. On one side of the paper there is a number in red, on the other side a number in blue. They show at the same time one side of the paper. If the two colors agree Pierluigi wins the number written in Carla's paper. Otherwise Carla wins what Pierluigi shows. One paper contains the number 7 in red and 3 in black, the other one 6 in red and 4 in black. The game looks fair, since the sum of the numbers in the two papers are the same. Which sheet of paper would you suggest Carla should choose?

7.3 Linear programming II

In this section we want to get some results on duality in linear programming by using a game theoretic approach. Our goal is to describe every possible situation for two linear programming problems in duality. We study the case when the primal problem presents nonnegativity constraints. At the end we shall see how to get the results also for the case when there are no nonnegativity constraints. Let us quickly recall the problems (see (7.1) and (7.2)).

We have an $m \times n$ matrix A and vectors b, c belonging to $\mathbb{R}^m$ and $\mathbb{R}^n$, respectively. The problem $(\mathcal{P})$ is

$$\begin{aligned} \text{minimize} \quad & \langle c, x \rangle \\ \text{such that} \quad & x \geq 0, Ax \geq b. \end{aligned} \tag{7.11}$$

and its dual problem is

$$\begin{aligned} \text{maximize} \quad & \langle \lambda, b \rangle \\ \text{such that} \quad & \lambda \geq 0, A^T \lambda \leq c. \end{aligned} \tag{7.12}$$

First, let us agree to call a problem *feasible (unfeasible)* if the constraint set is nonempty (empty), and call the minimum (maximum) problem *unbounded* if its value is $-\infty$ (∞). Since the value of the minimum problem always dominates the value of the maximum problem (in the case of our problems this can be seen with a one line proof, without appealing to previous results in duality theory), we immediately get that if one problem is unbounded, then the other one is necessarily unfeasible (remember that the value of a minimum (maximum) constrained problem such that no point satisfies the constraints is ∞ ($-\infty$)). It can also happen that both are unfeasible, as the following (trivial) example shows:

Example 7.3.1

$$A = \begin{pmatrix} -1 & 1 \\ 2 & -2 \end{pmatrix}; \quad b = (1, 0) \, c = (-1, 0).$$

What happens if one of the problems is unfeasible and the other one is feasible? We shall now show that the feasible problem must be unbounded. From the point of view of the values of the two problems, this means that it cannot happen that one is real, and the other one infinite.

Theorem 7.3.2 *Suppose the linear programming problem $\mathcal{P}$ is feasible, and its dual problem is unfeasible. Then the problem $\mathcal{P}$ is unbounded.*

Proof. Let us consider the game described by the following matrix:

$$\begin{pmatrix} a_{11} & \cdots & a_{m1} & -c_1 \\ \vdots & \vdots & \vdots & \vdots \\ a_{1n} & \cdots & a_{mn} & -c_n \end{pmatrix}.$$

Step 1. Let us see first that this game has value $v \geq 0$. Otherwise there would be a strategy $q = (q_1, \ldots, q_m, q_{m+1})$ for the second player such that it guarantees that she get a negative quantity against each row chosen by the first player. In formulas:

$$a_{1j}q_1 + \cdots + a_{mj}q_m - c_j q_{m+1} < 0, \quad j = 1, \ldots, n.$$

This will lead to a contradiction. For, if $q_{m+1} > 0$, setting $z_i = \frac{q_i}{q_{m+1}}$, $z = (z_1, \ldots, z_m)$, we get that

$$A^T z < c, \quad z \geq 0,$$

against the assumption that the dual problem is unfeasible. On the other hand, if $q_{m+1} = 0$, this implies that calling $z = (q_1, \ldots, q_m)$, then $A^T z \ll 0$ (the notation $a \ll b$ means $a_i < b_i$ for all i). But then, for a sufficiently large k, kz is feasible for the dual problem, which is impossible.

Step 2. We see now that if the value of the game is zero, then necessarily, for any optimal strategy $q = (q_1, \ldots, q_m, q_{m+1})$ of the second player, we must have $q_{m+1} = 0$. Otherwise, with a similar argument as before we see that

$$a_{1j}q_1 + \cdots + a_{mj}q_m - c_j q_{m+1} \leq 0, \quad j = 1, \ldots, n,$$

and setting $z_i = \frac{q_i}{q_{m+1}}$, $z = (z_1, \ldots, z_m)$, we get that

$$A^T z \le c, \quad z \ge 0,$$

and this is impossible.

Step 3. Let us now consider the first player. I claim that he has a strategy $x = (x_1, \ldots, x_n)$ such that

$$Ax \ge 0, \quad \langle x, c \rangle < 0.$$

This is obvious if the value of the game is positive, as he will be able to get a positive payoff against each column. If the value of the game is 0, the claim is intuitive from the point of view of the interpretation of the game, since we know from step 2 that it is never optimal for the second player to play the last column. Thus there must be a strategy x for the first player guaranteeing the he get at least zero (so that $Ax \ge 0$) and forcing her to avoid the last column (i.e., such that $\langle c, x \rangle < 0$). However, to show this mathematically is not immediate, and it will be shown in Lemma 7.3.5.

Step 4. As the minimum problem is feasible, there exists $\hat{x}$ such that $\hat{x} \ge 0$ and $A\hat{x} \ge b$. Consider now $x_t = \hat{x} + tx$, $t \ge 0$. Clearly, x_t satisfies $x_t \ge 0$ and $Ax_t \ge b$, for all $t > 0$. And from $\langle c, x \rangle < 0$ we get that $\langle c, x_t \rangle \to -\infty$, so that the problem is unbounded, and this ends the proof.   □

In other words, Theorem 7.3.2 implies that it cannot happen that one problem has finite value and the other one infinite. We shall see soon that the result can be considerably improved.

Exercise 7.3.3 Prove the following Lemma.

Lemma 7.3.4 *Suppose there are $p + 1$ vectors $v^1, \ldots, v^{p+1}$ in $\mathbb{R}^n$ such that for $z = (z_1, \ldots, z_n)$,*

$$\langle z, v^k \rangle \ge 0 \text{ for } 1 \le k \le p \implies \langle z, v^{p+1} \rangle \ge 0.$$

Then v^{p+1} lies in the convex cone C generated by $v^1, \ldots, v^p$: there are $\alpha_1 \ge 0, \ldots, \alpha_p \ge 0$ such that

$$v^{p+1} = \sum_{j=1}^{p} \alpha_j v^j.$$

Hint. Otherwise, separate C from v^{p+1} (C is closed, see Proposition 1.1.22). Thus there are $0 \ne z \in \mathbb{R}^n$ and c such that

$$\langle z, x \rangle \ge c > \langle z, v^{p+1} \rangle, \forall x \in C.$$

Observe that it must be $c \le 0$, and that if $c < 0$, the same string of inequalities above holds for $c = 0$ as well.

Lemma 7.3.5 *Given a game described by an $n \times m$ matrix P and with value v, either the second player has an optimal strategy $\bar{q} = (\bar{q}_1, \ldots, \bar{q}_m)$ such that $\bar{q}_m > 0$, or the first player has an optimal strategy $\bar{x} = (\bar{x}_1, \ldots, \bar{x}_n)$ such that $\langle \bar{x}, p_m \rangle > v$, where p_m is the m-th column of the matrix P.*

Proof. Without loss of generality we can assume $v = 0$. Otherwise, we could subtract v from each entry of the matrix, without clearly changing the optimal strategies of the players. Now consider the $n + m$ vectors

$$e_1 = (1, 0, \ldots, 0), \ldots, e_n = (0, \ldots, 0, 1), p_1, \ldots, p_{m-1}, -p_m.$$

It can happen that $-p_m$ is in the convex cone C generated by the other vectors, or it is not. We shall show that in the first case, the second player has an optimal strategy with the last component positive, while in the second case, the first player has an optimal strategy guaranteeing him positive payoff against the last column. In the first case, there are nonnegative numbers $\rho_1, \ldots, \rho_n, \lambda_1, \ldots, \lambda_{m-1}$ such that

$$-p_m = \sum_{j=1}^{n} \rho_j e_j + \sum_{j=1}^{m-1} \lambda_j p_j.$$

This implies

$$\sum_{j=1}^{m-1} \lambda_j p_{ij} + p_{im} = -\rho_i \leq 0,$$

for all i. Setting $\bar{q}_j = \frac{\lambda_j}{1 + \sum \lambda_i}, j = 1, \ldots, m-1, \bar{q}_m = \frac{1}{1 + \sum \lambda_i}, \bar{q} = (\bar{q}_1, \ldots, \bar{q}_m)$, then $\bar{q}$ is the optimal strategy we seek for the second player (remember, $v = 0$). Suppose now $-p_m \notin C$. Then there are numbers $\lambda_1, \ldots, \lambda_n$ such that setting $\lambda = (\lambda_1, \ldots, \lambda_n)$,

$$\langle e_j, \lambda \rangle \geq 0, j = 1, \ldots, n, \quad \langle p_j, \lambda \rangle \geq 0, j = 1, \ldots, m-1, \quad \langle -p_m, \lambda \rangle < 0.$$

The first inequality guarantees that $\lambda_i \geq 0$ for all i and the third one that they cannot be all zero. Setting $\bar{x}_i = \frac{\lambda_i}{\sum \lambda_i}, \bar{x} = (\bar{x}_1, \ldots, \bar{x}_n)$, we finally conclude that $\bar{x}$ is an optimal strategy for the first player with the required properties. $\square$

The previous analysis does not tell us what happens when both problems are feasible. In the next result we show that in this case both problems have solutions and there is no duality gap.

Theorem 7.3.6 *Suppose the two problems are both feasible. Then there are solutions $\bar{x}, \bar{\lambda}$ of the two problems, and $\langle c, \bar{x} \rangle = \langle b, \bar{\lambda} \rangle$.*

Proof. Again, we prove the theorem by appealing to a suitable game. Consider the following $(m + n + 1)$ square matrix:

$$
\begin{pmatrix}
0 & \cdots & 0 & -a_{11} & \cdots & -a_{1n} & b_1 \\
\vdots & \vdots & \vdots & \vdots & \vdots & \vdots & \vdots \\
0 & \cdots & 0 & -a_{m1} & \cdots & -a_{mn} & b_m \\
a_{11} & \cdots & a_{m1} & 0 & \cdots & 0 & -c_1 \\
\vdots & \vdots & \vdots & \vdots & \vdots & \vdots & \vdots \\
a_{1n} & \cdots & a_{mn} & 0 & \cdots & 0 & -c_n \\
-b_1 & \cdots & -b_m & c_1 & \cdots & c_n & 0
\end{pmatrix}
=
\begin{pmatrix}
0 & -A & b \\
A^T & 0 & -c \\
-b & c & 0
\end{pmatrix}.
$$

Observe that the above matrix is skew symmetric, so its value is zero and the optimal strategies for the players are the same. Let us call $(p, q, t) = (p_1, \ldots, p_m, q_1, \ldots, q_n, t)$ an optimal strategy for the first player. He will get a nonnegative payoff by playing the above strategy against any column chosen by the second player. Thus,

$$
Aq - tb \geq 0, \quad -A^T p + tc \geq 0, \quad \langle p, b \rangle - \langle q, c \rangle \geq 0.
$$

Suppose $t = 0$, for every optimal strategy for the first player. In such a case, there must be an optimal strategy for the second player guaranteeing a strictly negative result against the last row (see Lemma 7.3.5). Moreover, at every optimal strategy of the second player, she will play the last column with probability zero, because the first one plays the last row with probability zero. This amounts to saying that

$$
-Aq \leq 0, \quad A^T p \leq 0, \quad -\langle b, p \rangle + \langle c, q \rangle < 0.
$$

As both problems are feasible, there are $\hat{p} \geq 0, \hat{q} \geq 0$, such that $A\hat{q} \geq b$, $A^T \hat{p} \leq c$. As $\langle c, q \rangle < \langle b, p \rangle$, if $\langle c, q \rangle < 0$, then $\langle c, \hat{q} + rq \rangle \to -\infty$, for $r \to \infty$. But this is impossible, as the dual problem is feasible. Thus $\langle c, q \rangle \geq 0$, and so $\langle b, p \rangle > 0$. Again this leads to a contradiction, because it would imply that the dual problem is unbounded, against the assumption that the primal problem is feasible. Thus we must have $t > 0$ for at least an optimal strategy for the first player. Then, setting $\bar{x} = \frac{q}{t}$, $\bar{\lambda} = \frac{p}{t}$ from the above relations we get

$$
A\bar{x} \geq b, \quad A^T \bar{\lambda} \leq c, \quad \langle \bar{\lambda}, b \rangle \geq \langle \bar{x}, c \rangle.
$$

The first two conditions just say that $\bar{x}$ and $\bar{\lambda}$ are feasible for the problem and its dual respectively, while the third one is the required optimality condition, just remembering that the opposite inequality must hold at every pair of feasible vectors.   □

Summarizing the previous results, we have seen that if we exclude the (rather uninteresting) case when both problems are unfeasible, if one of the two is unfeasible, then necessarily the other one is unbounded, and if both are feasible, then they both have solutions and there is no duality gap, i.e., they are both regular.

We stated the previous results for a linear programming problem and its dual problem, enclosing nonnegativity conditions for the variable. But they are also valid in the case when the problem does not include this type of condition. Consider the problem without nonnegativity constraints,

$$\text{minimize} \quad \langle c, x \rangle$$
$$\text{such that} \quad Ax \geq b.$$

With a little trick, it is possible to find an equivalent problem, with nonnegativity constraint. Consider the problem

$$\text{minimize} \quad \langle \hat{c}, y \rangle$$
$$\text{such that} \quad y \geq 0, \hat{A}y \geq b.$$

where $y = (z, w)$, $\hat{c} = (c_1, \ldots, c_n, -c_1, \ldots, -c_n)$, $\hat{A} = (A, -A)$ and we put $x = z - w$. It is is straightforward to see that it is equivalent to the given one, and also its dual is equivalent to the dual of the initial one. Thus we can draw the same conclusions as before, even if the initial problem does not enclose nonnegativity conditions.

Let us finally observe that the study of regularity of the linear programming problem was more complicated than in the result obtained for the general mathematical programming problem, as there we made an assumption guaranteeing existence of a solution for the minimum problem, and a (strong) constraint qualification assumption, not required here. For instance, our analysis here allows having equality constraints in the problem. But in such a case the qualification condition required in Theorem 6.6.1 *never* applies.

7.4 Cooperative game theory

Cooperative game theory deals with a group of people, the players, trying to form coalitions in order to get advantages in some decision processes. For instance, companies providing connections to networks could be interested in sharing connection lines, people living in one city and working in another could be interested in car pooling, and so on. Cooperative game theory is interested in providing models in order to efficiently analyze such situations. It is outside the scope of this book to give here a complete picture of cooperative theory. There are books dedicated entirely to the subject (for instance, a classical and beautiful one is [Ow]). However, there are some parts of the theory with connections with linear programming, in particular to problems of the type described by Theorem 7.1.2, and thus we find it interesting to present some results here.

Thus, let us start by quickly describing the setting. We have a set N, called the set of players (usually we set $N = \{1, 2, \ldots, n\}$). They can form coalitions, which are simply subsets of N. To each coalition S is attached a real number, say $v(S)$, which establishes how much the members of the coalition can gain (globally), by staying together. So, here is the first definition.

Definition 7.4.1 A *side payment cooperative game* is a set N of players, together with a function

$$v \colon 2^N \to \mathbb{R},$$

with the property that $v(\{\emptyset\}) = 0$.

In recent books, this is the definition of a side payment game. Less recently, some extra condition was imposed. For instance, the so-called *superadditivity* condition could be required, i.e., for any two disjoint coalitions S, T, the following holds:

$$v(\{S \cup T\}) \geq v(\{S\}) + v(\{T\}).$$

This reflects the idea that making coalitions is convenient, and quite often is an assumption fulfilled in the applications. However, it does not seem to be necessary to include it in the very definition of the game.

Now, let us illustrate the definition by means of a simple example.

Example 7.4.2 A very good professional soccer player is playing forward in a low level team, and his salary is 100,000 Euros per year. A very good team needs an outstanding forward to win the Champions League, gaining 500,000 Euros. Let us agree that player one is the team, player two is the forward. How we define the function v? We can set $v(\{1\}) = 0$, $v(\{2\}) = 100,000$, $v(N) = 500,000$. It is likely that the two players will agree to "play together", but an interesting question is how they will share the 500,000 obtained by working together. (Perhaps we can conclude that game theory is not very realistic as no solution will foresee a salary of 2,000,000 for the good forward, the most likely result in recent years, at least in Italy, even if the Champion League was actually not guaranteed at all.)

There are very many (maybe too many) solution concepts for such games. Here we focus our attention on the so called *core* of the game. A solution for the game is a vector $x = (x_1, \ldots, x_n)$, where x_i represents what is assigned to the player i. Every reasonable solution will satisfy at least two minimal conditions: $x_i \geq v(\{i\})$ for all i, and $\sum x_i = v(N)$ (a vector x fulfilling these two conditions is called an *imputation*). Namely, the first condition simply says that x is refused if one player can get more by acting alone than with the distribution provided by x. This is reasonable, since the players will not participate in the grand coalition N, unless they get at least what they are able to get by acting alone. Surely, to come back to our example, the soccer player will gain more than 100,000 Euros when playing for the new team. The second condition says two things at the same time. First, it cannot happen that $\sum x_i > v(N)$, as the players cannot share *more* than they can actually get. At the same time, it would be stupid to distribute less (this is a big difference with the noncooperative theory, where it can happen that a rational solution (Nash equilibrium) does not distribute the whole utility available to the players). But we can make one more step. Suppose, for example, that x is proposed, and $x_1 + x_n < v(\{1, n\})$. Is x likely to be the solution? Actually, it is not, as the players labeled 1 and n will *refuse* such an agreement (thus making the distribution x impossible), as they can do better by acting together and without other guests. Thus, it makes sense to think that x will be a solution of the game provided no coalition will object to what is assigned to its players.

Definition 7.4.3 Let $v: 2^N \to \mathbb{R}$ be a side payment game. The *core* of the game, denoted by $C(v)$, is the set

$$C(v) = \left\{ x \in \mathbb{R}^n : \sum_{i=1}^{n} x_i = v(N) \quad \text{and} \quad \sum_{i \in S} x_i \geq v(S) \ \ \forall S \subset N \right\}.$$

Let us observe that the definition of core is not particularly meaningful for a two player game. All imputations belong to the core, and vice-versa.

Exercise 7.4.4 Let v be the following three player game: $v(S) = 1$ if $|S| \geq 2$, otherwise $v(S) = 0$. Prove that the core of v is empty. Let v be the following three player game: $v(\{i\}) = 0 = v(\{1, 2\})$, otherwise $v(S) = 1$. Prove that the core is the vector $(0, 0, 1)$.

In the first case the fact that the core is empty provides evidence that the coalitions of two players are too strong. They can all get the whole booty. This is the typical situation when the prize (e.g., a large amount of money) is assigned to one player if he has the majority of votes. It can be allowed that the player makes an agreement to share part of it with whomever votes for him (this explains the name of side payment game). But it can be easily imagined that no agreement is stable (if I promise you 50% of the money if you vote for me, then Maria can promise you 51% to get a vote, but I can react and so on). In the second game, the core highlights (perhaps rather brutally) the power of the third player with respect to the other ones.

Exercise 7.4.5 There are one seller and two potential buyers for an important, indivisible good. Let us agree that the player one, the seller, evaluates the good at a. Players two and three evaluate it b and c, respectively. We assume that $b \leq c$ (this is not a real assumption) and that $a < b$ (this is just to have a real three player game). Build up the corresponding cooperative game, and prove that the core $C(v)$ is given by

$$C(v) = \{(x, 0, c - x) : b \leq x \leq c\}.$$

The result of Exercise 7.4.5 is not surprising at all. The good will be sold to the buyer evaluating it higher, at a price which can vary from the price offered by the person evaluating it lower to the maximum possible price. This is quite reasonable. The price cannot be less than b, otherwise the second player could offer more. On the other hand, it cannot be more than c, as the third player would not buy a good for a price higher than the value he assigns to the good itself. This is not completely satisfactory as an answer. We would prefer to have more precise information. There are other solution concepts suggesting a single vector in this case (precisely, the price will be $(b + c)/2$, for the so-called *nucleolus*).

Exercise 7.4.4 shows that the core of a game can be empty. Thus it is of great interest to find conditions under which we can assure nonemptiness of the core. A smart idea is to characterize the core as the solution set of a particular linear programming problem, and then to look at its dual problem. This is what we are going to illustrate.

Now, observe that $C(v) \neq \emptyset$ if and only if the following linear programming problem:

$$\text{minimize} \quad \sum_{i=1}^{n} x_i$$

$$\text{such that} \quad \sum_{i \in S} x_i \geq v(S) \quad \text{for all } S \subset N, \tag{7.13}$$

has a minimum $\bar{x}$ such that $\sum_{i=1}^{n} \bar{x}_i \leq v(N)$. This is clear as such an element actually lies in the core, and vice-versa.

Just to familiarize ourselves with this, let us write the above linear programming problem for the three player game.

$$\begin{aligned} \text{minimize} \quad & x_1 + x_2 + x_3 \\ \text{such that} \quad & x_i \geq v(\{i\}), \quad i = 1, 2, 3, \\ & x_1 + x_2 \geq v(\{1, 2\}), \\ & x_1 + x_3 \geq v(\{1, 3\}), \\ & x_2 + x_3 \geq v(\{2, 3\}), \\ & x_1 + x_2 + x_3 \geq v(N). \end{aligned} \tag{7.14}$$

In matrix form,

$$\begin{aligned} \text{minimize} \quad & \langle c, x \rangle \\ \text{such that} \quad & Ax \geq b. \end{aligned}$$

where c, A, b are the following objects:

$$c = (1, 1, 1), \quad b = (v(\{1\}), v(\{2\}), v(\{3\}), v(\{1, 2\}), v(\{1, 3\}), v(\{2, 3\}), v(N))$$

and A is the following 7×3 matrix:

$$\begin{pmatrix} 1 & 0 & 0 \\ 0 & 1 & 0 \\ 0 & 0 & 1 \\ 1 & 1 & 0 \\ 1 & 0 & 1 \\ 0 & 1 & 1 \\ 1 & 1 & 1 \end{pmatrix}$$

The dimension of the matrix A is given by the number n of players, as far as the number of columns is concerned, and by the number $2^n - 1$, corresponding to the number of coalitions (except the empty set). Thus in the dual problem the variable will have $2^n - 1$ components, and a good idea is to use the letter S, denoting a coalition, for its index. Thus, in our example a generic dual variable is denoted by $(\lambda_{\{1\}}, \lambda_{\{2\}}, \lambda_{\{3\}}, \lambda_{\{1,2\}}, \lambda_{\{1,3\}}, \lambda_{\{2,3\}}, \lambda_N)$ and the dual problem (see Theorem 7.1.2) becomes

$$\begin{aligned}
\text{maximize} \quad & \lambda_{\{1\}} v(\{1\}) + \lambda_{\{2\}} v(\{2\}) + \lambda_{\{3\}} v(\{3\}) + \lambda_{\{1,2\}} v(\{1,2\}) \\
& + \lambda_{\{1,3\}} v(\{1,3\}) + \lambda_{\{2,3\}} v(\{2,3\}) + \lambda_N v(N) \\
\text{such that} \quad & \lambda_S \geq 0, \ \forall S, \\
& \lambda_{\{1\}} + \lambda_{\{1,2\}} + \lambda_{\{1,3\}} + \lambda_N = 1, \\
& \lambda_{\{2\}} + \lambda_{\{1,2\}} + \lambda_{\{2,3\}} + \lambda_N = 1, \\
& \lambda_{\{3\}} + \lambda_{\{1,3\}} + \lambda_{\{2,3\}} + \lambda_N = 1.
\end{aligned}$$

For the general case, thus we shall write the dual problem in the following way:

$$\begin{aligned}
\text{maximize} \quad & \sum_{S \subset N} \lambda_S v(S) \\
\text{such that} \quad & \lambda_S \geq 0, \text{ and } \sum_{S:i \in S \subset N} \lambda_S = 1, \ \forall i = 1, \ldots, n.
\end{aligned} \tag{7.15}$$

It is quite clear that both problems are feasible and bounded. Thus the maximum value of the dual problem agrees with the minimum value of the initial one. We can then claim:

Theorem 7.4.6 *The core $C(v)$ of the game v is nonempty if and only if every vector $(\lambda_S)_{S \subset N}$ fulfilling the conditions*

$$\lambda_S \geq 0, \qquad \forall S \subset N$$
$$\sum_{S:i \in S \subset N} \lambda_S = 1 \quad \forall i = 1, \ldots, n,$$

also satisfies

$$\sum_{S \subset N} \lambda_S v(S) \leq v(N).$$

At a first reading the above result could look uninteresting. It is not clear why solving the dual problem should be easier than solving the initial one. However, as often in game theory, it has a very appealing interpretation, which can convince us to go further in the analysis. First of all, let us observe that we can give an interpretation to the coefficients λ_S. The conditions

$$\lambda_S \geq 0, \qquad \forall S \subset N$$

$$\sum_{S:i\in S\subset N} \lambda_S = 1 \quad \forall i = 1,\ldots,n,$$

suggest looking at these coefficients as a possible "percentage" of participation of the players in a coalition. $\lambda_{\{1,2\}}$ represents, for instance, the percentage of participation of players one and two in the coalition $\{1,2\}$. Thus, in a sense, the theorem suggests that, no matter how the players decide their quota in the coalitions, the corresponding weighted values must not exceed the available amount of utility $v(N)$. It is clearly a way to control the power of the intermediate coalitions.

The geometry of the set of λ_S fulfilling the above constraints is quite clear. We have to intersect various planes with the cone made by the first orthant. As a result we get a convex polytope, having a finite number of extreme points, which are the only interesting points when one must maximize a linear function. The very important fact is that the theory is able to characterize these points. We do not go into much detail here, but rather we just describe the situation. A family $(S_1,\ldots,S_m)$ of coalitions (i.e. a subset of 2^N) is called *balanced* provided there exists $\lambda = (\lambda_1,\ldots,\lambda_m)$ such that $\lambda_i > 0 \ \forall i = 1,\ldots,m$ and, for all $i \in N$,

$$\sum_{k:i\in S_k} \lambda_k = 1.$$

λ is called a *balancing* vector.

Example 7.4.7 A *partition* of N (i.e., any family of disjoint sets covering N) is a balancing family, with balancing vector made up of all 1's. Let $N = \{1,2,3,4\}$; the family $(\{1,2\},\{1,3\},\{2,3\},\{4\})$ is balanced, with vector $(1/2,1/2,1/2,1)$. Let $N = \{1,2,3\}$, and consider the family $(\{1\},\{2\},\{3\},N)$. It is balanced, and every vector of the form $(1-p,p,p,p)$, $0 < p < 1$, is a balancing vector. The family $(\{1,2\},\{1,3\},\{3\})$ is not balanced.

Observe that in the case of a partition the balancing vector is unique, while it is not in the third example above. There is a precise reason for this. It is clear that in the third example we could erase some members of the collection (e.g., N) and still have a balanced family. However, it is not possible to erase a coalition from, for example, a partition, without destroying balancedness. Thus we can distinguish between *minimal* and nonminimal balancing families. The minimal ones are characterized by the fact that the balancing vector is unique. It can be shown that the extreme points of the constraint set in (7.15) are exactly the balancing vectors of the minimal balanced coalitions. Thus the following theorem, which we state without proof, holds:

Theorem 7.4.8 *The cooperative game v has a nonempty core if and only if, for every minimal balanced collection of coalitions, with balancing vector $\lambda = (\lambda_1,\ldots,\lambda_m)$,*

$$\sum_{k=1}^{m} \lambda_k v(S_k) \leq v(N).$$

Now, the (absolutely nontrivial) task is to see how many minimal balanced collections an N person game has. And also, in order to facilitate our job, to observe that partitions which are minimal and balanced can be ignored if we assume that the game is superadditive, because in such a case the condition required in the theorem is automatically fulfilled. Let us fully develop the case of a three player game. Let us put

$$\lambda_{\{1\}} = a, \lambda_{\{2\}} = b, \lambda_{\{3\}} = c,$$
$$\lambda_{\{1,2\}} = x, \lambda_{\{1,3\}} = y, \lambda_{\{2,3\}} = z,$$
$$\lambda_N = w.$$

The system of inequalities becomes

$$a + x + y + w = 1,$$
$$b + x + z + w = 1,$$
$$c + y + z + w = 1.$$

Taking into account the nonnegativity conditions, we have the following extreme points (we conventionally assign zero to a coalition not involved in the balanced family):

$(1, 1, 1, 0, 0, 0, 0)$ corresponding to the balanced family $(\{1\}, \{2\}, \{3\})$,

$(1, 0, 0, 0, 0, 1, 0)$ corresponding to the balanced family $(\{1\}, \{2, 3\})$,

$(0, 1, 0, 0, 1, 0, 0)$ corresponding to the balanced family $(\{2\}, \{1, 3\})$,

$(0, 0, 1, 1, 0, 0, 0)$ corresponding to the balanced family $(\{3\}, \{1, 2\})$,

$(0, 0, 0, 0, 0, 0, 1)$ corresponding to the balanced family (N),

and $(0, 0, 0, (1/2), (1/2), (1/2), 0)$

corresponding to the balanced family $(\{1, 2\}, \{1, 3\}, \{2, 3\})$.

Only the last one corresponds to a balanced family not being a partition of N. Thus, if the game is superadditive, we have just one condition to check: the core is nonempty provided

$$v(\{1, 2\}) + v(\{1, 3\}) + v(\{2, 3\}) \leq 2v(N).$$

This is not difficult. The situation however quickly becomes much more complicated when augmenting the number of players. For instance, in the case of four players, after some simplification, it can be shown that 11 inequalities must be checked to be true in order to have a nonempty core.

8

Hypertopologies, hyperconvergences

Life is not what we have experienced, but what we remember
and how we remember it in order to narrate it.
(G.G. Marquez, "Vivir para contarla")

8.1 Definitions and examples

One of the aims of these notes is to investigate the stability of a minimum problem. Roughly speaking, stability means that small changes in the data of the problem (the objective function, the constraint set) cause small changes in the basic objects of the problem itself, such as the inf value and the set of the minimizers. Clearly, one can give different meanings to the concept of small changes, but in any case every such meaning requires a topological structure on spaces of sets (for instance, to evaluate the changes of the set of minimum points of a given function) and on spaces of functions. The classical convergence notions for functions (for instance, pointwise convergence) do not work in the stability setting, as we shall deduce by means of an example. Rather, it will be more convenient to identify a function with its epigraph, and consequently, to define convergence of functions by means of convergence on spaces of sets. Thus, we are led to consider convergences/topologies on the set $c(X)$ of closed subsets of a metric space (X, d), and this chapter serves as an introduction to this topic. We shall focus only on the topologies on $c(X)$ *respecting* the topological structure of (X, d), in the following sense. The points of X are closed subsets of X, and thus elements of $c(X)$. Then X can be considered as embedded in $c(X)$ by identifying the point x with the singleton $\{x\}$. We are thus interested in those topologies/convergences in $c(X)$ such that the embedding of X is a bicontinuous bijection on its image. In other words, the sequence $\{x_n\}$ in X will converge to x if and only if the sequence $\{\{x_n\}\}$ will converge, in $c(X)$, to $\{x\}$. These topologies are usually called *hypertopologies*, though we shall often omit the prefix hyper in what follows. Together with $c(X)$, we shall also consider some of its important subsets. For instance $c_0(X)$, the family of *nonempty* closed subsets of X, or when X is a linear space, the family $C(X)$ $(C_0(X))$ of closed (nonempty) convex subsets of X.

Some of the topologies/convergences require only a topological structure on X, while others require that X be at least a metric space. In any case, being mainly interested in the convex case, we shall assume X to be at least a metric space, so several results presented in this chapter are not given in full generality.

Let me point out that, when introducing topologies/convergences, I shall emphasize the behavior of converging *sequences*, rather than *nets*. I do this also in the cases when the topology is not first countable, i.e., it cannot be described by sequences. This choice is motivated by the fact that in optimization one usually focuses on sequences. I direct the reader interested in the topological aspects to Appendix B.

The first topology we want to define on $c(X)$ is related to the definition of upper/lower semicontinuity we gave for a multifunction (see Definition 3.5.2). We start with it mainly for historical reasons. Being generally too fine, it is actually not much used in our setting.

So, let (X, d) be a metric space. We have already set $c(X)$ to be the collection of the closed subsets of X; if X is a linear space, denote by $C(X)$ the set of the closed convex subsets of X. Given sets $G \subset X$ and $V \subset X$, let us define

$$V^- := \{A \in c(X) : A \cap V \neq \emptyset\}$$

and

$$G^+ := \{A \in c(X) : A \subset G\}.$$

It is easy to verify that if G and V range over some subfamily $\mathcal{F}$ of open sets in X, then G^+ is a basis for a topology (if $\mathcal{F}$ is closed for the operation of intersection of a finite number of sets), called *an upper topology*, while V^- is a subbasis for another topology, called *a lower topology*.

Definition 8.1.1 We shall call *lower Vietoris topology* on $c(X)$ the topology having as a subbasis of open sets the family $\{V^- : V \text{ is open in } X\}$. We shall call *upper Vietoris topology* on $c(X)$ the topology having as a basis of open sets the family $\{G^+ : G \text{ is open}\}$. Finally, the *Vietoris topology* is the smallest topology finer than both the lower and upper topologies. A basis for it is given by the family of sets

$$G^+ \cap V_1^- \cap \cdots \cap V_n^-,$$

with $G, V_1, \ldots, V_n$ open in X and $n \in \mathbb{N}$.

We shall denote by V^-, V^+, V the lower and upper Vietoris topologies and the Vietoris topology, respectively. Hereafter, given a hypertopology τ, we shall use the notation $A_n \xrightarrow{\tau} A$ to denote that the sequence $\{A_n\}$ converges to A in the τ topology.

Of course, given a multifunction $F\colon X \to Y$ that is closed-valued, then it can be seen as a *function* $F\colon X \to c(Y)$. It is then easy to see that the multifunction F is upper semicontinuous at x if and only if $F\colon X \to (c(Y), V^+)$ is continuous at x; lower semicontinuity is instead related to continuity for the lower Vietoris topology.

Example 8.1.2 In $\mathbb{R}$ the sequence $\{A_n\} = \{\{0, n\}\}$ has a lower limit $\{0\}$ in $c_0(X)$ (and the empty set in $c(X)$). However, $\{0\}$ is not an upper limit. The sequence $\{A_n\} = \{[0, n]\}$ has Vietoris limit $[0, \infty]$. Let X be a Banach space and let $A_n = nB = \{x : \|x\| \le n\}$. Then X is the limit of $\{A_n\}$.

Example 8.1.3 In $\mathbb{R}^2$ let $A_n = \{(x, y) : x \ge 0, y \ge -\frac{1}{n}\}$. Then $\{A_n\}$ *does not* converge to $A = \{(x, y) : x \ge 0, y \ge 0\}$.

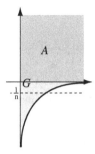

Figure 8.1. $A \in G^+$, $A_n \notin G^+$.

A basic neighborhood of an element A for the upper Vietoris topology contains sets that cannot be too big with respect to the set A, in the sense that they must be contained in an open set containing A. A dual argument can be used for the lower Vietoris topology. Put differently, let us observe that the upper Vietoris topology guarantees that a limit A of a sequence $\{A_n\}$ cannot be too small with respect to the sets A_n (and vice-versa for the lower topology). More precisely, if A is an upper (lower) limit of $\{A_n\}$ and $c(X) \ni B \supset A$ $(c(X) \ni B \subset A)$, then B is also an upper (lower) limit of $\{A_n\}$. This fact is a common feature of all upper and lower topologies we shall consider in the sequel.

The Vietoris topology is usually called a *hit and miss* topology since a typical basic open set for its lower part consists in a family of sets *hitting* a finite number of open sets V_i, while a typical basic open set for its upper part consists in a family of sets *missing* the closed set G^c. Several topologies are built up by following this pattern. For instance, if we want to get a topology coarser than the Vietoris topology, we can reduce the number of open sets as far as the upper part is concerned:

Definition 8.1.4 The *lower Fell topology* on $c(X)$ is the topology having the family $\{V^- : V \text{ is open}\}$ as subbasis of open sets. The *upper Fell topology* on $c(X)$ is the topology having the family $\{(K^c)^+ : K \text{ is compact}\}$ as a basis of open sets. *The Fell topology* has a basis of open sets the family

$$(K^c)^+ \cap V_1^- \cap \cdots \cap V_n^-,$$

where $V_1, \ldots, V_n$ are open sets, K is compact and $n \in \mathbb{N}$.

Let us denote by V^-, F^+, F, respectively, the lower Fell, upper Fell and Fell topologies.

Remark 8.1.5 It is obvious that the Fell topology is coarser than the Vietoris topology as the lower parts are the same, while the upper Vietoris is by definition finer than the upper Fell (strictly, unless the space (X, d) is compact). So in general we shall have more Fell converging sequences than Vietoris converging sequences.

Example 8.1.6 In $\mathbb{R}$ the sequence $\{A_n\} = \{\{0, n\}\}$ has Fell limit $\{0\}$. In $\mathbb{R}^2$ let $A_n = \{(x, y) : x \geq 0, y \geq -\frac{1}{n}\}$. $A_n \xrightarrow{F} A = \{(x, y) : x \geq 0, y \geq 0\}$. In $\mathbb{R}^2$ let $A_n = \{(x, y) : y = \frac{1}{n}x\}$. $A_n \xrightarrow{F} A = \{(x, y) : y = 0\}$. The sequence $\{A_{2n} = \{n\}, A_{2n+1} = \{-n\}\}$ has Fell limit the empty set in $c(X)$, but does not have a limit in $c_0(X)$.

We introduce now a metric on $c(X)$, which is one of the best known and most used way to measure distance between closed sets. It is the so-called *Hausdorff metric topology*.

Definition 8.1.7 Given two nonempty sets $A, C \in c(X)$, we define the *excess of A over C* :

$$e(A, C) := \sup_{a \in A} d(a, C) \in [0, \infty],$$

where, as usual, $d(a, C) := \inf_{c \in C} d(a, c)$.

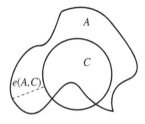

Figure 8.2. The excess of A over C, $e(A, C)$.

When $C = \emptyset$ and $A \neq \emptyset$, we set $e(A,C) = \infty$ (this is motivated by the fact that we shall always work in a linear setting. In arbitrary metric spaces, if the distance d is bounded, this definition could be revised).

Finally, set

$$h(A,C) := \max\{e(A,C), e(C,A)\}.$$

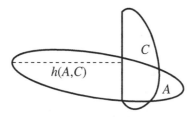

Figure 8.3. The Hausdorff distance between A and C, $h(A,C)$.

It is not hard to prove (see Exercise 8.1.9) that h defines an (extended) metric on $c(X)$, called the *Hausdorff metric topology* .

We have the following proposition, whose proof is left to the reader:

Proposition 8.1.8 *A sequence $\{A_n\}$ of elements of $c(X)$ converges in the Hausdorff sense to $A \in c(X)$ if*

$$e(A_n, A) \to 0 \quad \text{and} \quad e(A, A_n) \to 0.$$

The condition $e(A_n, A) \to 0$ will be called *upper* Hausdorff convergence, whereas the condition $e(A, A_n) \to 0$ will be called *lower* Hausdorff convergence.

Exercise 8.1.9 Verify that h defines a metric (valued in $[0, \infty]$) on $c(X)$.

Hint. The only nontrivial thing is the triangle inequality. Show that

$$e(A, B) \leq e(A, C) + e(C, B),$$

by noticing that $\forall a, c$

$$d(a, B) \leq d(a, c) + d(c, B) \leq d(a, c) + e(C, B)$$

whence

$$d(a, B) \leq d(a, C) + e(C, B), \forall a.$$

Exercise 8.1.10 Verify that

$$e(A, C) = \inf\{\varepsilon > 0 : A \subset S_\varepsilon[C]\},$$

where $S_\varepsilon[C] := \{x \in X : d(x, C) < \varepsilon\}$ (see Figure 8.4).

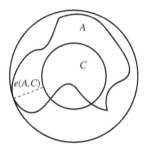

Figure 8.4. The excess of A over C, $e(A,C)$.

Example 8.1.11 In $\mathbb{R}$ the sequence $\{A_n\} = \{\{0,n\}\}$ has lower Hausdorff limit $\{0\}$, which is not the Hausdorff limit, while it is the Fell limit. The sequence $\{A_n\} = \{[0,n]\}$ does not have limit $[0,\infty]$ (it is only an upper limit). For, $e(A, A_n) = \infty, \forall n$. In $\mathbb{R}^2$ let $A_n = \{(x,y) : x \geq 0, y \geq -\frac{1}{n}\}$. Then $A_n \xrightarrow{\text{H}} A = \{(x,y) : x \geq 0, y \geq 0\}$. The empty set is isolated in $(c(X), h)$.

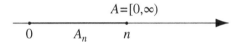

Figure 8.5. $e(A, A_n) = \infty$, A is the Vietoris limit of $\{A_n\}$.

Remark 8.1.12 The examples above show that the Vietoris and Hausdorff topologies are not comparable. This is due to the fact that the lower Vietoris topology is coarser than the analogous Hausdorff, while the opposite happens with the upper parts. The Fell topology is coarser than the Hausdorff topology. (Prove these statements as an exercise.)

Remark 8.1.13 If one is bothered by having a metric taking value ∞, there is a (standard) way to define a real valued (even bounded) metric equivalent to the former one. We can, for instance, consider

$$\hat{h}(A, c) = \min\{h(A, C), 1\}.$$

Then $\hat{h}$ is equivalent to h on $c(X)$.

Remark 8.1.14 Let $X = [0, \infty)$ be endowed with the metric ρ defined as $\rho(x, y) = |\frac{x}{1+x} - \frac{y}{1+y}|$. Then the sequence $\{A_n\} = \{[0,n]\}$ Hausdorff converges to X, as $e(X, A_n) = 1 - \frac{n}{1+n}$. On the other hand, (X, ρ) is topologically equivalent to (X, d), where d is the distance induced by the Euclidean metric on $\mathbb{R}$. Since $e_d(X, A_n) = \infty$ for all n, we can deduce that equivalent metrics

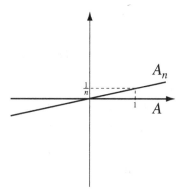

Figure 8.6. A is neither Vietoris nor Hausdorff limit of $\{A_n\}$; it is the Fell limit.

on X usually do not induce the same Hausdorff topology on the hyperspace. More precisely, one can show that two metrics on X induce the same Hausdorff convergence on the hyperspace if and only if they are uniformly equivalent. Thus there are topologies on $c(X)$, like the Vietoris and Fell, depending only on the *topology* of X, and others depending instead on the specific metric given on X. We have noted this fact just as an observation, and we shall not pursue this issue for the other hypertopologies that we shall introduce in the sequel.

Let us introduce now another structure on $c(X)$, by defining it in terms of convergence of sequences. Define the following sets:

$$\operatorname{Li} A_n := \{x \in X : x = \lim x_k, x_k \in A_k \text{ eventually}\}$$

and

$$\operatorname{Ls} A_n := \{x \in X : x = \lim x_k, x_k \in A_{n_k}, n_k \text{ a subsequence of the integers}\}.$$

Definition 8.1.15 The sequence $\{A_n\}$ is said to converge to A in the *Kuratowski* sense if
$$\operatorname{Ls} A_n \subset A \subset \operatorname{Li} A_n.$$

Condition $\operatorname{Ls} A_n \subset A$ relates to the upper part of the convergence, while $A \subset \operatorname{Li} A_n$ is the lower part. The set $\operatorname{Ls} A_n$ is called the *Limsup* of the sequence $\{A_n\}$, while the set $\operatorname{Li} A_n$ is called the *Liminf* of the sequence $\{A_n\}$.

Example 8.1.16 In $\mathbb{R}$ let us consider the sequence $\{A_n\} = \{\{0, n\}\}$. Then $\operatorname{Li} A_n = \operatorname{Ls} A_n = \{0\}$, hence $A_n \overset{\text{K}}{\to} A = \{0\}$. The sequence $\{[0, n]\}$ converges to $[0, \infty]$, the sequence $\{[n, \infty)\}$ converges to the empty set (in $c(X)$, in $c_0(X)$

it converges to nothing). Let X be a linear space and let $\{A_{2n} = B[0;n]$, $A_{2n+1} = B[0;\frac{1}{n}]\}$. Then $X = \operatorname{Ls} A_n$ while $\{0\} = \operatorname{Li} A_n$. In $\mathbb{R}^2$ let $A_n = \{(x,y) : x \geq 0, y \geq -\frac{1}{n}\}$. Then $A_n \xrightarrow{K} A = \{(x,y) : x \geq 0, y \geq 0\}$.

Exercise 8.1.17 Prove that

$$\operatorname{Li} A_n = \{x : \limsup d(x, A_n) = 0\},$$
$$\operatorname{Ls} A_n = \{x : \liminf d(x, A_n) = 0\}.$$

Suppose the sequence $\{A_n\} \subset c(X)$ is such that $d(\,\cdot\,, A_n) \to f(\cdot)$, where f is a continuous function. Then $A_n \to A$ in the Kuratowski sense, where $A = \{x : f(x) = 0\}$.

Exercise 8.1.18 Prove that $x \in \operatorname{Ls} A_n$ if and only for every open set W containing x there is a subsequence $\{n_k\}$ of the integers such that $A_{n_k} \cap V \neq \emptyset$. Prove that $x \in \operatorname{Li} A_n$ if and only if for every open set W containing x, $A_n \cap V \neq \emptyset$, eventually.

We shall see later that, if X is a Euclidean space, then Kuratowski convergence of sequences is compatible with the sequential convergence for a first countable topology.

8.2 Relations among topologies, new topologies

So far, we have introduced on $c(X)$ the Vietoris and Fell topologies, the Hausdorff metric topology, and Kuratowski convergence (for sequences). We now want to study some properties of the above structures. Later on, we shall define new hypertopologies. First, we see that the Limsup and Liminf of sequences of sets $\{A_n\}$ (not necessarily in $c(X)$), are closed, possibly empty, sets.

Proposition 8.2.1 $\operatorname{Li} A_n$ and $\operatorname{Ls} A_n$ are closed sets. Moreover,

$$\operatorname{Ls} A_n = \bigcap_{n \in \mathbb{N}} \overline{\bigcup_{k \geq n} A_k}.$$

Proof. Let us start by showing the statements concerning $\operatorname{Ls} A_n$. To show that it is closed, it is enough to prove that the above formula holds. So, let $A := \bigcap_{n \in \mathbb{N}} \overline{\bigcup_{k \geq n} A_k}$. Let us show that $\operatorname{Ls} A_n \subset A$. If $x \in \operatorname{Ls} A_n$, then $\exists x_k \to x$ such that $x_k \in A_{n_k} \forall k \in \mathbb{N}$. Then $x \in \overline{\bigcup_{k \geq n} A_k} \forall n \in \mathbb{N}$, whence $x \in A$. Now let $x \in A$. Then

$$\forall n \in \mathbb{N}, \exists x_{jn} \in \bigcup_{k \geq n} A_k \text{ such that } \lim_{j \to \infty} x_{jn} = x.$$

Let $x_1 = x_{j_1 1} \in A_{n_1}$ be such that $d(x_{j_1 1}, x) < 1$. Now, for $n = n_1 + 1$, it is possible to find $j_2 > n_1$, and $x_{j_2 n_1+1} \in A_{n_2}$ such that $d(x_{j_2 n_1+1}, x) < \frac{1}{2}$.

Setting $x_2 = x_{j_2 n_1 + 1}$, and proceeding in this way, we get at step k, $x_k = x_{j_k n_{k-1}+1} \in A_{n_k}$ $(n_k > n_{k-1})$ such that $d(x_k, x) \leq \frac{1}{k}$. So a subsequence n_k and a sequence $\{x_k\}$ are found such that $x_k \in A_{n_k}$ and $x_k \to x$. Thus $x \in \operatorname{Ls} A_n$.

We prove now that $\operatorname{Li} A_n$ is a closed set. In order to do this, we use the characterization seen in Exercise 8.1.18. Suppose $x \in \operatorname{cl} \operatorname{Li} A_n$. This means that for every open set O containing x we have $\operatorname{Li} A_n \cap O \neq \emptyset$. Take $z \in \operatorname{Li} A_n \cap O$ and an open set W such that $z \in W \subset O$. By definition, eventually $W \cap A_n \neq \emptyset$, and thus $O \cap A_n \neq \emptyset$ eventually.   □

The next exercise gives an alternative, more "constructive" proof of the fact that $\operatorname{Li} A_n$ is a closed set.

Exercise 8.2.2 Give an alternative proof of the fact that $\operatorname{Li} A_n$ is a closed set.

Hint. Let $x_j \in \operatorname{Li} A_n$ be such that $x_j \to x$. We must find a sequence $\{y_n\}$ such that $y_n \to x$ and $y_n \in A_n, \forall n \in \mathbb{N}$.

As $x_j \in \operatorname{Li} A_n$,
$$\exists x_{jk} \in A_k, \lim_{k \to \infty} x_{jk} = x_j.$$
Hence $\forall j \in \mathbb{N}$,
$$\exists N_j \in \mathbb{N}, \forall k > N_j, \, d(x_{jk}, x_j) < \frac{1}{j},$$
and we can assume that $N_1 < N_2 < \cdots < N_{j-1} < N_j < \cdots$. Set $N_0 = 0$, and define
$$y_n = x_{ln} \in A_n, \text{if } N_l < n \leq N_{l+1}.$$

$$
\begin{array}{ccccccccc}
A_1 & \cdots & \cdots & \cdots & A_n & \cdots & \cdots & \cdots & \operatorname{Li} A_n \\
\\
\cup & \cdots & \cdots & \cdots & \cup & \cdots & \cdots & \cdots & \cup \\
\\
x_{11} & \cdots & y_1 & \cdots & \cdots & \cdots & \cdots & \cdots & x_1 \\
\vdots & \vdots & \cdots & \cdots & \vdots & \vdots & \vdots & \vdots & \vdots \\
x_{j1} & \cdots & \cdots & x_{jN_j} & \cdots & y_n & \cdots & x_{jN_{j+1}} & \cdots & x_j \\
\vdots & \vdots & \cdots & \cdots & \vdots & \vdots & \vdots & \vdots & \vdots \\
 & & & & & & & & x
\end{array}
$$

Let $\varepsilon > 0$ and let $j \in \mathbb{N}$ be such that $\frac{1}{j} + d(x_l, x) < \varepsilon$, for all $l \geq j$. Then $\forall n > N_j$, if $n \in (N_l, N_{l+1}]$, $l \geq j$, we have
$$d(y_n, x) \leq d(y_n, x_l) + d(x_l, x) < \varepsilon.$$

The next propositions show some connections among the introduced convergences.

Proposition 8.2.3 *Let* $\{A_n\} \subset c(X)$ *be a sequence. Then the following are equivalent:*

(i) $\{A_n\}$ *converges to* A *for the lower Vietoris topology;*
(ii) $\{A_n\}$ *converges to* A *for lower Kuratowski convergence;*
(iii) $\forall x \in X,$

$$\limsup d(x, A_n) \leq d(x, A).$$

Proof. First, let us show that the condition $A \subset \operatorname{Li} A_n$ (lower Kuratowski) implies $\limsup d(x, A_n) \leq d(x, A), \forall x \in X$, i.e., that (ii) implies (iii). Let $\varepsilon > 0$, and let $a \in A$ be such that $d(x, A) \geq d(x, a) - \varepsilon$. For all (large) n, there is $a_n \in A_n$ such that $a_n \to a$. Then

$$\limsup d(x, A_n) \leq \limsup d(x, a_n) = d(x, a) \leq d(x, A) + \varepsilon.$$

We conclude, as $\varepsilon > 0$ is arbitrary.

We prove now that (iii) implies (i). To do this, let V be an open set such that $V \cap A \neq \emptyset$. We must show that $\{A_n\}$ meets V eventually. Let $a \in V \cap A$. Without loss of generality, we can suppose V is of the form $B(a; \varepsilon)$ for some $\varepsilon > 0$. Then

$$0 = d(a, A) \geq \limsup d(a, A_n).$$

This implies $d(a, A_n) < \varepsilon$ eventually, and thus $A_n \cap V \neq \emptyset$ eventually.

Finally, let us show that (i) implies (ii). To do this, let us suppose that A is a lower Vietoris limit of A_n, and let us show that $A \subset \operatorname{Li} A_n$. Let $a \in A$ and set $V_k = B(a; \frac{1}{k})$, the open ball centered at a with radius $\frac{1}{k}$. As $A \cap V_k \neq \emptyset$ for all k, there is N_k such that $\forall n > N_k, A_n \cap V_k \neq \emptyset$. We can also suppose that $\forall k \geq 1, N_k > N_{k-1}$, and we can set $N_0 = 1$. Let $a_n \in A_n \cap V_k$, if $N_k < n \leq N_{k+1}$. We have built up a sequence $\{a_n\}$ such that $a_n \in A_n$ for all n and $a_n \to a$. □

Proposition 8.2.4 *Let* $\{A_n\} \subset c(X)$ *be a sequence. Then* $A_n \xrightarrow{\text{K}} A$ *if and only if* $A_n \xrightarrow{\text{F}} A$.

Proof. In view of Proposition 8.2.3, we need to show the statement only for the upper parts. Suppose that for every compact set K such that $A \cap K = \emptyset$ then $A_n \cap K = \emptyset$ eventually, and let us show that $\operatorname{Ls} A_n \subset A$. Let $x_k \to x$, $x_k \in A_{n_k}$ and let us prove that $x \in A$. If for a subsequence $x_k \in A$, then $x \in A$. Otherwise $x_k \notin A$ for all large k. Let $K = \{x\} \cup \{x_k : k \in N\}$. Clearly K is a compact set and $K \cap A_n \neq \emptyset$ is true for all the elements of the subsequence n_k. Then $A \cap K \neq \emptyset$, whence $x \in A$. Conversely, let us assume that $\operatorname{Ls} A_n \subset A$ and $A \cap K = \emptyset$, for a compact set K. Suppose, by contradiction, that for a subsequence n_k, $A_{n_k} \cap K \neq \emptyset$. Let $x_k \in A_{n_k} \cap K$. Then there is a limit point x of x_k such that $x \in A \cap K$. But this is impossible, whence $A_n \cap K = \emptyset$ eventually. □

Thus the Kuratowski convergence of sequences describes the way sequences converge in the Fell topology. We emphasize here once again that not all the topologies we shall consider here can be described in terms of sequences. Furthermore, when introducing convergence of sequences, it is conceivable to do the same with nets. But it is important to understand that in this case it is not automatic to have the same relations with a given topology, for sequences and nets. We do not pursue this question here, as we believe that at first glance a reader interested in optimization is more concerned with the behavior of sequences, and less with topological questions. We make this remark because convergence of *nets* for the Fell and the Kuratowski convergences do not agree in general.

We saw in the previous proposition that the condition $\limsup d(x, A_n) \leq d(x, A), \forall x \in X$ is connected to lower Vietoris convergence. It is then natural to consider the dual condition $\liminf d(x, A_n) \geq d(x, A), \forall x \in X$, that one can expect to be related to an upper convergence. So, the following definition sounds quite natural:

Definition 8.2.5 The sequence $\{A_n\}$ is said to converge to A in the *Wijsman* sense if

$$\lim d(x, A_n) = d(x, A), \forall x \in X.$$

In the next proposition we see that if X is separable, sequences converging in the Wijsman sense are the same as sequences converging for a metric topology on $c_0(X)$ (an analogous result can be provided on $c(X)$).

Proposition 8.2.6 *Let X be separable and denote by $\{x_n : n \in \mathbb{N}\}$ a dense countable family in X. Then*

$$d(A, B) = \sum_{n=0}^{\infty} 2^{-n} \frac{|d(x_n, A) - d(x_n, B)|}{1 + |d(x_n, A) - d(x_n, B)|},$$

is a distance on $c(X)$ compatible with Wijsman convergence.

Proof. The family of functions

$$\{x \mapsto d(x, A) : A \in c(X)\}$$

is equilipschitz (with Lipschitz constant 1; prove it). Hence the condition $d(x_n, A_j) \to d(x_n, A), \forall n \in \mathbb{N}$, where $\{x_n : n \in \mathbb{R}\}$ is dense in X, actually is equivalent to $d(x, A_j) \to d(x, A), \forall x \in X$, i.e., to Wijsman convergence. The result now follows from Lemma 8.2.7 below. $\qquad\square$

Lemma 8.2.7 *For all n let us be given a sequence $\{a_{jn}\}_{j \in \mathbb{N}}$. Suppose there is $a > 0$ such that $|a_{jn}| \leq a$, for all n, j. Then for all n,*

$$\lim_{j \to \infty} \sum_{n=1}^{\infty} 2^{-n} a_{nj} = 0 \iff \lim_{j \to \infty} a_{nj} = 0.$$

Exercise 8.2.8 Prove Lemma 8.2.7.

We see now the connections between the Fell and Wijsman convergences.

Proposition 8.2.9 *The following relations are true:*

(i) $\liminf d(x, A_n) \geq d(x, A), \forall x \in X$ *implies that the sequence* $\{A_n\}$ *upper Fell converges to* A;

(ii) *if* X *is a Euclidean space, then the converse also holds true, and the two convergences are the same.*

Proof. Let $\liminf d(x, A_n) \geq d(x, A), \forall x \in X$ and let K be a compact set such that $K \cap A = \emptyset$. We must show that $K \cap A_n = \emptyset$ eventually. If for a subsequence it happens that $A_{n_k} \cap K \neq \emptyset$, then there is a limit point $K \ni x$ of $x_k \in A_{n_k} \cap K$. Then $\liminf d(x, A_n) = 0$, implying $d(x, A) = 0$, whence $x \in K \cap A$, which is a contradiction. To conclude, suppose that X is a Euclidean space, that $\{A_n\}$ upper Fell converges to A and that there are $x \in X$ and n_k such that $d(x, A_{n_k}) < r < d(x, A)$. Then $A \cap B[x; r] = \emptyset$, while $A_{n_k} \cap B[x; r] \neq \emptyset$, and this contradiction shows the claim.   □

Remark 8.2.10 The second relation in the above proposition holds, more generally, in every metric space X with the property that the closed balls are compact in it. The proof is the same.

From the above propositions it follows that Wijsman convergence is finer than Fell convergence, and that they agree in Euclidean spaces.

Example 8.2.11 Let X be a separable Hilbert space with $\{e_n : n \in \mathbb{N}\}$ as an orthonormal basis. Let $A_n = \{2e_1 \cup e_n\}$ and $A = \{2e_1\}$. Then $\{A_n\}$ Fell converges to A, but it does not converge in the Wijsman sense, as $2 = d(0, A) > \lim d(0, A_n) = 1$. This example shows that if the balls of X are not compact, usually the Wijsman convergence is strictly finer than the Fell convergence.

The next result offers a useful characterization of the Hausdorff convergence, remembering that Wijsman convergence amounts to pointwise convergence of the distance functions $f_A(\cdot) = d(\cdot, A)$.

Theorem 8.2.12 $A_n \overset{\mathrm{H}}{\to} A$ *if and only if*

$$\sup_{x \in X} \{|d(x, A_n) - d(x, A)|\} \to 0.$$

Proof. It is enough to show that

$$\sup\{|d(x, A) - d(x, B)| : x \in X\} = h(A, B)$$

for every $A, B \in c(X)$. Let $x \in X$ and let us show that

$$d(x, B) \leq d(x, A) + e(A, B).$$

Let $\varepsilon > 0$ and $a \in A$ be such that $d(x, a) \leq d(x, A) + \varepsilon$. Then, $\forall b \in B$,

$$d(x, b) \leq d(x, a) + d(a, b),$$

whence

$$d(x, B) \leq d(x, a) + d(a, B) \leq d(x, A) + \varepsilon + e(A, B).$$

Therefore, $\forall x \in X$,

$$d(x, B) - d(x, A) \leq e(A, B)$$

and, by interchanging the roles of the sets A and B,

$$d(x, A) - d(x, B) \leq e(B, A),$$

implying

$$\sup\{|d(x, A) - d(x, B)| : x \in X\} \leq h(A, B).$$

On the other hand,

$$e(A, B) = \sup\{d(a, B) - d(a, A) : a \in A\} \leq \sup\{d(x, B) - d(x, A) : x \in X\},$$

and we conclude. $\square$

From the previous result we get that Wijsman convergence is coarser than Hausdorff convergence. It is also coarser than Vietoris convergence, for the lower parts are the same, while the upper Vietoris is finer than the upper Hausdorff (and thus finer than the upper Wijsman).

So far we have seen two convergences on $c(X)$ that can be characterized by two different types of convergence of the family of functions

$$\{f_A(\cdot) = d(\cdot, A) : A \in c(X)\}.$$

More precisely, when using uniform convergence, we generate the Hausdorff metric topology, while pointwise convergence generates the Wijsman topology. It is natural at this point to ask what happens if we consider a third natural convergence mode on the family of functions $\{f_A(\cdot) = d(\cdot, A) : A \in c(X)\}$, namely uniform convergence on bounded sets. The convergence we shall define now as a "localization" of the Hausdorff convergence provides the right answer to the question. Let $x_0 \in X$, where X is a metric space. If A, C are nonempty sets, define

$$e_j(A, C) := e(A \cap B[x_0; j], C) \in [0, \infty),$$
$$h_j(A, C) := \max\{e_j(A, C), e_j(C, A)\}$$

If C is empty and $A \cap B[x_0; j]$ nonempty, set $e_j(A, C) = \infty$.

Definition 8.2.13 The sequence $\{A_n\}$ is said to *Attouch–Wets converge* to A if

$$\lim_{n \to \infty} h_j(A_n, A) = 0 \text{ for all large } j.$$

It is easy to verify that the above convergence is independent of the point x_0, and that the sequence of balls $B[x_0; j]$ can be replaced by any sequence of nested closed bounded sets covering X. In the sequel, when X is a linear space, we always choose $x_0 = 0$.

Theorem 8.2.14 $A_n \overset{\text{AW}}{\to} A$ if and only if $\sup_{x \in B[x_0;j]} \{|d(x, A_n) - d(x, A)|\} \to 0, \forall j$.

Proof. It is left as an exercise.

Remark 8.2.15 As we have already noticed, the family of functions

$$\{x \mapsto d(x, A) : A \in c(X)\}$$

is equilipschitz. Hence, if one has Wijsman convergence of $\{A_n\}$ to A, i.e., $d(x, A_n) \to d(x, A), \forall x \in X$, by the Ascoli–Arzelà theorem it also holds that $d(\,\cdot\,, A_n) \to d(\,\cdot\,, A)$ uniformly on the compact sets, since $\{d(\,\cdot\,, A_n), d(\,\cdot\,, A)\}$ is an equibounded family (on the bounded sets). This means that one has also AW convergence, when the bounded sets are compact. Thus, if X is a finite-dimensional space, Wijsman and Attouch–Wets convergences coincide.

8.3 A convergence for sequences of convex sets

The Fell topology in infinite-dimensional spaces is often too weak to produce interesting results. On the other hand, for several purposes AW convergence is too restrictive. So, it is useful to introduce a new convergence, intermediate between the two, which will be, as we shall see, useful in reflexive (infinite-dimensional) Banach spaces and in a convex setting. We shall restrict our attention to the set $C(X)$ of closed convex subsets of a *reflexive* Banach space X. The basic idea in constructing this new convergence, called *Mosco convergence*, is to exploit the two natural topologies with which X can be endowed.

Definition 8.3.1 Given $A_n, A \in C(X)$, $n = 1, \ldots$, we say that $A_n \overset{\text{M}}{\to} A$ if

$$\text{w-Ls}\, A_n \subset A \subset \text{s-Li}\, A_n,$$

where w-Ls A_n indicates that in the definition of Ls A_n we use the *weak* topology on X, while s-Li A_n indicates that in the definition of Li A_n we use the *norm* topology in X.

It is easy to verify that $A_n \overset{\text{M}}{\to} A$ if and only if $A_n \overset{\text{K}}{\to} A$ both in the norm and the weak topologies on X. For, it always holds that

$$\text{w-Ls}\, A_n \supset \text{s-Ls}\, A_n \supset \text{s-Li}\, A_n,$$
$$\text{w-Ls}\, A_n \supset \text{w-Li}\, A_n \supset \text{s-Li}\, A_n,$$

whence Mosco convergence of $\{A_n\}$ to A implies that A is the Kuratowski limit in the weak and in the norm topologies at the same time. Moreover, let us remark that in the definition we could consider weakly closed sets, rather than closed and convex sets, and also in a nonreflexive setting. However, only in reflexive spaces is the Mosco convergence compatible with a topology with good properties. And we give the notion with convex sets, as we shall use it later only in a convex setting.

Exercise 8.3.2 Let X be a separable Hilbert space with orthonormal basis $\{e_n : n \in \mathbb{N}\}$. Show that $[0, e_n] \overset{M}{\to} \{0\}$ and also $\mathrm{sp}\{e_n\} \overset{M}{\to} \{0\}$ ($\mathrm{sp}\{e_n\}$ means the linear space generated by e_n). On the other hand, we have $\mathrm{sp}\{\bigcup_{k \le n} e_n\} \overset{M}{\to} X$.

Exercise 8.3.3 Let us define a topology τ on $C(X)$ with the following basis of open sets:

$$(wK^c)^+ \cap V_1^- \cap \cdots \cap V_n^-,$$

where $V_1, \ldots, V_n$ are norm open sets, wK is a weakly compact set and $n \in \mathbb{N}$. Following what we did with the Fell topology and the Kuratowski convergence, prove that a sequence $\{A_n\}$ in $C(X)$ τ converges to A if and only if $\{A_n\}$ Mosco converges to A. Observe also that, in the definition of τ, wK weakly compact can be substituted by wK weakly compact and convex, as Exercise 8.3.4 shows.

Exercise 8.3.4 The upper τ topology defined in Exercise 8.3.3 is generated also by the family

$$\{(wC^c)^+ : wC \text{ weakly compact and convex}\}.$$

Hint. Let $A \in C(X)$ and suppose $A \cap K = \emptyset$, with K weakly compact. Every $x \in K$ can be strictly separated from A by means of a hyperplane. This generates an open halfspace containing x and whose closure does not intersect A, and K is contained in a finite number $S_1, \ldots, S_n$ of closed halfspaces not intersecting A. Then

$$A \in (\mathrm{cl\,co}(S_1 \cap K)^c)^+ \cap \cdots \cap (\mathrm{cl\,co}(S_n \cap K)^c)^+ \subset (K^c)^+.$$

Proposition 8.3.5 *Let X be a reflexive Banach space. Mosco convergence in $C(X)$ is finer than Wijsman convergence. If X is a Hilbert space, the two convergences coincide.*

Proof. The lower parts of the two topologies always coincide as Proposition 8.2.3 shows. So, let us concentrate on the upper parts. To begin with, let us suppose that $A \supset \text{w-Ls}\, A_n$ and prove that $\liminf d(x, A_n) \ge d(x, A), \forall x \in X$. If $\liminf d(x, A_n) = \infty$, there is nothing to prove. Otherwise, let $l = \liminf d(x, A_n)$ and let n_k be such that $d(x, A_{n_k}) \to l$. Fix $\varepsilon > 0$, and let $a_k \in A_{n_k}$ be such that eventually $d(x, a_k) \le l + \varepsilon$. Then, for a subsequence k_j,

a_{k_j} weakly converges to a, which we denote hereafter with the usual notation: $a_{k_j} \rightharpoonup a$. Moreover $a \in A$, by assumption. Then

$$d(x, A) \leq \|x - a\| \leq \liminf \|x - a_{k_j}\| \leq l + \varepsilon,$$

and we conclude, as $\varepsilon > 0$ was arbitrary.

Suppose now X is a Hilbert space. We must then show that the condition $\liminf d(x, A_n) \geq d(x, A)$ (upper Wijsman) implies that, given $x_k \in A_{n_k}$ such that $x_k \rightharpoonup x$, then $x \in A$. As both convergences are unaffected by translations of sets with a fixed element, we can suppose $x = 0$ and, by contradiction, that $0 \notin A$. Let p be the projection of 0 on A (see Exercise 4.1.4), $p \neq 0$ and let $\lambda \geq 0$. Then, by assumption

$$\liminf \|x_k + \lambda p\| \geq \liminf d(-\lambda p, A_{n_k}) \geq d(-\lambda p, A) = (1 + \lambda)\|p\|.$$

Setting $a = \frac{\lambda}{\lambda+1} \in [0, 1)$, we then have

$$\liminf \|(1 - a)x_k + ap\| \geq \|p\|, \forall a \in [0, 1).$$

This leads to a contradiction, as the Exercise 8.3.6 shows.   □

Exercise 8.3.6 Show that if $x_k \rightharpoonup 0$ and if $p \neq 0$, then there is $a \in [0, 1)$ such that
$$\liminf \|(1 - a)x_k + ap\| < \|p\|.$$

Hint. Let M be such that $M \geq \|x_k\|$ for all k. Then

$$\liminf \|(1 - a)x_k + ap\|^2 \leq (1 - a)^2 M^2 + a^2 \|p\|^2 < \|p\|^2$$

if $a > \frac{M^2 - \|p\|^2}{M^2 + \|p\|^2}$.

The previous result can be refined. Mosco and Wijsman convergences coincide if and only if X is a reflexive Banach space whose dual space X^* enjoys the property that the weak and norm topologies coincide on the boundary of the unit ball [BF].

8.4 Metrizability and complete metrizability

In this section we want to give some (partial) results on metrizability of $c_0(X)$ and $C_0(X)$, endowed with some hyperspace topology. We shall focus on selected hypertopologies, mainly those which will be used more often in the sequel. Analogous results can be given for $c(X)$ and $C(X)$, but we want to avoid the empty set here, since it is not necessary to consider it in future results and in this way we avoid some technicalities.

We have seen in Proposition 8.2.6 that when X is separable, it is possible to define a metric d on $c_0(X)$ such that d-converging sequences are the sequences converging in the Wijsman sense. It has been proved that $(c_0(X), d)$

is complete if and only if the closed balls of X are compact [LL], for instance, in the finite dimensional case. An interesting result shows that when X is separable, $(c_0(X), d)$ is *topologically complete*, i.e., there is another distance ρ, generating the same open sets as d, such that $(c(X), \rho)$ is complete [Be].

We now want to see that the Hausdorff metric topology h defines a complete distance on $c_0(X)$. The same proof applies to $c(X)$.

Theorem 8.4.1 *Let (X, d) be a complete metric space. Then $c_0(X)$, endowed with the Hausdorff metric topology, is a complete metric space.*

Proof. Let $\{A_n\}$ be a Cauchy sequence in $(c(X), h)$, and fix $\varepsilon > 0$. We shall see that $\{A_n\}$ converges to the set $A = \limsup A_n$. For every $k > 0$, there is n_k such that $\forall n, m \geq n_k$, $h(A_n, A_m) < (\varepsilon/2^k)$. Without loss of generality, we can suppose $\{n_k\}$ to be a strictly increasing sequence. Let $n \geq n_1$ and let $x_1 \in A_{n_1}$. Then there is $x_2 \in A_{n_2}$ such that $d(x_1, x_2) < (\varepsilon/2)$. By induction, we can find, for all k, $x_k \in A_{n_k}$ such that $d(x_k, x_{k-1}) \leq (\varepsilon/2^k)$. Thus $\{x_k\}$ is a Cauchy sequence in (X, d) and so it has a limit, say x. Clearly, $x \in A$. Moreover $d(x_1, x) \leq \varepsilon$. Summarizing, we found n_1 such that, for all $n > n_1$, $e(A_n, A) \leq \varepsilon$. We now show the other required inequality. Let $x \in A$. Then there is $x_j \in A_{n_j}$ such that $x_j \to x$. Thus, we can take j so large that $d(x_j, x) \leq (\varepsilon/2)$ and $n_j \geq n_1$. If $m \geq n_1$, $e(A_{n_j}, A_m) \leq (\varepsilon/2)$, and so $d(x, A_m) < \varepsilon$. Thus $e(A, A_m) \leq \varepsilon$, and the proof is complete. $\square$

The next result deals with the Attouch–Wets convergence. Let X be a normed linear space (for simplicity, the result holds in any metric space).

Theorem 8.4.2 *Define, on $c_0(X)$,*

$$\mathrm{aw}(A, C) = \sum_{n=0}^{\infty} 2^{-n} \sup_{\|x\| \leq n} \frac{|d(x, A) - d(x, C)|}{1 + |d(x, A) - d(x, C)|}.$$

Then aw *is a distance on $c_0(X)$ compatible with the Attouch–Wets topology, and $(c(X), \mathrm{aw})$ is complete.*

Proof. Let $A_j, A \in c(X)$. Then, by Lemma 8.2.7, $aw(A_j, A) \to 0$ if and only if

$$\sup_{\|x\| \leq n} \frac{|d(x, A_j) - d(x, A)|}{1 + |d(x, A_j) - d(x, A)|} \to 0$$

if and only if

$$m_n := \sup_{\|x\| \leq n} |d(x, A_j) - d(x, A)| \to 0$$

and this shows that aw is a distance compatible with the Attouch–Wets topology. Now, take a Cauchy sequence $\{A_j\}$. Observe that there is k such that $A_j \cap kB \neq \emptyset$ eventually. Otherwise for all k there would be j_k such that $A_{j_k} \cap kB = \emptyset$. Fix $x_j \in A_j$, for all j. Then there would be

$s \in \{j_1, j_2, \ldots, j_k, \ldots\}$ such that $A_s \cap (\|x_j\|+1)B = \emptyset$. But then, for $n > \|x_j\|$ we would have

$$m_n(A_j, A_s) \geq |d(x_j, A_j) - d(x_j, A_s)| \geq 1,$$

and this would contradict the Cauchy character of $\{A_j\}$. Thus we can suppose, without loss of generality, that $A_j \cap kB \neq \emptyset$ for all j and for some k. Now, as $\{A_j\}$ is Cauchy, we get from Lemma 8.2.7 that

$$m_n(A_j, A_i) \to 0$$

for all n. Then

$$\sup_{\|x\| \leq n} d(x, A_j) - d(x, A_i) \geq \sup\{d(x, A_j) : x \in A_i \cap nB\} = e(A_i \cap nB, A_j).$$

This implies that $\{h(A_j \cap nB)\}$ is a Cauchy sequence for all n. Then from Theorem 8.4.1 we can conclude that $\{A_j \cap nB\}$ has a limit, call it C_n. Now it is easy to show that $A = \bigcup_n C_n$ is the AW-limit of the sequence $\{A_j\}$.   □

The following result deals with the Mosco topology. Thus X will be a reflexive and separable Banach space, and we focus on the set $C_0(X)$ of nonempty closed convex subsets of X. We need to appeal to a result that it is unnecessary to prove here. Thus we only remind the reader of it. The result claims that X can be equivalently renormed in such a way that Mosco convergence of $\{C_n\}$ to C is equivalent to Wijsman convergence of $\{C_n\}$ to C (thus extending Proposition 8.3.5) and this is also equivalent to the condition $p_{C_n}(x) \to p_C(x)$ for all x, where, as usual, $p_A(x)$ denotes the projection of x over the set A. Having this in mind, we can prove:

Theorem 8.4.3 *Let X be a reflexive separable Banach space. Then $(C(X), \tau_M)$ is topologically complete.*

Proof. Since the Mosco topology on $C_0(X)$ does not change if we renorm X in an equivalent way, we can suppose that Mosco convergence is equivalent to Wijsman convergence and to convergence of projections. Since X is separable, we can find a countable family $\{x_n : n \in \mathbb{N}\}$ which is dense in X. Now, define

$$m(A, C) = \sum_{n=1}^{\infty} 2^{-n} \frac{\|p_C(x_n) - p_A(x_n)\|}{1 + \|p_C(x_n) - p_A(x_n)\|}.$$

We want to show that m is a distance, compatible with the Mosco topology, such that $(C_0(X), m)$ is complete. First, m is a distance. The only thing we need to verify is that $m(A, C) = 0$ implies $A = C$. Now $m(A, C) = 0$ implies $p_C(x_n) = p_A(x_n)$ for all n. Suppose $x \in A \cap C^c$. Then $d(x, a) = 0$, $d(x, C) > 0$ and we can find n such that $d(x_n, A) < d(x_n, C)$. But this implies $p_C(x_n) \neq p_A(x_n)$. Thus m is a distance. Now, $m(C_j, C) \to 0$ implies $x_n - p_{C_j}(x_n) \to x_n - P_C(x_n)$ for all n, and this in turn implies $d(x_j, C_n) \to d(x_j, C)$ for all j. As we have seen, this implies also $d(x, C_n) \to d(x, C)$ for all $x \in X$, i.e., Wijsman convergence. This implies Mosco convergence of $\{C_n\}$ to C. Conversely, Mosco convergence of $\{C_n\}$ to C implies the convergence of projections, and thus $d(C_n, C) \to 0$. Now, consider a Cauchy sequence $\{C_n\}$ in $(C_0(X), m)$. We want to prove that it has a limit. From the definition of m, we get that $\{P_{C_n}(x_j)\}$ is a Cauchy sequence, for all j. Call c_j its limit and set $C = \mathrm{cl}\{\bigcup c_j\}$. First, we prove that $C = \mathrm{Li}\, C_n$. Let us show first $\mathrm{Li}\, C_n \subset C$. Fix $\varepsilon > 0$ and $c \in \mathrm{Li}\, C_n$. Then there are $y_n \in C_n$ for all n, such that $\|c - y_n\| < \varepsilon$ eventually, and j such that $\|x_j - c\| < \varepsilon$. This implies $d(x_j, C_n) < 2\varepsilon$ eventually, i.e., $\|p_{C_n}(x_j) - x_j\| < 2\varepsilon$ and thus $\|p_{C_n}(x_j) - c\| < 3\varepsilon$, eventually. This finally implies $\|c_j - c\| \leq 3\varepsilon$, and thus $\mathrm{Li}\, C_n \subset C$. For the other relation, as $\mathrm{Li}\, C_n$ is a closed set, it is enough to show that, given j, $c_j \in \mathrm{Li}\, C_n$. But this is obvious, as $c_j = \lim p_{C_n}(x_j) \ni C_n$. Thus $C \in C(X)$, as $\mathrm{Li}\, C_n$ is a closed convex set. We have shown that $C = \mathrm{Li}\, C_n$. To conclude, it remains to show that $C_n \to C$ in the sense of Mosco or, equivalently, in the Wijsman sense. We have

$$\limsup_n d(x_k, C_n) = \limsup_n \|x_k - p_{C_n}(x_k)\| \leq \limsup_n \|x_k - p_{C_n}(x_j)\|$$
$$= \|x_k - c_j\|.$$

As this is true for all j, this implies

$$\limsup_n d(x_k, C_n) \leq d(x_k, C).$$

On the other hand,

$$\liminf_n d(x_k, C_n) = \liminf_n \|x_k - p_{C_n}(x_k)\| = \|x_k - c_k\| \geq d(x_k, C).$$

We conclude by the density of $\{x_k : k \in \mathbb{R}\}$.   □

The next compactness result is very useful, since it implies, in particular, that a sequence of closed sets in a Euclidean space always admits convergent subsequences. To get this result, it is necessary to include the empty set in $c(X)$. Of course, a compactness result is at the same time a completeness result.

Theorem 8.4.4 Let X be a metric space. Then $c(X)$, endowed with the Fell topology, is a compact space.

Proof. The proof appeals to a theorem of Alexander, claiming that, if each covering of the space made by a family of open sets taken from a *subbasis*, has a finite subcovering, then the space is compact. Hence, let us consider a covering of $c(X)$ made by a family of open sets of the form

$$\bigcup_{i \in I} (K_i^c)^+ \cup \bigcup_{j \in J} V_j^-.$$

Both the sets of indices I and J must contain at least one element, for the empty set does not belong to any element of the form V_j^-, while X itself does not belong to K_i^c. Moreover we claim that there is a compact set $K_{\bar{i}}$ such that

$$K_{\bar{i}} \subset \bigcup_{j \in J} V_j.$$

Otherwise, for each i we could take $x_i \in K_i \setminus \bigcup_{j \in J} V_j$. Then the element of $c(X)\,\mathrm{cl}\{x_i : i \in I\}$ would not belong to the given initial covering, which is impossible. As $K_{\bar{i}}$ is a compact set and $K_{\bar{i}} \subset \bigcup_{j \in J} V_j$, there is a finite number of indices, say $1, \ldots, m$, such that

$$K_{\bar{i}} \subset \bigcup_{j=1,\ldots,m} V_j.$$

Now, it is easy to check that

$$(K_{\bar{i}}^c) \cup \bigcup_{j=1,\ldots,m} V_j^-$$

is a covering of $c(X)$. □

The result provided by the next exercise is useful for subsequent results.

Exercise 8.4.5 Let (X, d) be a complete metric space and let $A \subset X$ be a closed set. Then (A, d) is complete. Suppose instead A is an open set. Show that A is topologically complete.

Hint. The function $f \colon A \to \mathbb{R}$, $f(x) = (1/d(x, A^c))$ is continuous, and thus its graph G is a closed subset of $X \times \mathbb{R}$, a complete space. Observe that G is homeomorphic to A.

Remark 8.4.6 Clearly, if we remove the element $\{\emptyset\}$ from $c(X)$, endowed with the Fell topology, we get an open set, thus a topologically complete space, in view of Exercise 8.4.5. In the particular case of X being a Euclidean space, a complete metric is that one provided by the Attouch–Wets (or also the Wijsman) metric topology we considered above.

The next exercise provides a compactness criterion for the AW topology.

Exercise 8.4.7 Let X be a normed space and let $\mathcal{F} \subset c(X)$. If there is a family of compact sets K^n such that $F \in \mathcal{F}$ implies $F \cap nB_X \subset K^n$, except possibly for a finite number of sets in $\mathcal{F}$, then $\mathcal{F}$ is relatively compact in the AW topology.

Now we switch our attention to function spaces to show complete metrizability. Remember that we identify (lower semicontinuous) functions with their epigraphs. First, the following result holds:

Proposition 8.4.8 *Let τ be a convergence on $c(X)$ at least as fine as the Kuratowski convergence. Then*

$$\mathcal{F} := \{A \in c(X) : \exists f : A = \text{epi } f\},$$

is τ closed.

Proof. Let us take a sequence in $\mathcal{F}$ (with some abuse of notation we call it $\{f_n\}$) converging to a set A, and prove that there is f such that $A = \text{epi } f$. To show this, we take $(x, r) \in A$ and we must show that $(x, s) \in A$ for all $s > r$. For each n, there is $(x_n, r_n) \in \text{epi } f_n$ with the property that $x_n \to x$, $r_n \to r$. Thus $r_n \leq s$ eventually, and $(x_n, s) \in \text{epi } f_n$. Thus its limit (x, s) must belong to A. □

We stated the above result for Kuratowski convergence of sequences, since we are interested in results with the Mosco and Attouch–Wets convergences, which are finer. The above proof also holds, without any change, when X is a Banach space and we concentrate our attention on the subset $C(X)$ of $c(X)$.

From the previous result it follows that whenever τ is a topology for which $c(X)$ (or $C(X)$) is complete (or topologically complete), then the subset made by the epigraphs is also (topologically) complete, as it is a closed set. But actually we are interested in a subset of $\mathcal{F}$, namely the proper functions, i.e., those never assuming value $-\infty$. Remembering that a convex, lower semicontinuous function assuming value $-\infty$ cannot have finite values, thus $\Gamma(X)$ can be characterized as the subset of A of the functions assuming a finite value at a point.

Proposition 8.4.9 *Let X be a reflexive Banach space. Then the set of functions f such that there exists x with $|f(x)| < \infty$ is open in the Mosco topology.*

Proof. Take f and $\bar{x}$ such that $|f(\bar{x})| < \infty$. Since $f \in \Gamma(X)$, it is lower bounded in $B[\bar{x}; 1]$, there is a such that $f(x) > a$ for all $x \in B[\bar{x}; 1]$. Now, consider the sets

$$V := B\big((x, f(x)); 1\big), \quad wK := B[x; 1] \times \{a\}.$$

Clearly, V is open and wK is weakly compact. Thus $\mathcal{W} = V^- \cap (K^c)^+$ is an open neighborhood of f. Take $g \in \mathcal{W}$. Then there must be x_g such that $a < g(x_g) < f(x) + 1$, and this ends the proof. □

Thus, with the help of Exercise 8.4.5 we can conclude the following:

Theorem 8.4.10 *Let X be a Banach space. Then $(\Gamma(X), \text{aw})$ is topologically complete. Let X be a reflexive and separable Banach space. Then $(\Gamma(X), m)$ is topologically complete.*

Proof. It is enough to observe that an open set of a topologically complete space is topologically complete as well. □

8.5 A summary of the topologies when X is a normed space

In this section we summarize some results we have seen above, under the assumption that X is a normed linear space. This is the most interesting case from the perspective of this book, and can also serve as a useful starting point for the study of the convex case.

Besides the Vietoris topology, described mainly for historical reasons (and for its connections with the ideas of lower and upper semicontinuity of a multifunction), but too fine to be used in problems of optimization, we have introduced the Hausdorff metric topology. This one too is very fine, but it serves as an introduction to coarser topologies. For instance, one can declare a sequence $\{A_n\}$ converging to A if $A_n \cap kB$ converges in the Hausdorff sense to $A \cap kB$ for all (large) k. The Hausdorff metric topology can also be seen as the topology characterized by a uniform convergence of distance functions, i.e., A_n converges to A for the Hausdorff metric topology if and only if $d(\,\cdot\,, A_n) \to d(\,\cdot\,, A)$ uniformly. Thus it is natural to define convergence of a sequence $\{A_n\}$ to a set A by requiring $d(\,\cdot\,, A_n) \to d(\,\cdot\,, A)$ in *different ways*. Naturally enough, we can consider pointwise convergence and uniform convergence on bounded sets. It is not strange that uniform convergence on bounded sets is equivalent to the "localized" Hausdorff convergence we mentioned before. This is the Attouch–Wets convergence, thus characterized in two different useful ways. Moreover, as $d(\,\cdot\,, A)$ is a family of equilipschitz functions (i.e., with constant 1), pointwise convergence (giving raise to Wijsman convergence) and uniform convergence on bounded sets coincide whenever the bounded sets are compact, i.e., in the finite dimensional case. In this case, Wijsman convergence is convergence for a topology, the Fell topology, described in terms of a hit and miss topology, as the Vietoris topology. Moreover, this topology makes $c(X)$ compact. Summarizing, in the finite dimensional case we essentially have a (unique) useful topology, that we describe in several ways, which are useful in different contexts. These are the Fell, Wijsman, and Attouch–Wets topologies. Moreover, convergence in this completely metrizable topology can be described by Kuratowski convergence of sets.

In the infinite dimensional case we have introduced a new convergence (on $C(X)$), finer than Kuratowski convergence (which is too weak for many purposes), and coarser than Attouch–Wets convergence. This is the Mosco convergence that exploits both the norm and the weak topologies on X, and thus not surprisingly enjoys good properties only when X is reflexive. Thus, in infinite dimensions, we can essentially consider, in increasing order, the following convergences: Wijsman, Mosco, and Attouch–Wets.

To conclude, I want to remark that quite possibly the introduction (probably a little brutal) of several convergences/topologies can in some sense leave the reader feeling annoyed and skeptical about the value of introducing so many topological structures. Instead, I strongly believe that having several convergences is quite useful, I would even say necessary, when dealing, for instance, with stability properties in optimization. This is due to the fact that several times we apply standard approximation procedures (e.g., penalization, Riesz–Galerkin, etc.) and it is important to know for which topologies they generate converging sequences. Knowing under which assumptions these topologies guarantee some form of stability automatically provides us with the conditions under which these methods allow sequences approximating the solutions to be constructed. For this reason, we are not yet satisfied having only the topologies introduced before. So, to end this section, we introduce, without many comments, some new topologies, which will be mentioned later in connection with a stability problem. One of them is particularly important. It is the so-called slice topology. The reader who is not really interested in going into further detail, can also skip these definitions (except for the *slice* topology, Definition 8.5.4) and the connected results later on. However, the reader who is interested in knowing more about these topologies is directed to the section in Appendix B.

It is simple to explain how to define new topologies dedicated to this. We have seen that natural convergences arise when considering (different modes of) convergences of distance functions. $A_n \to A$ in some sense if and only if $d(\,\cdot\,, A_n) \to d(\,\cdot\,, A)$ in an appropriate sense. For instance, pointwise convergence gives rise to Wijsman convergence. The idea then is to consider other geometric functionals related to sets, for instance, the gap functional.

The gap functional $D(A, B)$ between two (closed) sets A, B is defined in the following way:

$$D(A, B) := \inf\{d(a, b) : a \in A, b \in B\}.$$

Thus new hypertopologies can be defined in the following way: $A_n \to A$ in a certain sense if $D(A_n, C) \to D(A, C)$ for all C in a prescribed class of closed sets. Wijsman convergence is exactly such a kind of convergence, provided we take $C = \{x\}$, with x ranging in X (or just a dense subset, as we know). This is what we intend to do now.

Definition 8.5.1 We say that the sequence $\{A_n\}$ in $c(X)$ converges for the *proximal* topology if
$$D(A_n, C) \to D(A, C),$$
for all closed sets C.

Definition 8.5.2 We say that the sequence $\{A_n\}$ in $c(X)$ converges for the *bounded proximal* topology if

$$D(A_n, C) \to D(A, C),$$

for all (closed) bounded sets C.

The next definitions apply to the convex case. Thus X is a Banach space.

Definition 8.5.3 We say that the sequence $\{A_n\}$ in $C(X)$ converges for the *linear* topology if

$$D(A_n, C) \to D(A, C),$$

for all closed convex sets C.

Definition 8.5.4 We say that the sequence $\{A_n\}$ in $C(X)$ converges for the *slice* topology if

$$D(A_n, C) \to D(A, C),$$

for all (closed) convex bounded sets C.

As a final comment, I note that all these topologies have the lower Vietoris topology as a lower part.

8.6 Epiconvergence of functions and a first stability result

In this section we see how set convergence provides a useful tool in the study of stability. We shall define convergence of functions in terms of convergence of their epigraphs. The choice of the epigraph is motivated by the fact that we focus on minimum problems; by using hypographs, a symmetric theory can be pursued for maxima. One of the reasons for introducing and studying the class of the convex and lower semicontinuous functions by emphasizing their geometric properties (via epigraphs and level sets) is indeed related to the introduction of convergence of functions by means of set convergence of their epigraphs.

So, given a metric space X, we shall identify a lower semicontinuous function $f \colon X \to (-\infty, \infty]$ with its epigraph, a closed subset of $X \times \mathbb{R}$. This space is naturally topologized with the product topology. Usually, we shall take $\bar{d}[(x, r), (y, s)] := \max\{d(x, y), |r - s|\}$ for the product metric (often we shall write d instead of $\bar{d}$, when no confusion occurs).

Finally, given a convergence, or a topology τ on $c(X)$, we shall use the notation $f_n \overset{\tau}{\to} f$ to indicate that epi $f_n \overset{\tau}{\to}$ epi f in $c(X \times \mathbb{R})$.

Proposition 8.6.1 *Let* $f, f_1, \ldots \colon X \to [-\infty, \infty]$ *be lower semicontinuous functions. Then the following are equivalent:*

(i)   Li epi $f_n \supset$ epi f;
(ii) $\forall x \in X$, *such that* $\exists x_n \to x$, $\limsup f_n(x_n) \leq f(x)$.

Proof. Let $x \in X$, $-\infty < f(x) < \infty$. As $(x, f(x)) \in$ epi f, there exists $(x_n, r_n) \in$ epi f_n such that $x_n \to x, r_n \to f(x)$. Hence $\limsup f_n(x_n) \leq \lim r_n = f(x)$. If $f(x) = -\infty$, we substitute $f(x)$ with an arbitrary real number and we proceed in the same way. Then (i) implies (ii). To see the opposite implication, let $(x, r) \in$ epi f. We must find $(x_n, r_n) \in$ epi f_n such that $x_n \to x$, $r_n \to r$. From (ii), $\exists x_n \to x$, with $\limsup f_n(x_n) \leq f(x) \leq r$. It is then enough to choose $r_n = \max\{f_n(x_n), r\}$ to conclude. $\square$

Proposition 8.6.2 *Let $f, f_1, \ldots : X \rightarrow [-\infty, \infty]$ be lower semicontinuous functions. The following are equivalent:*

(i) $\operatorname{Ls} \operatorname{epi} f_n \subset \operatorname{epi} f$;
(ii) $\forall x \in X, \forall x_n \rightarrow x, \; \liminf f_n(x_n) \geq f(x)$.

Proof. Suppose $\operatorname{Ls} \operatorname{epi} f_n \subset \operatorname{epi} f$; let $x_n \rightarrow x$ and let $\liminf f_n(x_n) < \infty$. Fix $r > \liminf f_n(x_n)$. Then there exists an increasing sequence $\{n_k\}$ of indices such that $f_{n_k}(x_{n_k}) < r$. As $(x_{n_k}, r) \in \operatorname{epi} f_{n_k}$ it follows that $(x, r) \in \operatorname{epi} f$. This means $f(x) \leq r$ and so the second condition holds, as $r > \liminf f_n(x_n)$ is arbitrary. To see the opposite implication, let $(x, r) \in \operatorname{Ls} \operatorname{epi} f_n$. Then there exists $(x_k, r_k) \in \operatorname{epi} f_{n_k}$ such that $x_k \rightarrow x$, $r_k \rightarrow r$. We must show that $r \geq f(x)$. Set $x_n = x$ if $n \notin \{n_1, n_2, \ldots\}$. Then

$$r = \lim r_k \geq \limsup f_{n_k}(x_k) \geq \liminf f_n(x_n) \geq f(x),$$

and this concludes the proof of the proposition. $\square$

Observe that when dealing with convergence of functions, there is no need to appeal to subsequences to characterize the condition $\operatorname{Ls} \operatorname{epi} f_n \subset \operatorname{epi} f$.

It is now very easy to collect together the two previous propositions. The result is a fundamental one, and thus we establish it in the form of theorems.

Theorem 8.6.3 *Let $f, f_1, \ldots : X \rightarrow [-\infty, \infty]$ be lower semicontinuous functions. Then the following are equivalent:*

(i) $f_n \xrightarrow{\mathrm{K}} f$;
(ii) (a) $\forall x \in X, \forall x_n \rightarrow x, \; \liminf f_n(x_n) \geq f(x)$,
      (b) $\forall x \in X, \exists x_n \rightarrow x, \; \limsup f_n(x_n) \leq f(x)$.

Theorem 8.6.4 *Let X be a reflexive Banach space and let $f, f_1, \ldots : X \rightarrow [-\infty, \infty]$ be weakly lower semicontinuous functions. Then the following are equivalent:*

(i) $f_n \xrightarrow{\mathrm{M}} f$;
(ii) (a) $\forall x \in X, \forall x_n \rightharpoonup x, \; \liminf f_n(x_n) \geq f(x)$,
      (b) $\forall x \in X, \exists x_n \rightarrow x, \; \limsup f_n(x_n) \leq f(x)$.

Now, suppose we have a sequence $\{f_n\}$ of functions converging in some sense to a limit function f. We are interested in what happens to the basic parameters of a minimum problem, i.e., the inf value and the set of the minima. In other words, does $\inf f_n$ converge to $\inf f$? Does a selection of the minima of f_n converge to a minimum point of f? Actually these questions are a little naive as such. We shall make them more precise in the sequel. But we start by seeing that pointwise convergence is absolutely *not* adequate to provide a form of stability.

Example 8.6.5 Let $f_n : [0, 1] \rightarrow \mathbb{R}$ be defined as

$$f_n(x) = \begin{cases} 0 & 0 \le x \le 1 - \frac{2}{n}, \\ -nx + n - 2 & 1 - \frac{2}{n} \le x \le 1 - \frac{1}{n}, \\ 2nx - 2n + 1 & 1 - \frac{1}{n} \le x \le 1. \end{cases}$$

We have that $\inf f_n = -1$, while $\inf f = 0$. Lower semicontinuity of the value function is missing. Moreover the sequence $\{1 - \frac{1}{n}\}$ of the minima of f_n converges to a point which is not a minimum for f. A nightmare from the point of view of stability! But there is an even deeper reason for considering pointwise convergence not adequate in this setting. The approximating functions are (lower semi) continuous, while their pointwise limit is not. As we know, dealing with abstract minimum problems, the requirement that a function be at least lower semicontinuous is mandatory. Thus pointwise convergence has this negative feature, too. On the other hand, defining convergence of functions via convergence of epigraphs does not cause any problem. First of all, a sequence of epigraphs will converge (in *any* hypertopology) to an epigraph. Moreover, a lower semicontinuous function is characterized by closedness of its epigraph. Thus convergence in $c(X)$ of epigraphs ensures that a limit (in $c(X)$) of a sequence of epigraphs is a closed epigraph, i.e. a lower semicontinuous function. In other words, variational convergences of (lower semicontinuous) functions will always provide a lower semicontinuous limit.

The above example exhibits a sequence of functions having a pointwise limit which is different, at one point, from the epi limit (find this last one!). Actually, it is not difficult to produce an example of a sequence of (continuous) functions on $[0, 1]$ converging to the zero function in the sense of the epigraphs, and not converging to zero at any point.

We start with the first stability result.

Theorem 8.6.6 *Let $f, f_1, \dots : X \to [-\infty, \infty]$ be lower semicontinuous functions and suppose $f_n \xrightarrow{\text{K}} f$.*
Then

(i) $\limsup \inf f_n \le \inf f$;
(ii)   *if x_k minimizes f_{n_k}, n_k a subsequence of the integers, and if x_k converges to x, then x minimizes f and $\lim(\inf f_{n_k}) = \inf f$.*

Proof. We suppose $\inf f$ a real number; the case $\inf f = -\infty$ can be handled in the same way. Let $\varepsilon > 0$ and x be such that $f(x) \le \inf f + \varepsilon$. As there exists x_n such that $\limsup f_n(x_n) \le f(x)$, we get

$$\limsup \inf f_n \le \limsup f_n(x_n) \le f(x) \le \inf f + \varepsilon,$$

and this proves the first statement. As to the second one,

$$\inf f \le f(x) \le \liminf f_{n_k}(x_k) = \liminf \inf f_{n_k}$$
$$\le \limsup \inf f_{n_k} \le \limsup \inf f_n \le \inf f,$$

hence all inequalities above are actually equalities.   □

Let us observe that the upper semicontinuity of the value function inf follows from a weak lower convergence of the functions, i.e., the Vietoris convergence. Conversely, upper Kuratowski convergence of the functions does not automatically provide lower semicontinuity of the inf function, but it is necessary to have also some compactness condition. Upper convergence of the epigraphs actually guarantees that every limit point, *if any*, of minima of the approximating functions is indeed a minimum point for the limit function. Observe also that we stated the theorem for the Kuratowski convergence of the functions, but a similar statement holds for the Mosco convergence too. In such a case, for (weakly lower semicontinuous) functions defined on a reflexive Banach space, to get existence of a minimum point for the limit function it is enough to show that a subsequence of approximating functions all have a minimum point in a fixed bounded set.

This is the first basic and abstract result, and it is necessary to translate it when we deal with more concrete situations. For instance, when we have constrained problems, it will be necessary to take into account how the constraint sets affect the convergence of the objective functions.

We conclude the chapter with several exercises on convergence of sets and functions. The reader should be aware that the proofs of some of them are much more than simple exercises.

Exercise 8.6.7 Show that given $A_n, A \in c(X)$, then

$$I_{A_n} \xrightarrow{\tau} I_A$$

if and only if

$$A_n \xrightarrow{\tau} A,$$

where τ is Kuratowski or AW or M convergence.

Exercise 8.6.8 Let f_n, f be real valued lower semicontinuous functions. Show that if $f_n \to f$ uniformly on bounded sets, then $f_n \xrightarrow{K} f$.

Exercise 8.6.9 Let A_n be closed convex subsets of a normed space. Show that $\operatorname{Li} A_n$ is (closed and) convex (possibly empty).

Exercise 8.6.10 Let $f_n \colon \mathbb{R} \to \mathbb{R}$ be defined as

$$f_n(x) = \begin{cases} 1 - \frac{2n}{n-1}x & 0 \le x \le 1 - \frac{1}{n}, \\ 2nx + 1 - 2n & 1 - \frac{1}{n} \le x \le 1, \\ \infty & \text{otherwise.} \end{cases}$$

Then f_n is a convex continuous function for each n. Find the pointwise and Kuratowski limits of the sequence $\{f_n\}$.

Now, let

$$f_n(x) = \begin{cases} -\frac{x}{n} & x \le n, \\ -1 & x \ge 1. \end{cases}$$

Find the pointwise and Kuratowski limits of the sequence $\{f_n\}$.

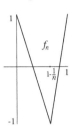

Figure 8.7.

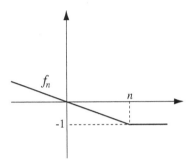

Figure 8.8. inf f_n *does not converge to* inf f.

Exercise 8.6.11 Let X be a separable Hilbert space and let $\{e_n : n \in \mathbb{N}\}$ be an orthonormal basis. Let

$$f(x) = \|x\|^2, \, f_n(x) = \sum_{i \neq n} (x, e_i)^2.$$

Does $f_n \overset{\text{M}}{\to} f$? $f_n \overset{\text{AW}}{\to} f$?

Exercise 8.6.12 Show that $A_n \overset{\text{M}}{\to} A \iff d(\cdot, A_n) \overset{\text{M}}{\to} d(\cdot, A)$.

Exercise 8.6.13 Find $A_n \overset{\text{K}}{\to} A$, $C_n \overset{\text{K}}{\to} C$, all nonempty closed convex subsets of $\mathbb{R}^k$, and such that $\{A_n \cap C_n\}$ does not Kuratowski converge to $A \cap C$.

Exercise 8.6.14 Show that if $A \subset \operatorname{Li} A_n$, $C \subset \operatorname{Li} C_n$, are nonempty closed convex subsets of $\mathbb{R}^k$, and if int $A \cap C \neq \emptyset$, then $A \cap C \subset \operatorname{Li}(A_n \cap C_n)$.

Hint. Show at first that $(\operatorname{int} A) \cap C \subset \operatorname{Li}(A_n \cap C_n)$. If $x \in (\operatorname{int} A) \cap C$ there exists $c_n \in C_n$ such that $c_n \to x$. As $x \in \operatorname{int} A$, then $c_n \in \operatorname{int} A$ eventually. Show that eventually $c_n \in A_n$. Then show that $A \cap C = \overline{\operatorname{int} A \cap C}$. See Figure 8.9.

Exercise 8.6.15 Show that if $f_n \overset{\text{M}}{\to} f, f, f_n, f \in \Gamma(X)$, X a reflexive Banach space, then $(f^a)_n \overset{\text{M}}{\to} f^a$ for all $a > \inf f$. What if $a = \inf f$?

Hint. If $a > \inf f$, $f^a = \overline{\{x \in X : f(x) < a\}}$.

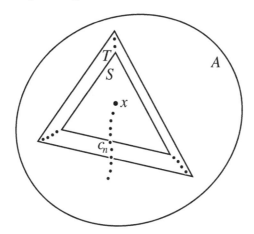

Figure 8.9. (Hint for Exercise 8.6.14) A contains the simplex T. A smaller simplex S is eventually contained in A_n.

Exercise 8.6.16 If A_n, A are closed convex subsets of a normed space, $A_n \overset{\text{AW}}{\to} A$, A bounded, show $A_n \overset{\text{H}}{\to} A$.

Hint. Let $A \subset rB$, and let $x \in A$ be such that $\|x\| < r$. Then $A_n \subset (r+1)B$. Otherwise, there would exist $a_k \in A_{n_k}$ such that $\|a_k\| > r+1$, and as there is $A_{n_k} \ni x_k \to x$, on the line segment $[x_k, a_k]$ there is a point $y_k \in A_{n_k}$ such that $\|y_k\| = r+1$. Then $d(y_k, A) \ge 1$, against the fact that $e(A_{n_k} \cap (r+1)B, A) \to 0$. Hence $e(A_n, A) = e(A_n \cap (r+1)B) \to 0$. Moreover $e(A, A_n) = e(A \cap rB, A_n) \to 0$.

Exercise 8.6.17 Given $A \in c(X)$, let us denote by

$$\text{diam}A := \sup\{d(a_1, a_2) : a_1, a_2 \in A\}.$$

If A_n, A are closed convex subsets of a normed space and if $A_n \overset{\text{AW}}{\to} A$, then $\text{diam}A_n \to \text{diam}A$. Is the same true with Mosco convergence? Is the same conclusion true for a convergence weaker than AW?

Exercise 8.6.18 Let $f_n, f \in \Gamma(X)$, let $x_n \to x$, and let $x_n^*, x^* \in X^*$ be such that $x_n^* \to x^*$. Setting

$$g_n(x) = f_n(x + x_n) + \langle x_n^*, x \rangle, g(x) = f(x + \bar{x}) + \langle x^*, x \rangle,$$

show that f_n Mosco (AW) converges to f if and only if g_n Mosco (AW) converges to g.

Continuity of some operations between functions

Adgnosco veteris vestigia flammae.
(P. Virgilio Marone, "Eneide")

In this section we shall investigate the behavior of the convergence of functions with respect to some important operations. We connect the convergence of a sequence of functions with the convergence of the conjugates, we study the convergence of the sum of two converging sequences, and we provide some results on the convergence of functions and of their subdifferentials. The study of these questions has relevance in optimization, as we shall see. We shall focus our attention on the AW, Mosco and slice convergences.

9.1 Continuity of the conjugation

The first topic we consider is the continuity of the Fenchel transform. We start by establishing a useful lemma.

Lemma 9.1.1 *Let $g \in \Gamma(X)$, $0 \in \operatorname{dom} g$, $x^* \in X^*$, $R > 0$, $s \in \mathbb{R}$, be such that*

$$g(x) > \langle x^*, x \rangle - s, \forall x, \|x\| \le R.$$

Then there are $y^* \in X^*$, $k \le s$ such that

(i) $g(x) \ge \langle y^*, x \rangle - k, \forall x \in X$;

(ii) $\|y^* - x^*\| \le \frac{s + g(0)}{R}$.

Proof. Let $A := \{(x, t) : \|x\| \le R, t \le \langle x^*, x \rangle - s\}$. Then epi g and int A can be separated. Thus there are $z^* \in X^*, a, c \in \mathbb{R}$ such that $(z^*, a) \ne (0^*, 0)$ and

$$\langle z^*, x \rangle + ar \ge c > \langle z^*, z \rangle + at, \forall (x, r) \in \operatorname{epi} g, (z, t) \in \operatorname{int} A. \qquad (9.1)$$

As usual, $a \ge 0$. Suppose $a = 0$. Applying the above inequalities to $x = 0$ and $(z, t) \in A$, with z such that $\|z\| \le R$ and suitable t, we get $0 \ge \langle z^*, z \rangle$, and this implies $z^* = 0$, which is impossible. Thus $a > 0$. Dividing by a, we get from (9.1) that if $(x, r) \in$ epi f,

$$\left\langle \frac{z^*}{a}, x \right\rangle + r \geq \frac{c}{a},$$

from which, setting $y^* = \frac{-z^*}{a}$, $-k = \frac{c}{a}$, we get

$$g(x) \geq \langle y^*, x \rangle - k, \ \forall x \in X,$$

implying, in particular, $g(0) \geq -k$. We then need to show that y^* fulfills condition (ii). From (9.1) we also get that $\langle y^*, x \rangle - k \geq t \ \forall (x, t) \in A$, whence, with the choice of $t = \langle x^*, x \rangle - s$,

$$\langle y^* - x^*, x \rangle \geq k - s, \forall x, \ \|x\| \leq R,$$

implying both $s \geq k$ (with the choice of $x = 0$) and

$$-R\|y^* - x^*\| \geq k - s.$$

Therefore

$$\|y^* - x^*\| \leq \frac{s - k}{R} \leq \frac{s + g(0)}{R}.$$

$\square$

Now we can state the first continuity result.

Theorem 9.1.2 *Let X be a reflexive Banach space. Let $f_n, f \in \Gamma(X)$. Then*

$$f_n \xrightarrow{M} f \Longleftrightarrow f_n^* \xrightarrow{M} f^*.$$

Proof. It is enough to show that

$$f_n \xrightarrow{M} f \Longrightarrow f_n^* \xrightarrow{M} f^*.$$

Namely, from this follows that

$$f_n^* \xrightarrow{M} f^* \Longrightarrow f_n^{**} \xrightarrow{M} f^{**}$$

and we conclude, since $f_n^{**} = f_n$, $f^{**} = f$. Let us then show that

$$f_n \xrightarrow{M} f \Longrightarrow f_n^* \xrightarrow{M} f^*.$$

At first, we prove that

$$\text{epi } f^* \supset \text{w-Ls epi } f_n^*,$$

or, equivalently, that, if $x_n^* \rightharpoonup x^*$, then

$$\liminf f_n^*(x_n^*) \geq f^*(x^*).$$

For every $x \in X$ there exists $x_n \to x$ such that $\limsup f_n(x_n) \le f(x)$. Hence

$$\liminf f_n^*(x_n^*) \ge \liminf(\langle x_n^*, x_n \rangle - f_n(x_n)) = \langle x^*, x \rangle - \limsup f_n(x_n)$$
$$\ge \langle x^*, x \rangle - f(x).$$

We conclude, as x is arbitrary.

We verify now that for each $x^* \in X^*$ there exists $x_n^* \in X^*$ such that

$$x_n^* \to x^* \text{ and } \limsup f_n^*(x_n^*) \le f^*(x^*).$$

To do this, let us suppose $f^*(x^*) < \infty$ and let s be such that $f^*(x^*) < s$, fix $\varepsilon > 0$ and let us seek for $x_n^* \in X^*$ such that $(x_n^*, s) \in \mathrm{epi}\, f_n^*$ and $\|x_n^* - x^*\| < \varepsilon$ eventually.

We divide the proof into two steps.

Step 1. We prove the result with the further assumption that $0 \in \mathrm{dom}\, f$ and that there is c such that $f_n(0) \le c$.

(a) From the definition of f^*, we have that

$$f(x) > \langle x^*, x \rangle - s, \forall x \in X.$$

(b) Let $R > 0$. We verify that there is N such that $\forall n > N$

$$f_n(x) > \langle x^*, x \rangle - s, \text{ for } \|x\| \le R.$$

Otherwise, there would exist a subsequence n_k and x_k such that $\|x_k\| \le R$ and $f_{n_k}(x_k) \le \langle x^*, x_k \rangle - s$. Along a subsequence, $x_k \to x$, whence

$$f(x) \le \liminf f_{n_k}(x_k) \le \lim \langle x^*, x_k \rangle - s = \langle x^*, x \rangle - s,$$

in contradiction with Step 1(a).

(c) From Step 1(b), we can apply Lemma 9.1.1, with $R > \frac{s+c}{\varepsilon}$, to the function $g(x) = f_n(x)$, $n > N$.

Thus there are $x_n^* \in X^*$, $k_n \le s$ such that

$$f_n(x) > \langle x_n^*, x \rangle - k_n, \forall x \in X \text{ and } \|x_n^* - x^*\| \le \frac{s+c}{R} < \varepsilon.$$

On the other hand $f_n(x) > \langle x_n^*, x \rangle - k_n \; \forall x \in X$ is equivalent to saying that

$$k_n \ge f_n^*(x_n^*),$$

and this is enough to conclude, recalling that $k_n \le s$.

Step 2. We now extend the proof to an arbitrary sequence of functions in $\Gamma(X)$. Let $\bar{x} \in \mathrm{dom}\, f$. There exists $x_n \to \bar{x}$ such that $\limsup f_n(x_n) \le f(\bar{x})$, hence there exists c such that $f_n(x_n) \le c$ eventually. Let us now consider the functions

$$g_n(x) = f_n(x_n + x), \quad g(x) = f(\bar{x} + x).$$

As Exercise 8.6.18 shows, $g_n \overset{\text{M}}{\to} g$. We can now apply the result proved in Step 1 to g_n, g, to conclude that $g_n^* \overset{\text{M}}{\to} g^*$. On the other hand

$$g_n^*(x^*) = f_n^*(x^*) - \langle x^*, x_n \rangle, g^*(x^*) = f^*(x^*) - \langle x^*, \bar{x} \rangle.$$

Appealing again to Exercise 8.6.18, we can conclude that $f_n^* \overset{\text{M}}{\to} f^*$.   □

Remark 9.1.3 We saw that the relation w-Ls epi $f_n^* \subset$ epi f^* (upper convergence condition for the conjugates) is simply implied by epi $f \subset$ s-Li epi f_n (lower convergence of the functions). It is then natural to ask if, dually, epi $f^* \subset$ s-Li epi f_n^* (lower convergence of the conjugates) is implied by the upper convergence condition w-Ls epi $f_n \subset$ epi f. Let us consider the following example. Let

$$f_n(x) = n, \quad f(x) = x.$$

Obviously epi $f \supset$ Ls epi f_n, but it is not true that epi $f^* \subset$ Li epi f_n^*. The above proof however shows that to guarantee that epi $f^* \subset$ Li epi f_n^*, it is enough to have upper convergence of the functions and the existence of $\bar{x} \in$ dom f, and $x_n \to \bar{x}$ such that lim sup $f_n(x_n) \leq f(\bar{x})$. This is clearly implied by the assumption that epi $f \subset$ Li epi f_n.

This observation might appear to be nothing but a boring and useless specification; on the contrary, it will be used to provide a proof of a subsequent theorem (Theorem 9.3.1) much simpler than the original one.

The same theorem holds for the slice convergence, and its proof is similar. Remember that for X a normed space, a sequence $\{A_n\}$ of closed convex sets converges for the slice topology to a (closed, convex) set A if $D(A_n, B) \to D(A, B)$ for all B closed convex and bounded (remember D is the gap functional $D(A, B) = \inf\{d(a, b) : a \in A, b \in B\}$). On the dual space X^* we consider the same gap functional, with B ranging over the family of weak* closed convex, bounded subsets. Let us state the theorem and see the necessary changes in the proof.

Theorem 9.1.4 *Let X be a normed space. Let $f_n, f \in \Gamma(X)$. Then*

$$f_n \overset{\text{sl}}{\to} f \Longleftrightarrow f_n^* \overset{\text{sl}^*}{\to} f^*.$$

Proof. The second part of the proof is the same, with the following remarks.

- If $f_n \overset{\text{sl}}{\to} f$ and if there are $s \in \mathbb{R}$, $x^* \in X^*$ such that $s > f^*(x^*)$, then for each $R > 0$, there exists N such that, $\forall n > N$,

$$f_n(x) > \langle x^*, x \rangle - s,$$

if $\|x\| \leq R$. For, there exists $\varepsilon > 0$ such that for all $x \in X$,

$$f(x) > \langle x^*, x \rangle - (s - \varepsilon),$$

implying

$$D(\text{epi } f, B) > 0,$$

where $B = \{(x, \alpha) : \|x\| \leq R, \alpha = \langle x^*, x \rangle - s\}$. Hence

$$D(\text{epi } f_n, B) > 0$$

eventually, and thus, as required,

$$f_n(x) > \langle x^*, x \rangle - s,$$

if $\|x\| \leq R$.

- If $x_n \to \bar{x}$, then $g_n(\cdot) = f_n(x_n + \cdot) \xrightarrow{\text{sl}} g(\cdot) = f(\bar{x} + \cdot)$;
- If $x_n^* \to \bar{x}^*$, then $g_n^*(\cdot) = f_n^*(\cdot) + \langle x_n^*, \cdot \rangle \xrightarrow{\text{sl}} g(\cdot) = f(\cdot) + \langle \bar{x}^*, \cdot \rangle$.

To conclude, let us see that the lower slice (Vietoris) convergence of $\{f_n\}$ to f implies the upper slice convergence of $\{f_n^*\}$ to f^*. Let $D(\text{epi } f^*, B) > 0$, where B is a convex, weak* closed and bounded set and prove that $D(\text{epi } f_n^*, B) > 0$ eventually. There is $R > 0$ such that $\|x^*\| \leq R$ for every x^* such that $(x^*, r) \in B$. There is some small $\varepsilon > 0$ such that $D(B, \text{epi } f^*) > 3\varepsilon$. Thus epi f^* and $B_{3\varepsilon}$ can be separated. As $B_{3\varepsilon}$ is a weak* closed convex bounded set, we have thus existence of a separating hyperplane, which as usual can be assumed to be nonvertical. Thus there exist $x \in X$, $a \in \mathbb{R}$ fulfilling

$$f^*(x^*) \geq \langle x, x^* \rangle - a \text{ for all } x^* \in X^*; \tag{9.2}$$

and

$$\langle x, x^* \rangle - a > r + 3\varepsilon \text{ for } (x^*, r) \in B. \tag{9.3}$$

(9.2) is equivalent to saying that $a \geq f(x)$. By the lower Vietoris convergence, there exists $x_n \to x$ such that

$$f_n(x_n) \leq a + \varepsilon,$$

eventually. This means that

$$f_n^*(x^*) \geq \langle x_n, x^* \rangle - a - \varepsilon$$

for all x^* and thus, if $\|x^*\| \leq R$ and n is so large that $\|x_n - x\| < \frac{\varepsilon}{R}$,

$$f_n^*(x^*) \geq \langle x, x^* \rangle - a - 2\varepsilon.$$

Let $(x^*, r) \in B$. From (9.3) and the inequality above we thus get

$$f_n^*(x^*) > r + 3\varepsilon - 2\varepsilon = r + \varepsilon.$$

Thus $D(\text{epi } f_n^*, B) \geq \varepsilon$, and this concludes the proof.   □

We now see the same result with the Attouch–Wets convergence. To do this, we start by proving the following technical lemma.

Lemma 9.1.5 *Let $f, g \in \Gamma(X)$, $r, R > 0$, $0 < t < 1$, $a \in \mathbb{R}$, $x^* \in X^*$ be such that*

$$\|x^*\| \leq r, |a| \leq r, f(x) \geq \langle x^*, x \rangle - a, \forall x \in X.$$

Moreover, let us suppose that

$$e(\operatorname{epi} g \cap sB_{X \times \mathbb{R}}, \operatorname{epi} f) \leq t,$$

with some $s > \max\{R, rR + 2r + 1\}$. Then

$$g(x) \geq \langle x^*, x \rangle - a - (r+1)t, \quad for \ \|x\| \leq R.$$

Proof. Let $\|x\| \leq R(< s)$ and $|g(x)| \leq s$. Then there are $(\bar{x}, \bar{a}) \in \operatorname{epi} f$ such that

$$\|x - \bar{x}\| \leq t, |\bar{a} - g(x)| \leq t$$

whence

$$g(x) \geq \bar{a} - t \geq f(\bar{x}) - t \geq \langle x^*, \bar{x} \rangle - a - t \geq \langle x^*, x \rangle - a - (r+1)t.$$

Now, let $\|x\| \leq R$ and $g(x) < -s$. Applying the previous formula to $(x, -s) \in \operatorname{epi} g$, we get

$$-s \geq \langle x^*, x \rangle - a - (r+1)t > -Rr - a - (r+1)t > -s.$$

Hence there does not exist x such that $\|x\| \leq R$ with $g(x) \leq -s$. Finally, if $\|x\| \leq R$ and $g(x) \geq s$, then

$$\langle x^*, x \rangle - a - (r+1)t \leq Rr + 2r + 1 < s \leq g(x).$$

Summarizing, for each x such that $\|x\| \leq R$, we have

$$g(x) \geq \langle x^*, x \rangle - a - (r+1)t.$$

$\square$

Theorem 9.1.6 *Let X be a normed space, let $f_n, f \in \Gamma(X)$. Then*

$$f_n \overset{\text{AW}}{\to} f \Longleftrightarrow f_n^* \overset{\text{AW}}{\to} f^*.$$

Proof. As in the case of the Mosco convergence, it is enough to show only that $f_n \overset{\text{AW}}{\to} f$ implies $f_n^* \overset{\text{AW}}{\to} f^*$; moreover, we can assume that $0 \in \operatorname{dom} f$ and that there exists c such that $f_n(0) \leq c$. We must show that given $\varepsilon, r > 0$, there exists N such that $\forall n > N$,

$$e(\operatorname{epi} f^* \cap rB_{X^* \times \mathbb{R}}, \operatorname{epi} f_n^*) < \varepsilon,$$
$$e(\operatorname{epi} f_n^* \cap rB_{X^* \times \mathbb{R}}, \operatorname{epi} f^*) < \varepsilon.$$

We prove the first formula, the second one being completely analogous. Let $(x^*, a) \in \operatorname{epi} f^* \cap rB_{X^* \times \mathbb{R}}$. Then

$$f(x) \geq \langle x^*, x \rangle - a, \ \forall x \in X.$$

By Lemma 9.1.5, with the choice of $t = \frac{\varepsilon}{2(r+1)}$, $R > \frac{a+\varepsilon+c}{\varepsilon}$ and $s = Rr+2r+2$, there exists N such that $\forall n > N$, $\forall x$ such that $\|x\| \leq R$,

$$f_n(x) \geq \langle x^*, x \rangle - a - \frac{\varepsilon}{2}.$$

By Lemma 9.1.1 there are $x_n^* \in X^*$, $k_n \leq a + \frac{\varepsilon}{2}$ such that

$$f_n(x) \geq \langle x_n^*, x \rangle - k_n, \ \forall x \in X$$

and

$$\|x_n^* - x^*\| \leq \frac{a + \varepsilon + c}{R} < \varepsilon.$$

Then $(x_n^*, a + \frac{\varepsilon}{2}) \in \operatorname{epi} f_n^*$ and $\mathrm{d}[(x_n^*, a + \frac{\varepsilon}{2}), (x^*, a)] < \varepsilon.$ $\qquad\square$

9.2 Continuity of the sum

With pointwise convergence or uniform convergence (on bounded sets) it is quite clear that the limit of a sum of convergent sequences is the sum of the limits. The same certainly is not obvious (nor even true, in general), for epiconvergences. Consider the following example:

Example 9.2.1 In $\mathbb{R}^2$ let us consider $A_n = \{(x,y) : 0 \leq x \leq 1, y = \frac{1}{n}x\}$ $A = \{(x,y) : 0 \leq x \leq 1, y = 0\}$, $B_n = B = A$. Then $A_n \overset{\mathrm{H}}{\to} A$, B_n converges to B in every topology, but $A_n \cap B_n$ converges to $A \cap B$ in no lower topology. This shows that a sum theorem does not hold in general, as a sequence of sets converges in some usual hypertopology if and only if the sequence of the indicator functions converges to the indicator function of the limit (see Exercise 8.6.7). On the other hand, the indicator function of the intersection is exactly the sum of the indicator functions, obviously.

On the other hand, Exercise 8.6.14 suggests that some extra condition, at least in the convex case, can provide a positive result. So, let us more carefully investigate this topic. Let us start with an example showing what happens with Mosco convergence.

Example 9.2.2 Let X be a separable Hilbert space, with basis $\{e_n, n \in \mathbb{N}\}$. Let

$$A_n = \operatorname{sp}\left\{\bigcup_{i \leq n} e_i\right\}, B = \left\{\sum_{n \in \mathbb{N}} \frac{1}{n} e_n\right\}.$$

Then $A_n \overset{\mathrm{M}}{\to} X$, but $A_n \cap B$ does not converge to B. This shows that Exercise 8.6.14 cannot simply be extended to infinite dimensions with Mosco convergence.

We intend now to show that it is possible to extend the result of Exercise 8.6.14 to infinite dimensions, just using Attouch–Wets convergence. To prepare the proof of the result, we first present two simple Lemmas.

Lemma 9.2.3 *Let X be a normed space, let A, B, C be closed convex sets such that B is bounded. If $A + B \subset C + B$, then $A \subset C$.*

Proof. Suppose there exists $a \in A$ such that $a \notin C$. Then there are $x^* \neq 0$, $\varepsilon > 0$ such that

$$\langle x^*, a \rangle > \langle x^*, c \rangle + \varepsilon, \ \forall c \in C.$$

As B is bounded, there exists $\bar{b} \in B$ such that $\langle x^*, \bar{b} \rangle \geq \langle x^*, b \rangle - \varepsilon, \forall b \in B$. Then

$$\langle x^*, a \rangle + \langle x^*, \bar{b} \rangle > \langle x^*, b + c \rangle, \ \forall b \in B, c \in C,$$

in contradiction with $a + \bar{b} \in C + B$. □

The next lemma is quite simple, and it shows the reason why the intersection theorem is true for AW convergence, but fails (in the same form) for Mosco convergence (see also Example 9.2.2).

Lemma 9.2.4 *Let X be a normed space, let A be a closed convex set such that $B[a; r] \subset A$. Moreover, let C be such that $e(A, C) \leq t < r$. Then $B[a; r - t] \subset C$.*

Proof. Since

$$(a + (r - t)B) + tB \subset A \subset C + tB,$$

we conclude, by appealing to Lemma 9.2.3. □

We are now able to prove the sum theorem.

Theorem 9.2.5 *Let $f_n, f, g_n, g \in \Gamma(X)$ be such that*

$$f = \text{AW-lim} f_n, \quad g = \text{AW-lim} g_n.$$

Moreover, suppose that

$$there\ exists\ \bar{x} \in \text{dom}\ f\ where\ g\ is\ continuous. \tag{9.4}$$

Then

$$f + g = \text{AW-lim}(f_n + g_n).$$

Proof. We shall show that $\forall \varepsilon > 0, \forall r > 0$ there exists N such that $\forall n > N$ the following holds:

$$e\big(\text{epi}(f + g) \cap r B_{X \times \mathbb{R}}, \text{epi}(f_n + g_n)\big) < \varepsilon, \tag{9.5}$$

which shows the lower part of the convergence. The proof of the upper part of the convergence is shown in the same fashion.

By assumption (9.4), there exists $(\bar{x}, \bar{a}) \in \text{int}(\text{epi}\,g) \cap \text{epi}\,f$. Then there exists $s > 0$ such that $B[(\bar{x}, \bar{a}); 2s] \subset \text{epi}\,g$. Lemma 9.1.5 provides existence of k such that

$$f, f_n, g, g_n \geq -k \text{ on } rB.$$

(In fact, f, g each have an affine minorizing function, so the functions f_n have a common affine minorizing function, and the same with the functions g_n. Thus they all are lower bounded on bounded sets. By the way, observe that the existence of a common minorizing function is a consequence of the upper part of the convergence, and it matters only for the proof of the upper part of the convergence.) Choose l such that $l \geq r + k$ and $P = B[(\bar{x}, \bar{a}); 2s] \subset lB_{X \times \mathbb{R}}$, suppose $\varepsilon < s$ and let N be so large that

$$e\big(\text{epi}\,f \cap lB_{X \times \mathbb{R}}, \text{epi}\,f_n\big) < \frac{\varepsilon s}{16l} \text{ and } e\big(\text{epi}\,g \cap lB_{X \times \mathbb{R}}, \text{epi}\,g_n\big) < \frac{\varepsilon s}{16l}. \quad (9.6)$$

Let us verify that if $n > N$, then (9.5) is fulfilled. Take $(x, a) \in \text{epi}(f + g) \cap rB_{X \times \mathbb{R}}$ and look for $(x_n, a_n) \in \text{epi}\,f_n + g_n$ such that $d[(x, a), (x_n, a_n)] < \varepsilon$. Observe that $\|x\| \leq r$ and that there exists $\hat{a} \geq f(x)$ such that $a = \hat{a} + g(x)$. Moreover $|a| \leq r$, $-k < a$, $-k < g(x)$ whence $\|x\| \leq l$, $|g(x)| \leq l$, $|\hat{a}| \leq l$. Let us now explain in qualitative terms what we are doing to do. For a better understanding, the reader could also look at Fig. 9.1. Then checking the calculations is only tedious. The line segment $[(\bar{x}, \bar{a}), (x, g(x)))$ belongs to the interior of epi g. We select a point c on it, sufficiently close to $(x, g(x))$, as a center of a suitable ball contained in epi g. Thus a ball of a small diameter is contained in epi g_n, eventually. Now, we proceed to operating on f. We consider the point $\bar{c}$ with the same first coordinate as c and lying in the line segment, contained in epi f, joining $(\bar{x}, \bar{a})$ and $(x, \hat{a})$. Since $\bar{c}$ is in epi f, we can approximate it as closely as we want by points $p_n = (x_n, \hat{a}_n)$ lying in epi f_n. Now we go back and select points in epi g_n, as follows. We consider points $q_n = (x_n, \alpha)$, whose first coordinate equals the first one of p_n, and α is the second coordinate of c. They are sufficiently close to c to fall in the prescribed ball contained in epi g_n eventually. The job is over. We only need to take $r_n = (x_n, \hat{a}_n + \alpha)$ to get the desired points as close as needed to (x, a). If this is not convincing, read the following calculations. Set $t = \frac{\varepsilon}{8l}$, and consider the ball $\hat{S}$ of center $c = (1 - t)(x, g(x)) + t(\bar{x}, \bar{a})$ and radius $2st$; the center lies on the line segment with endpoints $(x, g(x))$ and $(\bar{x}, \bar{a})$. Then

(i) $\hat{S} \subset \text{epi}\,g$;
(ii)   if $(z, m) \in \hat{S}$ then $d[(z, m), (x, g(x))] < \frac{\varepsilon}{4}$.

By (9.5) and Lemma 9.2.2, the ball S with same center and radius εt, is contained in epi g_n. The point $\bar{c} = (1 - t)(x, \hat{a}) + t(\bar{x}, \bar{a})$ belongs to epi f and thanks to (9.5), we find points $p_n = (x_n, \hat{a}_n) \in \text{epi}\,f_n$ with distance less than $\frac{\varepsilon s}{16l}$ from $\bar{c}$.

Thus the point $q_n = (x_n, (1 - t)g(x) + t\bar{a}) \in \text{epi}\,g_n$, for

$$d[(x_n, (1 - t)g(x) + t\bar{a}), c] = d(x_n, (1 - t)x + t\bar{x}) < \frac{\varepsilon s}{16l} < \varepsilon t,$$

whence
$$(x_n, (1-t)g(x) + t\bar{a}) \in S \subset \text{epi } g_n.$$

Hence the point $r_n = (x_n, (1-t)g(x) + t\bar{a} + \hat{a}_n)$ belongs to $\text{epi}(f_n + g_n)$ and is at a distance less than ε from $(x, g(x))$.   □

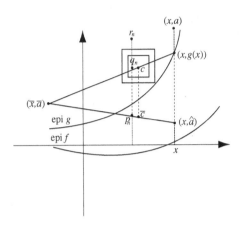

Figure 9.1.

The next theorem provides sufficient conditions in the case of the Mosco convergence. A similar result holds, in a normed space, for the slice convergence.

Theorem 9.2.6 *Let X be a reflexive Banach space and let $f_n, f, g_n, g \in \Gamma(X)$ be such that*
$$f_n \xrightarrow{\text{M}} f, \quad g_n \xrightarrow{\text{M}} g.$$
Moreover suppose there are $\bar{x} \in X$ and $r > 0$ such that the functions f_n are upper equibounded on $B[\bar{x}; r]$. Then
$$f_n + g_n \xrightarrow{\text{M}} (f + g).$$

Proof. The proof relies on the following facts:

(i) f is upper bounded on $B[\bar{x}; r]$;
(ii) $\text{epi }(f + g) = \text{cl}\{(x, \alpha) : x \in \text{int dom } f \cap \text{dom } g, \alpha \geq f(x) + g(x)\}$;
(iii) given $x \in \text{int dom } f \cap \text{dom } g$, f_n, f are lower and upper equibounded on a neighborhood of x.

Let us start by proving the previous claims.

(i) Let k be such that $f_n \leq k$ on $B[\bar{x}; r]$. Since $\liminf f_n(x) \geq f(x)$, for all $x \in X$, then $f \leq k$ on $B[\bar{x}, ; r)]$.

(ii) Take $(x, r) \in$ epi $f + g$. Then $r \geq f(x) + g(x)$. Take points of the form (y, s), where y lies on the line segment $[\bar{x}, x) \subset$ int dom $f \cap$ dom g, and $s = \max\{r, f(y) + g(y)\}$. It should be clear that every open set containing (x, r) also contain points of the form above and this shows (ii).

(iii) The lower equiboundedness of f_n, f (on the bounded sets) can be seen as in step 1(b) of the proof of Theorem 9.1.2. Since f_n are upper equibounded on $B[\bar{x}; r]$, the upper equiboundedness of f_n, f around a point $x \in$ int dom f can be seen as a simple variant of the proof of Theorem 2.1.2 (remember that there is a sequence $x_n \to x$ such that $f_n(x_n)$ is upper equibounded, by $f(x) + 1$, for instance).

All of this allows us to conclude that epi$(f + g) \subset$ Li epi$(f_n + g_n)$, which is the hard part of the convergence, since it does not work without specific ad hoc assumptions. Let us see how. (ii) implies that it is enough to verify that if (x, α) is such that $x \in$ int dom $f \cap$ dom g and $\alpha \geq f(x) + g(x)$, then there exists $(x_n, \alpha_n) \in$ epi$(f_n + g_n)$ such that $x_n \to x$, $\alpha_n \to \alpha$. As $g_n \overset{M}{\to} g$, there exists $x_n \to x$ such that $\limsup g_n(x_n) \leq g(x)$. Now (iii) above and Lemma 2.1.8 imply that the functions f_n, f are equilipschitz on a neighborhood of x, and thus $\lim f_n(x_n) \to f(x)$. So, it is enough to choose $\alpha_n = \max\{\alpha, f_n(x_n) + g_n(x_n)\}$. And now the easy task, the upper part of the convergence: if $x_n \rightharpoonup x$, then $\liminf f_n(x_n) \geq f(x)$, $\liminf g_n(x_n) \geq g(x)$, whence $\liminf(f_n(x_n) + g_n(x_n)) \geq f(x) + g(x)$ (you probably noticed we do not need *any* assumption to conclude this but convergence of the two sequences; no extra conditions are required). $\qquad \square$

It is also possible to provide theorems like those above in terms of other hypertopologies. For instance, the following theorem, whose proof is simple and left to the reader, holds:

Theorem 9.2.7 *Let $f_n, f, g_n, g \in \Gamma(X)$ be such that*

$$f_n \overset{AW}{\to} f, \quad g_n \overset{bp}{\to} g.$$

Suppose also

$$\exists \bar{x} \in \text{dom } g \text{ where } f \text{ is continuous.} \tag{9.7}$$

Then

$$(f_n + g_n) \overset{bp}{\to} (f + g).$$

Theorem 9.2.7 can be of interest because, in order to get some stability results, it is not necessary to assume that the sequence of the sums converges in the AW sense. The bounded proximal convergence suffices. Please observe also that the condition (9.7) is *not* symmetric. In other words, we could *not* assume the following:

$$\exists \bar{x} \in \text{dom } f \text{ where } g \text{ is continuous.}$$

The reason should be clear.

From the previous theorems we can get useful information for the convergence of sequences of intersecting sets.

Corollary 9.2.8 *Let $\{A_n\}, \{B_n\} \subset C(X)$, suppose $A_n \overset{AW}{\to} A$, $B_n \overset{AW}{\to} B$, and suppose moreover there exists $a \in \text{int } A \cap B$. Then $A_n \cap B_n \overset{AW}{\to} A \cap B$.*

Corollary 9.2.9 *Let $\{A_n\}, \{B_n\} \subset C(X)$, suppose $A_n \overset{AW}{\to} A$, $B_n \overset{bp}{\to} B$, and suppose moreover there exists $a \in \text{int } A \cap B$. Then $A_n \cap B_n \overset{bp}{\to} A \cap B$.*

Corollary 9.2.10 *Let $\{A_n\}, \{B_n\} \subset C(X)$, suppose $A_n \overset{M}{\to} A$, $B_n \overset{M}{\to} B$, and suppose moreover there are $a \in B$ and $r > 0$ such that $B[a; r] \subset A$. Then $A_n \cap B_n \overset{M}{\to} A \cap B$.*

The proofs of all these corollaries rely on the facts that $C_n \to C$ if and only if $I_{C_n} \to I_C$, where the convergence is intended in the sense of the three topologies above, and that the sum of the indicator functions is the indicator of the intersection.

Let us conclude this section with two useful results. They allow approximating arbitrary functions in $\Gamma(X)$ by regular functions.

Given $f \in \Gamma(X)$, the function $f_n = f \nabla n \| \cdot \|$ is called the *n-Lipschitz regularization* of f, the largest n-Lipschitz function minorizing f.

The following result holds:

Theorem 9.2.11 *Let X be a normed space, let $f \in \Gamma(X)$. Let $\{f_n\}$ be the sequence of n-Lipschitz regularizations of f. Then $f = AW\text{-}\lim f_n$.*

Proof. Since the sequence of the indicator functions $I_{nB^*} : X^* \to (-\infty, \infty]$ AW converges to the zero function, by the sum theorem we get

$$f^* + I_{nB^*} \overset{AW}{\to} f^*.$$

By the continuity theorem of the conjugation operation, we get

$$(f^* + I_{nB^*})^* \overset{AW}{\to} f^{**} = f.$$

From the Attouch–Brézis theorem then

$$(f^* + I_{nB^*})^* = f \nabla n \| \cdot \| = f_n,$$

and this allows us to conclude, a simple, beautiful proof relying on nontrivial previous results! □

The following could be an application with Mosco convergence.

Proposition 9.2.12 *Let X be a separable Hilbert space, with an orthonormal basis $\{e_n : n \in \mathbb{N}\}$. Let $X_n = \text{sp}\{\bigcup_{i \leq n} e_i\}$, the space generated by the first n vectors of the basis. Let $f \in \Gamma(X)$ be continuous at a point $x \in X$. Then the sequence $\{f_n\}$,*

$$f_n = f + I_{X_n} : X \to (-\infty, \infty],$$

converges in the Mosco sense to f.

Thus f can be approximated by functions with finite dimensional domain. Observe that actually the former convergence is stronger (bounded proximal).

9.3 Convergence of functions and of their subdifferentials

Everyone has studied in some calculus class, theorems relating convergence of regular functions with convergence of their derivatives. Thus, it is of interest to ask if there is any connection between the variational convergences of convex functions and set convergence of the graphs of their subdifferentials. In this final section of the chapter, we provide one typical result in this sense, and we give references for some others. It is well known that convergence of functions usually does not provide information on convergence of derivatives. Once again, the convex case provides an exception, as we have seen in Lemma 3.6.4.

Now, we provide a result dealing with Mosco convergence of functions.

Theorem 9.3.1 *Let X be a reflexive Banach space, let $f_n, f \in \Gamma(X)$. Then $f_n \xrightarrow{M} f$ if and only if one of the following (always equivalent) conditions holds:*

(i) $(x, x^*) \in \partial f \Rightarrow \exists (x_n, x_n^*) \in \partial f_n \colon x_n \to x, x_n^* \to x^*, f_n(x_n) \to f(x)$;

(ii) $(x^*, x) \in \partial f^* \Rightarrow \exists (x_n^*, x_n) \in \partial f_n^* \colon x_n \to x, x_n^* \to x^*, f_n^*(x_n^*) \to f^*(x)$.

Proof. First observe that(i) and (ii) above are equivalent. Suppose $(x, x^*) \in \partial f$. Then, from $f_n^*(x_n^*) = f_n(x_n) - \langle x_n^*, x_n \rangle$ and from(i) it immediately follows that $f_n^*(x_n^*) \to f^*(x)$, and vice-versa. Now suppose $f_n \xrightarrow{M} f$ and fix $\varepsilon > 0$ and $(x, x^*) \in \partial f$. Then we must find $(x_n, x_n^*) \in \partial f_n$, and N such that for $n \geq N$, the following hold:

- $\|x_n - x\| \leq \varepsilon$;
- $\|x_n^* - x^*\|_* \leq \varepsilon$;
- $f_n(x_n) \leq f(x) + \varepsilon$.

Fix $\eta > 0$ such that $\max\{\eta(1 + \|x^*\|_*), \eta(\eta + 1)\} < \frac{\varepsilon}{2}$. Let $\tau > 0$ be such that $3\tau < \eta^2$. There exist $u_n \to x$ and $u_n^* \to x^*$ such that, eventually,

$$f_n(u_n) \leq f(x) + \tau, f_n^*(u_n^*) \leq f^*(x^*) + \tau$$

(we are using the theorem about the continuity of the conjugation operation). Let N be so large that the following hold:

$$\|u_n - x\| < \frac{\varepsilon}{2}, \quad \|u_n^* - x^*\|_* < \frac{\varepsilon}{2},$$
$$f_n(u_n) \leq f(x) + \tau, \quad f_n^*(u_n^*) \leq f^*(x^*) + \tau, \quad \langle x^*, x \rangle \leq \langle u_n^*, u_n \rangle + \tau.$$

Then for $n \geq N$,

$$f_n(u_n) + f_n^*(u_n^*) \leq f(x) + f^*(x^*) + 2\tau = \langle x^*, x \rangle + 2\tau \leq \langle u_n^*, u_n \rangle + \eta^2,$$

and this implies that $u_n^* \in \partial_{\eta^2} f_n(u_n)$. Thus, from Theorem 4.2.10 (with the choice of $\sigma = 1 + \|x^*\|_*$), we can claim the existence of $(x_n, x_n^*) \in \partial f_n$ such that

$$\|x_n - u_n\| \leq \eta, \|x_n^* - u_n^*\| \leq \eta(1 + \|x^*\|_*), f_n(x_n) \leq f_n(u_n) + \eta(\eta + 1).$$

Thus

$$\|x_n - x\| \leq \|x_n - u_n\| + \|u_n - x\| \leq \eta + \frac{\varepsilon}{2} < \varepsilon.$$

The other two inequalities follow in exactly the same way, and thus one of the implications is proved. Let us now see the other one. First, we want to prove that for every sequence $\{x_n\}$ such that $x_n \rightharpoonup x$, then $\liminf f_n(x_n) \geq f(x)$. We provide the proof in the case when $f(x) = \infty$, the other case being completely analogous. Fix $k > 0$, let $x_n \rightharpoonup x$ and let us see that $\liminf f_n(x_n) \geq k$. From Theorem 4.2.17 we get that there is $(y, y^*) \in \partial f$ such that

$$f(y) + \langle y^*, x - y \rangle \geq k + 1.$$

By assumption, there exists $(y_n, y_n^*) \in \partial f_n$ such that $y_n \to y$, $y_n^* \to y^*$, $f_n(y_n) \to f(y)$. Thus, for large n,

$$f_n(x_n) \geq f_n(y_n) + \langle y_n^*, x_n - y_n \rangle \geq f(y) + \langle y^*, x - y \rangle - 1 \geq k.$$

Clearly, the same argument applies to conjugates, so we can conclude that for every $\{x_n^*\}$ such that $x_n^* \rightharpoonup x^*$, set $\liminf f_n^*(x_n^*) \geq f^*(x^*)$. Moreover, it is clear that there exist some $y^* \in X^*$ and $y_n^* \to y^*$ such that $\limsup f_n^*(y_n^*) \leq f^*(y^*)$. Thus, from Remark 9.1.3 we can also conclude that the second condition guaranteeing Mosco convergence, i.e., lower convergence of the epigraphs, holds true. The proof is complete. □

The previous result relates Mosco convergence of a sequence of functions to (Vietoris) lower convergence of their subdifferentials (plus a normalization condition). It is thus of interest to ask whether anything can be said about upper convergence of the subdifferentials. The next lemma will be useful in drawing some conclusions on this.

Lemma 9.3.2 *Let $A: X \to X^*$ be a maximal monotone operator, let $A_n: X \to X^*$ be a monotone operator for each $n \in \mathbb{N}$. Suppose moreover there is lower convergence of the graphs of A_n to the graph of A: $\mathrm{Li}\, A_n \supset A$. Then there is upper convergence of the graphs of A_n to the graph of A: $\mathrm{Ls}\, A_n \subset A$.*

Proof. Let $(x_k, y_k) \in A_{n_k}$ for all k and suppose $(x_k, y_k) \to (x, y)$. We must prove that $(x, y) \in A$. Take any $(u, v) \in A$. By assumption, there exist $(u_n, v_n) \in A_n$ such that $(u_n, v_n) \to (u, v)$. Next,

$$0 \leq \langle v_{n_k} - y_k, u_{n_k} - x_k \rangle \to \langle v - y, u - x \rangle.$$

Thus $\langle v - y, u - x \rangle \geq 0$ for all $(u, v) \in A$, and thus, by maximality, $(u, v) \in A$. □

Then Theorem 9.3.1 states the equivalence between Mosco convergence of functions and Kuratowski convergence of the associate subdifferentials plus a condition on convergence of the values. Actually the result can be given a sharper formulation as it can be seen that the following result holds:

Theorem 9.3.3 *Let X be a reflexive Banach space, let $f_n, f \in \Gamma(X)$. Then the following are equivalent:*

(i) $f_n \xrightarrow{\text{M}} f$;

(ii) $\partial f_n \xrightarrow{\text{K}} \partial f$ *and there exists* $(x_n, x_n^*) \in \partial f_n$ *such that* $(x_n, x_n^*) \to (x, x^*)$ $(\in \partial f)$ *and* $f_n(x_n) \to f(x)$.

To conclude, we observe that the same result holds for the slice convergence, in any Banach space (see, e.g., [Be, Corollary 8.3.8]), and that a similar result can be provided, relating Attouch–Wets convergence of functions and of their subdifferentials [BT].

10

Well-posed problems

Anyone aware of being the creator of his own reality
would equally be conscious of the possibility,
always immanent, to create it in a different way.
(P. Watzlawick, "Die erfundene Wirklichkeit")

When minimizing a function, usually we are not able to find in an analytic way the global minimizer(s) (if any!) of the function. For this reason, we introduced some algorithms in Chapter 4 in order to build up a sequence converging to the (or some) minimizer. Thus we are interested in finding some notion highlighting not only that a problem has a solution, but also that the solution is "easy to find", at least in principle. This topic is known in the literature under the name of "well-posedness" of a problem. Another, aspect related to this subject is to require that a problem be "stable under small perturbations". Of course, this is a very loose requirement, but it can be understood, at least from a qualitative point of view. When modeling a problem, we quite often make some simplifying assumptions, in order to better handle it. Thus, when the problem is expressed in the form of a minimum problem, we can argue that the performance function (the function to be minimized) is known up to some (small) error. Moreover, when solving the problem in practice, we are often led to approximate the performance function with a sequence of functions for which it is easier to find the minima. A typical situation is when the function is defined in an infinite-dimensional domain and we consider its projection on a sequence of finite dimensional spaces invading it (the so called Riesz–Galerkin method). Thus it appears clear that it is interesting to know if the sequence of minima found with this procedure actually approaches the true solution of the problem, and this is exactly what we intend by stability of the initial problem. This chapter analyzes some of the numerous results in this setting, and is focused on showing that the two aspects – well-posedness and stability – are deeply related. The chapter begins by considering some different notions of well-posedness.

10.1 Tykhonov, Levitin–Polyak and strong well-posedness

We shall consider a metric space (X, d), and we suppose a function $f \colon X \to (-\infty, \infty]$ is given which is at least lower semicontinuous, so that its epigraph and its level sets are all closed sets.

Definition 10.1.1 Let (X, d) be a metric space, and let $f \colon X \to \mathbb{R}$ be lower semicontinuous. Then (X, f) (or simply f) is said to be *Tykhonov well-posed* if

(i)   there exists a unique $\bar{x} \in X$ such that $f(\bar{x}) \le f(x), \forall x \in X$;
(ii)   every sequence $\{x_n\}$ such that $f(x_n) \to \inf f$ is such that $x_n \to \bar{x}$.

Let us observe that requiring uniqueness in condition (i) is actually redundant, as it is implied by (ii). Sequences $\{x_n\}$ as in condition (ii) are called, as we well know, *minimizing sequences*. So the definition declares well-posed those functions such that points with values close to the minimum value are actually close to the solution point.

Sometimes the uniqueness of the solution is a too restrictive assumption.

Definition 10.1.2 f is said to be *Tykhonov well-posed in the generalized sense* if

(i)   there exists $\bar{x} \in X$ such that $f(\bar{x}) \le f(x), \forall x \in X$;
(ii)   every minimizing sequence $\{x_n\}$ has a subsequence converging to a minimum point.

Then, if f is well-posed in the generalized sense, $\arg \min f$ is a nonempty compact set.

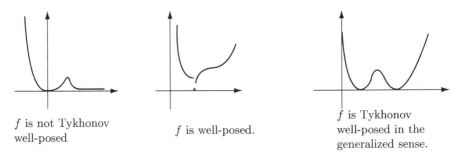

f is not Tykhonov well-posed

f is well-posed.

f is Tykhonov well-posed in the generalized sense.

Figure 10.1.

Example 10.1.3 In $\mathbb{R}$ the function $x^2 e^{-x}$ has a unique minimum point, but it is not Tykhonov well-posed. If (X, d) is a compact space, then f is Tykhonov well-posed in the generalized sense. More generally, if there exists $a > \inf f$ such that f^a is compact, then f is Tykhonov well-posed in the generalized sense. So each time it is possible to apply the Weierstrass theorem, we actually face a Tykhonov well-posed problem.

Example 10.1.4 Let $f\colon \mathbb{R}^n \to (-\infty, \infty]$ be a convex, lower semicontinuous function with a unique minimum point. Then f is Tykhonov well-posed. If $\operatorname{Min} f$ is nonempty and compact, then f is Tykhonov well-posed in the generalized sense. This is a consequence of Proposition 4.3.1. Now let X be a separable Hilbert space with orthonormal basis $\{e_n : n \in \mathbb{R}\}$. Let $f(x) = \sum_{n=1}^{\infty} \frac{(x, e_n)^2}{n^2}$. Then f is continuous, convex and it has a unique minimum point, but it is not Tykhonov well-posed.

Example 10.1.5 Let X be a Hilbert space, $L\colon X \to X$ a symmetric linear bounded operator. Suppose there is $a > 0$ such that $\langle Lx, x \rangle \geq a\|x\|^2$. Then $f(x) = \frac{1}{2}\langle Lx, x \rangle - \langle x^*, x \rangle$ is Tykhonov well-posed for all $x^* \in X$. Conversely, if the problem of minimizing f has one and only one solution for all x^*, then f is Tykhonov well-posed for all $x^* \in X$. This last statement relies on the fact that f is differentiable, with derivative $Lx - x^*$. Having one and only one solution for the problem of minimizing f, means that the equation $Lx = x^*$ has one and only one solution, i.e., L is invertible, and thus there is $a > 0$ such that $\langle Lx, x \rangle \geq a\|x\|^2$.

The next proposition provides a useful characterization of Tykhonov well-posedness. It is called the Furi–Vignoli criterion.

Proposition 10.1.6 *Let X be a complete metric space and let $f\colon X \to (-\infty, \infty]$ be a lower semicontinuous function. The following are equivalent:*

(i)   *f is well-posed;*
(ii) $\inf_{a > \inf f} \operatorname{diam} f^a = 0$.

Proof. If (ii) does not hold, then it is possible to find $\varepsilon > 0$ and two minimizing sequences $\{x_n\}$ and $\{y_n\}$ such that $d(x_n, y_n) \geq \varepsilon, \forall n$. This implies that at least one of them does not converge to the minimum point, and this is impossible. Conversely, let $\{x_n\}$ be a minimizing sequence. Then (ii) implies that $\{x_n\}$ is a Cauchy sequence, and thus it converges to a minimum point, as f is lower semicontinuous. This point is also unique, because it belongs to f^a, for all $a > \inf f$. $\square$

Well-posedness in the generalized sense can be characterized in a similar fashion.

Proposition 10.1.7 *Let X be a complete metric space and let $f\colon X \to (-\infty, \infty]$ be a lower semicontinuous function. Then*

- *If f is Tykhonov well-posed in the generalized sense, then Min f is compact and*

$$\forall \varepsilon > 0, \exists a > \inf f \text{ such that } f^a \subset B_\varepsilon[\text{Min } f]. \qquad (10.1)$$

- *If $\forall \varepsilon > 0, \exists a > \inf f$, such that $f^a \subset B_\varepsilon[\text{Min } f]$, and if Min f is a compact set, then f is Tykhonov well-posed in the generalized sense.*

Proof. Suppose f is Tykhonov well-posed in the generalized sense. Every sequence from Min f has a subsequence converging to some point of Min f, and this means that Min f is compact. Now suppose (10.1) does not hold. Then there is $\varepsilon > 0$ such that for each n it is possible to find x_n such that $f(x_n) \leq \inf f + \frac{1}{n}$ and $d(x_n, \text{Min } f) \geq \varepsilon$. Thus $\{x_n\}$ is a minimizing sequence with no subsequences converging to a minimum point, which is impossible.

Let us now see the second claim. Let $\{x_n\}$ be a minimizing sequence. Then $\forall a > \inf f$, $x_n \in f^a$ eventually. Thus $d(x_n, \text{Min } f) \to 0$. This means that for all $n \in \mathbb{N}$ there is $y_n \in \text{Min } f$ such that $d(x_n, y_n) \to 0$. Now we conclude by exploiting the compactness of Min f. □

The above proposition shows that setting

$$\text{Lev}: \mathbb{R} \to X, \text{Lev}(a) = f^a,$$

then Lev is upper Vietoris (or, equivalently, Hausdorff) continuous at $a = \inf f$. The converse is true provided Min f is a compact set. Other notions of well-posedness have been considered in the literature, for instance by requiring upper continuity of the multifunction Lev at the level $\inf f$, but without compactness assumption (in this case upper Vietoris and Hausdorff give rise to different notions). We do not pursue this issue here.

Let us see another characterization of well-posedness.

Definition 10.1.8 Let $T \subset [0, \infty)$ be a set containing the origin. A function $c: T \to [0, \infty)$ is said to be *forcing*, provided it is increasing, $c(0) = 0$ and $c(t) > 0$ implies $t > 0$.

Then the following holds:

Proposition 10.1.9 *Let (X, d) be a metric space and $f: X \to \mathbb{R}$. Then f is Tykhonov well-posed if and only if there are $\bar{x}$ and a forcing function c such that*

$$f(x) \geq f(\bar{x}) + c(d(x, \bar{x})).$$

In case X is a normed linear space and f convex, then c can be chosen convex too.

Proof. If there are $\bar{x}$ and a forcing function c as in the statement, then f is Tykhonov well-posed, with solution $\bar{x}$. For, if $f(x_n) \to f(\bar{x})$, then $c(d(\bar{x}, x_n)) \to 0$ and, since c is increasing and positive for $t > 0$, this implies $d(\bar{x}, x_n) \to 0$. Conversely, let us suppose f Tykhonov well-posed, with solution $\bar{x}$. Set

$$c(t) = \inf_{x:d(x,\bar{x})\geq t} \{f(x) - f(\bar{x})\}.$$

It is clearly increasing, moreover we cannot have $c(t) = 0$ for some $t > 0$ because, in such a case, we would have a minimizing sequence $\{x_n\}$ fulfilling the condition $d(x_n, \bar{x}) \geq t(> 0)$, against Tykhonov well-posedness. Now, suppose f convex and, without loss of generality, $\bar{x} = 0$ and $f(0) = 0$. Let $b > a > 0$. Suppose $\|x\| \geq b$. Then $c(a) \leq f(\frac{a}{b}x) \leq \frac{a}{b}f(x)$. Since this is true for all x such that $\|x\| \geq b$, this implies $\frac{c(a)}{a} \leq \frac{c(b)}{b}$, which means that the function $\frac{c(\cdot)}{\cdot}$ is increasing, and thus $c(\cdot)$ is convex (see Proposition 1.2.11). □

In the proposition above, when f is convex, a forcing function is

$$c(t) = \inf_{\|x-\bar{x}\|\geq t} \{f(x) - f(\bar{x})\} = \inf_{\|x-\bar{x}\|=t} \{f(x) - f(\bar{x})\}.$$

Example 10.1.10 Let (X, d) be a metric space, let $f: X \to \mathbb{R}$. Suppose moreover f has a minimum point $\bar{x}$. Then, for all $a > 0$, the function $g(\cdot) = f(\cdot) + ad(\cdot, \bar{x})$ is Tykhonov well-posed. This remark allows us to get a little improvement in the statement of the Ekeland variational principle, as follows.

Let (X, d) be a complete metric space and let $f: X \to (-\infty, \infty]$ be a lower semicontinuous, lower bounded function. Let $\varepsilon > 0$, $r > 0$ and $\bar{x} \in X$ be such that $f(\bar{x}) < \inf_X f + r\varepsilon$. Then, there exists $\hat{x} \in X$ enjoying the following properties:

(i) $d(\hat{x}, \bar{x}) < r$;
(ii) $f(\hat{x}) < f(\bar{x}) - \varepsilon d(\bar{x}, \hat{x})$;
(iii) the function $f(\cdot) + \varepsilon d(\hat{x}, \cdot)$ is Tykhonov well-posed.

We have just seen that Tykhonov well-posedness is related to the existence of a forcing function. On the other hand, the existence of a forcing function for f provides important information on the smoothness of f^* at 0^*, and conversely. Thus Tykhonov well-posedness of a function is related to Fréchet differentiability of its conjugate at 0^*. Let us see this important result, a particular case of a famous theorem by Asplund–Rockafellar.

Theorem 10.1.11 *Let X be a reflexive Banach space and $f \in \Gamma(X)$. If f^* is Fréchet differentiable at a point $p \in X^*$, with $\nabla f^*(p) = \bar{x}$, then there is a forcing function c such that $f(x) - \langle p, x \rangle \geq f(\bar{x}) - \langle p, \bar{x} \rangle + c(\|x - \bar{x}\|)$. Conversely, if there are a forcing function c and a point $\bar{x}$ such that $f(x) - \langle p, x \rangle \geq f(\bar{x}) - \langle p, \bar{x} \rangle + c(\|x - \bar{x}\|)$, then f^* is Fréchet differentiable at p with $\nabla f^*(p) = \bar{x}$. Thus f^* is Fréchet differentiable at p if and only if $f(\cdot) - \langle p, \cdot \rangle$ is Tykhonov well-posed.*

Proof. Step 1. From Fréchet differentiability of f^* to Tykhonov well-posedness of f: let f^* be Fréchet differentiable at p, with $\nabla f^*(p) = \bar{x}$. Consider the function $g(x) = f(x + \bar{x}) - \langle p, x \rangle + f^*(p)$. Observe that $g^*(x^*) = f^*(x^* + p) - \langle x^*, \bar{x} \rangle - f^*(p)$. Thus $g^*(0^*) = 0$, g^* is differentiable at the origin, and

$\nabla g^*(0^*) = 0$. Suppose we have proved the statement for g. Then, for all $u \in X$, $g(u) \geq g(0) + c(\|u\|)$. Setting $u = x - \bar{x}$, we then get

$$f(x) - \langle p, x - \bar{x} \rangle \geq f(\bar{x}) + c(\|x - \bar{x}\|).$$

In other words, it is enough to show the claim in the special case when $f^*(0^*) = 0$, f^* is differentiable at the origin and $\nabla f^*(0^*) = 0$. By way of contradiction, let us suppose there is $\bar{t} > 0$ such that $\inf_{\|x\|=\bar{t}} f(x) = 0$. Thus there is $\{x_n\}$ such that $\|x_n\| = \bar{t} \ \forall n$ and $f(x_n) \to 0$. Fix any $a > 0$ and take x_n^* such that $\|x_n^*\|_* = 1$ and $\langle x_n^*, x_n \rangle = \bar{t}$. Then

$$\sup_{\|x^*\|_*=a} f^*(x^*) \geq f^*(ax_n^*) \geq \langle ax_n^*, x_n \rangle - f(x_n) \geq \bar{t}\|x^*\|_* - f(x_n).$$

As the above relation holds for each n and $f(x_n) \to 0$, we get

$$\sup_{\|x^*\|_*=a} f^*(x^*) \geq \bar{t}\|x^*\|_*.$$

This contradicts the fact that 0 is the Fréchet derivative of f^* at 0^*.

Step 2. From Tykhonov well-posedness of f to Fréchet differentiability of f^*: A similar argument as before allows us to take $\bar{x} = 0$, $f(0) = 0$, $p = 0^*$. Now, fix $\varepsilon > 0$ and observe that $f(x) \geq \frac{c(\varepsilon)}{\varepsilon}\|x\|$ if $\|x\| \geq \varepsilon$. Let

$$g(x) = \begin{cases} 0 & \text{if } \|x\| \leq \varepsilon, \\ \frac{c(\varepsilon)}{\varepsilon}\|x\| - c(\varepsilon) & \text{otherwise.} \end{cases}$$

Then $f(x) \geq g(x)$ for all x and thus $f^*(x^*) \leq g^*(x^*)$ for all x^*. Let us evaluate $g^*(x^*)$, if $\|x^*\|_* \leq \frac{c(\varepsilon)}{\varepsilon}$. We have

$$\sup_{\|x\|\geq\varepsilon} \left\{ \langle x^*, x \rangle - \frac{c(\varepsilon)\|x\|}{\varepsilon} + c(\varepsilon) \right\} \leq \sup_{\|x\|\geq\varepsilon} \left\{ \left(\|x^*\|_* - \frac{c(\varepsilon)}{\varepsilon} \right)\|x\| + c(\varepsilon) \right\}$$
$$= \varepsilon\|x^*\|_*.$$

On the other hand,

$$\sup_{\|x\|\leq\varepsilon} \{ \langle x^*, x \rangle - g(x) \} \leq \varepsilon\|x^*\|_*.$$

Thus $g^*(x^*) \leq \varepsilon\|x^*\|_*$ and so

$$0 \leq f^*(x^*) \leq \varepsilon\|x^*\|_*,$$

provided $\|x^*\|_* \leq \frac{c(\varepsilon)}{\varepsilon}$. This means that 0 is the Fréchet derivative of f^* at 0^*, and this ends the proof. $\qquad \square$

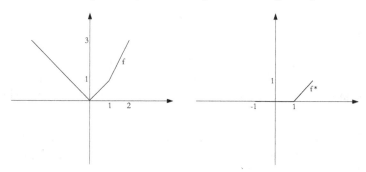

Figure 10.2. $f(\cdot) - \langle p, \cdot \rangle$ Tykhonov well-posed, f^* Fréchet differentiable, for $p \in (-1, 1) \cup (1, 2)$.

Now, we focus in particular on constrained problems. In principle, the concept of Tykhonov well-posedness can be used in constrained optimization too, but only in an abstract way. For, if (X, d) is a metric space, $A \subset X$, $f : X \to (-\infty, \infty]$ and one must minimize f over A, one can consider the restriction of f to A (obviously a metric space with the distance induced by X), and apply the definition to the problem of minimizing $f : A \to (-\infty, \infty]$. This procedure however does not take into account the fact that several algorithms for constrained problems use methods providing approximate solutions which do not lie in the constraint set, but get closer and closer to it. It is thus useful also to consider minimizing sequences "close" to the constraint set. So, suppose we are given the problem (A, f) of minimizing f over the set A.

Definition 10.1.12 A sequence $\{x_n\}$ is said to be a *Levitin–Polyak* minimizing sequence if $\lim f(x_n) = \inf_A f$ and $d(x_n, A) \to 0$. A sequence $\{x_n\}$ is a *strongly* minimizing if $\limsup f(x_n) \le \inf_A f$ and $d(x_n, A) \to 0$.

Definition 10.1.13 The problem (A, f) is said to be *Levitin–Polyak*, (respectively, *strongly*) well-posed if every Levitin–Polyak minimizing, (respectively, strongly minimizing) sequence converges to the minimum point of f over A.

Exercise 10.1.14 In $\mathbb{R}^2$ consider $f(x, y) = x^2 - x^4 y^2$ to be minimized over $A := \{(x, y) : y = 0\}$. Then (A, f) is Tykhonov, but not Levitin–Polyak, well-posed.

It is obvious that strong well-posedness implies Levitin–Polyak well-posedness (there are more strongly minimizing sequences than Levitin–Polyak minimizing sequences). On the other hand, the two notions coincide in several situations, as is suggested in the following exercise.

Exercise 10.1.15 Show that (A, f) Levitin–Polyak well-posed implies (A, f) strongly well-posed in each of the following situations:

(i)   either f is uniformly continuous, or f is uniformly continuous on the bounded sets and A is bounded (in this case both definitions agree with Tykhonov well-posedness);

(ii) f is continuous and X is a normed space;

(iii) X is a reflexive Banach space, $f \in \Gamma(X)$ and A is a closed convex set.

The next proposition shows how the notion of strong well-posedness can be helpful in algorithms using penalization methods. It should be noticed that the same result does not hold for problems that are merely Tykhonov well-posed.

Proposition 10.1.16 *Let X be a Banach space, $g \in \Gamma(X)$, and let $f \colon X \to \mathbb{R}$ be lower semicontinuous and lower bounded. Suppose $\lim_{\|x\| \to \infty} f(x) = \infty$ and that there is $\bar{x}$ such that $g(\bar{x}) < 0$. Let $A := \{x \in X : g(x) \le 0\}$ and suppose that the problem of minimizing f over A is strongly well-posed. Finally, setting*

$$f_n(x) := f(x) + n \max\{g(x), 0\},$$

let $\{\varepsilon_n\}$ be such that $\varepsilon_n \downarrow 0$ and $x_n \in X$ be such that

$$f_n(x_n) \le \inf_{x \in X} f_n(x) + \varepsilon_n, \ \forall n \in \mathbb{N}.$$

Then $x_n \to a$, where a is the solution of the problem.

Proof. First let us remark that

$$-\infty < \inf_X f \le \inf_X f_n \le \inf_A f,$$

providing $\limsup_{n \to \infty} g(x_n) \le 0$. Moreover $f(x_n) \le f_n(x_n)$ implies that $\{x_n\}$ is a bounded sequence. Let us now show that $d(x_n, A) \to 0$. Define a sequence $\{y_n\} \subset A$ as follows: if $g(x_n) \le 0$, let $y_n := x_n$. Otherwise, let

$$y_n := \frac{g(x_n)}{g(x_n) - g(\bar{x})} \bar{x} + \left(1 - \frac{g(x_n)}{g(x_n) - g(\bar{x})}\right) x_n.$$

Then $y_n \in A$ for all large n, and $d(x_n, y_n) = \frac{g(x_n)}{g(x_n) - g(\bar{x})} \|x_n - \bar{x}\| \to 0$. As

$$\limsup_{n \to \infty} f(x_n) \le \limsup_{n \to \infty} f_n(x_n) \le \inf_A f,$$

therefore, $\{x_n\}$ is a strongly minimizing sequence, and this completes the proof.   □

The Tykhonov well-posedness criterion provided in Proposition 10.1.6 can be easily extended to strong well-posedness as follows:

Proposition 10.1.17 *Let X be a complete metric space, let $A \subset X$ be a closed set and let $f \colon X \to (-\infty, \infty]$ be a lower semicontinuous function. Then the following are equivalent:*

(i)   *The minimum problem (A, f) is strongly well-posed.*

(ii) $\inf_{\epsilon > 0, a > \inf_A f} \operatorname{diam}\{x \in X : f(x) \le a \text{ and } d(x, A) \le \epsilon\} = 0$.

Proof. See Exercise 10.1.18.   □

Exercise 10.1.18 Prove Proposition 10.1.17.

An analogous proposition holds for the generalized strong well-posedness (see Proposition 10.1.7).

Example 10.1.19 Let $A \subset \mathbb{R}^n$ be a closed convex set, and let $f : \mathbb{R}^n \to \mathbb{R}$ be convex, lower semicontinuous and with a unique minimum point over A. Then the problem (A, f) is Levitin–Polyak (and strongly) well-posed. This is not difficult to see; the result also follows from subsequent theorems.

10.2 Stability

In this section we focus on the stability of various constrained minimization problems. To start with, we shall consider perturbations acting only on the constraint set. This is the setting. We are given a function $f : X \to \mathbb{R}$ to be minimized over a set $A \subset X$, and we are interested in the continuity of the function

$$v : (c(X), \tau) \to [-\infty, \infty], \ v(A) = \inf\{f(x) : x \in A\},$$

where τ is a hypertopology to be specified.

As was already argued, the upper semicontinuity of the function v will follow under weak assumptions. In this setting it is actually enough to assume the lower Vietoris convergence of the sets (remember that in sequential terms this can be expressed by the condition $\limsup d(x, A_n) \le d(x, A), \forall x \in X$), and the upper semicontinuity of the function f. On the other hand, it is useless to take a finer lower convergence for the sets in order to get lower semicontinuity of the function v. The game must be played with the upper part of the convergence of the sets. Thus, we shall appeal to Theorem B.4.6 to provide the result. So, we shall deal with a real valued and continuous function f, to be minimized over sets in $c(X)$ or over particular subfamilies of $c(X)$, such as the convex sets. Moreover, we suppose a family Ω of subsets of $c(X)$ is given, containing at least the singletons of X. This last condition is necessary to provide the upper semicontinuity of the value function. We shall provide the results in terms of convergence of sequences, noticing that the same results hold, with the same proofs, for nets, too. Convergences in $c(X)$ and/or $C(X)$ are defined in the following fashion: $A_n \to A$ for some hypertopology if and only if

$$D(A_n, F) \to D(A, F) \ \text{for all } F \in \Omega, \tag{10.2}$$

where Ω is a prescribed subfamily of $c(X)$ and, as usual, D is the gap functional: $D(A, F) = \inf\{d(a, b) : a \in A, b \in F\}$. This is the way to characterize various hyperspace topologies, as can be seen in Appendix B dedicated to this topic. Here in any case we recall the needed result.

The basic abstract result is the following:

Theorem 10.2.1 *Let $A, A_n \in c(X)$ and suppose the following hold:*

(i) *if $a \in \mathbb{R}$ is such that $f^a = \{x \in X; f(x) \le a\} \ne \emptyset$, then $f^a \in \Omega$;*
(ii) *$D(A, E) = \lim D(A_n, E), \forall E \in \Omega$;*
(iii) *$\inf\{f(x) : x \in A\} = \inf_{\varepsilon > 0} \inf\{f(x) : x \in B_\varepsilon[A]\}$.*

Then $v(A) := \inf_{x \in A} f(x) = \lim_{n \to \infty} v(A_n)$.

Proof. We must check lower semicontinuity of v. Suppose, by contradiction, the existence of $a \in \mathbb{R}$ such that

$$v(A_n) < a < v(A),$$

for some subsequence of the sets A_n that we still label by n. Then $A_n \cap f^a \ne \emptyset$ whence, from (ii), $D(A, f^a) = 0$. Then there is a sequence $\{a_k\} \in f^a$ such that $d(a_k, A) \to 0$. It follows that, $\forall \varepsilon > 0$, $a_k \in B_\varepsilon[A]$ eventually, but by (iii), $f(a_k) > a$, providing the desired contradiction.   □

Condition (iii) in Theorem 10.2.1 is a technical one, and it is necessary to show some more concrete cases when it is fulfilled. Here are some examples.

Proposition 10.2.2 *Each of the following conditions implies condition* (iii) *of Theorem* 10.2.1.

- *There is $\varepsilon > 0$ such that f is uniformly continuous on $B_\varepsilon[A]$.*
- *The problem (A, f) is strongly well-posed, in the generalized sense.*

Proof. Let us show the proof in the case when (A, f) is strongly well-posed; the other one is left to the reader. Let $a = \inf_{\varepsilon > 0} \inf\{f(x) : x \in B_\varepsilon[A]\}$. There exists a sequence $\{x_n\} \subset X$ such that $f(x_n) < a + \frac{1}{n}$ and $d(x_n, A) < \frac{1}{n}$. The sequence $\{x_n\}$ is a strongly minimizing sequence, as in the Definition 10.1.13, hence it has a subsequence converging to a point $x \in A$ minimizing f on A. Then $\inf_{x \in A} f(x) \le a$, and this ends the proof.   □

Observe that the conclusion of Theorem 10.2.1 also holds under the assumption $D(A, E) = 0$ implies $A \cap E \ne \emptyset$ for each $E \in \Omega$, without assuming condition (iii).

We now see some possible applications of the previous results.

Corollary 10.2.3 *Let X be a metric space. Let $f : X \to \mathbb{R}$ be continuous and with compact level sets. Let $\{A_n\}$ be a sequence of closed sets converging in Wijsman sense to A. Then $v(A) = \lim_{n \to \infty} v(A_n)$.*

Proof. Apply Theorem 10.2.1, remembering that the Wijsman topology (see Example B.4.14) can be characterized by convergence of the gap functionals, with Ω in (10.2) the family of compact subsets of X. Moreover, compactness of the level sets of f ensures that the problem (A, f) is strongly well-posed, in the generalized sense (alternatively, the first condition of Proposition 10.2.2 holds).   □

The compactness assumption of the level sets of f is not too strong if X is, for instance, a Euclidean space; a sufficient condition, for instance, is that f be coercive, so that $f(x) \to \infty$ if $\|x\| \to \infty$. Moreover, even if our formulation of the result does not cover this case, the result is true in infinite dimensions too, provided f is weakly lower semicontinuous, with weakly compact level sets. On the other hand, if X is a separable Hilbert space with basis $\{e_n : n \in \mathbb{N}\}$, and $f(x) = \max \{-\|x\|, \|x\| - 2\}$, $A_n = [0, e_n]$ and $A = \{0\}$, we have that $v(A_n) = -1$, $v(A) = 0$, showing that coercivity of f is not enough in infinite dimensions to get continuity of v. But if f is radial, which means that there exists a function $g : [0, \infty) \to \mathbb{R}$ such that $f(x) = g(\|x\|)$, with $g(t) \to \infty$ if $t \to \infty$, then the result is true. It is enough to remember that in the normed spaces the Wijsman topology is generated also by the family Ω of balls.

Corollary 10.2.4 *Let X be a metric space. Let $f : X \to \mathbb{R}$ be continuous. Let $\{A_n\}$ be a sequence of closed sets converging to A for the proximal topology. Suppose moreover the problem (A, f) is strongly well-posed in the generalized sense. Then $v(A) = \lim_{n \to \infty} v(A_n)$.*

Proof. The proximal topology can be characterized by convergence of the gap functionals, with Ω in (10.2) the family of the closed subsets of X.   □

Example 10.2.5 In $\mathbb{R}^2$, let $A_n = \{(x, \frac{1}{n}) : x \in \mathbb{R}, n \in \mathbb{N}\}$, $A = \{(x, 0) : x \in \mathbb{R}\}$, $f(x, y) = \max \{x^2 - x^4 y, -1\}$. As $\{v(A_n)\}$ does not converge to $v(A)$, we see that the assumption that (A, f) is (only) Tykhonov well-posed is not enough to get the result.

Corollary 10.2.6 *Let X be a metric space. Let $f : X \to \mathbb{R}$ be continuous and with bounded level sets. Let $\{A_n\}$ be a sequence of closed sets converging to A for the bounded proximal topology. Suppose moreover the problem (A, f) strongly well-posed in the generalized sense. Then $v(A) = \lim_{n \to \infty} v(A_n)$.*

Proof. The bounded proximal topology can be characterized by convergence of the gap functionals, with Ω in (10.2) the family of bounded subsets of X.   □

Example 10.2.7 In a separable Hilbert space with basis $\{e_n : n \in \mathbb{N}\}$, let $A_n = (1 + \frac{1}{n})B_X$, $A = B_X$, $f(x) = \max \{-\sum(x, e_n)^{2n}, \|x\| - 10\}$. Observe that if $\|x\|^2 \le 1$, then $(x, e_n)^2 \le 1$ for all n, and so $\sum(x, e_n)^{2n} \le \|x\|^2 \le 1$. Thus $v(A) = -1$, while $v(A_n) < -4$. Thus $\{v(A_n)\}$ does not converge to $v(A)$, and we see that the assumption of having bounded level sets is not enough to guarantee the result, unless either f is uniformly continuous, or the problem (A, f) is strongly well-posed.

Corollary 10.2.8 *Let X be a normed space. Let $f : X \to \mathbb{R}$ be continuous and convex. Let $\{A_n\}$ be a sequence of closed convex sets converging to A for the linear topology. Suppose moreover that the problem (A, f) is strongly well-posed in the generalized sense. Then $v(A) = \lim_{n \to \infty} v(A_n)$.*

Proof. The linear topology can be characterized by convergence of the gap functionals, with Ω in (10.2) the family of the closed convex subsets of X. □

Corollary 10.2.9 *Let X be a normed space. Let $f: X \to \mathbb{R}$ be continuous, convex and with bounded level sets. Let $\{A_n\}$ be a sequence of closed convex sets converging to A for the slice topology. Suppose moreover the problem (A, f) is strongly well-posed in the generalized sense. Then $v(A) = \lim_{n\to\infty} v(A_n)$.*

Proof. The slice topology can be characterized by convergence of the gap functionals, with Ω in (10.2) the family of the closed bounded convex subsets of X.   □

The previous result applies to Mosco convergence too, in reflexive Banach spaces. Later we shall see a direct proof of it.

We have established continuity of the value function under various assumptions of the convergence of the constraint sets. What about the behavior of the minimizers? The following exercise suggests a result in this direction.

Exercise 10.2.10 *Let $f: X \to \mathbb{R}$ be continuous. Suppose LS $A_n \subset A$ and $v(A_n) \to v(A)$. Then Ls Min $A_n \subset$ Min A.*

I conclude this part by mentioning that the above stability results are taken from [LSS].

We now begin to study convex problems in more detail. The first result is of the same type as the previous ones, but it tackles the problem from a different point of view. Namely, we analyze stability, not of a single given problem, but of a whole class of problems at the same time.

The setting is the following. Let X be a reflexive Banach space, let $f: X \to \mathbb{R}$ be (at least) convex and lower semicontinuous, and let us consider the problem of minimizing f over a closed convex set C. As before, we are interested to the stability of the problem with respect to the constraint set C. We start with some auxiliary results. The first one highlights a property of those convex functions having at least a minimum on every closed convex set of a reflexive Banach space.

Proposition 10.2.11 *Let X be a reflexive Banach space, and let $f \in \Gamma(X)$. Suppose f has a minimum on every closed convex set C of X. Then, one and only one of the following alternatives holds:*

* Min f *is an unbounded set;*
* f^a *is a bounded set $\forall a \in \mathbb{R}$.*

Proof. As f has a minimum point on X, with a translation of the axes we can suppose, without loss of generality, that $f(0) = 0 = \inf f$. Suppose, by contradiction, that there are $\bar{r} > 0$ such that Min $f \subset \bar{r}B$ and a sequence $\{x_k\}$ such that

$$f(x_k) \to \inf f = 0 \text{ and } \|x_k\| > k!.$$

We shall build up a closed convex set C such that f does not have a minimum point on C, and this will provide the desired contradiction. Let $v_k := \frac{x_k}{\|x_k\|}$. Then $\{v_k\}$ is a minimizing sequence and along a subsequence, $v_k \rightharpoonup v_0$. As $\{rv_k\}$ is a minimizing sequence for all $r > 0$ and as $rv_k \rightharpoonup rv_0$, then rv_0 is a minimum point for f for all $r > 0$, implying $v_0 = 0$. Let $v_1^* \in X^*$ be such that

$$\|v_1^*\|_* = 1 \text{ and } \langle v_1^*, v_1 \rangle = 1.$$

Since $v_k \rightharpoonup 0$, there exists $n_1 \in \mathbb{N}$ such that $\forall n \geq n_1$,

$$|\langle v_1^*, v_n \rangle| < 1.$$

Let $v_2^* \in X^*$ be such that

$$\|v_2^*\|_* = 1 \text{ and } \langle v_2^*, v_{n_1} \rangle = 1.$$

There exists $n_2 > n_1$ such that $\forall n \geq n_2$,

$$|\langle v_1^*, v_n \rangle| < \frac{1}{2!} \text{ and } |\langle v_2^*, v_n \rangle| < \frac{1}{2!}.$$

By induction we find, $\forall j \in \mathbb{N}$, $v_j^* \in X^*$ and a subsequence n_j such that

$$\|v_j^*\|_* = 1, \langle v_j^*, v_{n_{j-1}} \rangle = 1, |\langle v_i^*, v_n \rangle| < \frac{1}{j!},$$

for $n \geq n_j$, and $i = 1, \ldots, j$. Now, let $z_j = v_{n_j} \|x_{n_j}\|$ and let $v^* := \sum_{j=1}^{\infty} \frac{v_j^*}{3^j} \in X^*$. Observe that $\|v^*\|_* \leq \frac{1}{2}$. Finally, set

$$C := \{x \in X : \langle v^*, x \rangle \geq \bar{r}\}.$$

If $x \in C$, then $\bar{r} \leq \langle v^*, x \rangle \leq \frac{1}{2}\|x\|$, whence $\|x\| \geq 2\bar{r}$. Therefore $f(x) > 0, \forall x \in C$. To conclude, it is enough to show that $\inf_C f = 0$; to get this, we shall see that $z_j \in C$ eventually. Namely,

$$\langle v^*, z_j \rangle = \sum_{m=1}^{\infty} \frac{\langle v_m^*, \|x_{n_j}\| v_{n_j} \rangle}{3^m}$$

$$= \|x_{n_j}\| \left\{ \sum_{m=1}^{j} \frac{\langle v_m^*, v_{n_j} \rangle}{3^m} + \frac{1}{3^{j+1}} \langle v_{j+1}^*, v_{n_j} \rangle + \sum_{m=j+2}^{\infty} \frac{\langle v_m^*, v_{n_j} \rangle}{3^m} \right\}$$

$$\geq j! \left\{ -\frac{j}{j!} + \frac{1}{3^{j+1}} - \sum_{m=j+2}^{\infty} \frac{1}{3^m} \right\} = j! \left\{ \frac{1}{2 \cdot 3^{j+1}} - \frac{1}{(j-1)!} \right\} \to \infty.$$

$\square$

Let us start with a first stability result, which could however also be deduced from previous statements[*] (Theorem 8.6.6).

Proposition 10.2.12 *Let $f: X \to \mathbb{R}$ be convex and continuous. Moreover, suppose f has bounded level sets and only one minimum point over each closed convex set. Let $C_n, C \in C(X)$ be such that $C_n \overset{\text{M}}{\to} C$. Then*

$$\inf_{C_n} f \to \inf_{C} f,$$
$$\underset{C_n}{\text{Min}}\, f \rightharpoonup \underset{C}{\text{Min}}\, f.$$

Proof. From the assumption made on f it follows that $\text{Min}_A\, f$ is a singleton for each closed convex set A. Let $c_n = \text{Min}_{C_n} f$, $c = \text{Min}_C f$. It is clear that $\limsup f(c_n) \le f(c)$. For, there exists $x_n \in C_n$ such that $x_n \to c$. Then

$$\limsup f(c_n) \le \limsup f(x_n) = f(c).$$

Moreover, $\{c_n\}$ is a bounded sequence, as it is contained (eventually) in the level set of height $\inf_C f + 1$. Thus there are a subsequence c_{n_j} and a point $\bar{c} \in C$ such that $c_{n_j} \to \bar{c}$. Moreover,

$$\inf_{C} f \ge \limsup f(c_{n_j}) = f(\bar{c}).$$

This shows that $\bar{c}$ minimizes f over C and, by uniqueness of the minimum point of f over C, $\bar{c} = c$. Uniqueness of the limit point guarantees that $c_n \rightharpoonup c$ (not only along a subsequence). ☐

Exercise 10.2.13 f is said to be *quasi convex* if $\forall x, \forall y \ne x, \forall t \in (0,1)$,

$$f(tx + (1-t)y) \le \max\{f(x), f(y)\},$$

and *strictly quasi convex* if in the above inequality the symbol $<$ is substituted for $\le$. Show that a sufficient condition for a function f to have at most a minimum point on every closed convex set is that f is strictly quasi convex.

We want to now show how, in the above setting, adding an assumption of well-posedness increases the property of stability in minimum problems. In particular, we shall see that if we add the assumption that the problems of minimizing f over A are Tykhonov well-posed for every closed convex set A, in Proposition 10.2.12 we are able to draw a much stronger conclusion, since we show that there actually is strong convergence of the minimum points, rather than only weak convergence. Here is the result.

Theorem 10.2.14 *Let X be a reflexive Banach space, let $f: X \to \mathbb{R}$ be convex, continuous, bounded on the bounded sets and such that the problem (A, f) is Tykhonov well-posed for each closed convex subset $A \subset X$. Let $C_n, C \subset X$ be closed convex subsets of X. Then*

$$C_n \overset{\text{M}}{\to} C \implies \underset{C_n}{\text{Min}}\, f \to \underset{C}{\text{Min}}\, f.$$

Proof. First observe that $\mathrm{Min}_{C_n} f$, $\mathrm{Min}_C f$ are singletons. Let us denote them by c_n, c respectively, and by $\bar{x}$ the minimum point of f over X. Proposition 10.2.12 shows that $c_n \rightharpoonup c$. Now, if $c = \bar{x}$, then $c_n \to c$, as $\{c_n\}$ is a minimizing sequence for the problem (X, f), a well-posed one, by assumption. In this case the theorem is proved. If $c \neq \bar{x}$, set $a = f(c)$. The closed convex sets f^a and C can be weakly separated by a hyperplane as, by the continuity of f, $\bar{x} \in \mathrm{int}\, f^a$. Hence there are $0^* \neq x^* \in X^*$, $r \in \mathbb{R}$ such that

$$C \subset H^+ := \{x \in X : \langle x^*, x \rangle \geq r\},$$

and

$$f^a \subset H^- := \{x \in X : \langle x^*, x \rangle \leq r\}.$$

Denote by

$$H := \{x \in X : \langle x^*, x \rangle = r\},$$

and by

$$H_0 := \{x \in X : \langle x^*, x \rangle = 0\}.$$

There exists $l \in X$ such that $\langle x^*, l \rangle \neq 0$ and

$$X = H_0 \oplus \mathrm{sp}\{l\};$$

indeed, for every element $x \in X$ there are (unique) $x_0 \in H_0$ and $m \in \mathbb{R}$ such that $x = x_0 + ml$. Hence $c_n - c = x_n + m_n l$, $x_n \in H_0$. Observe that $m_n \to 0$. For $0 = \lim \langle x^*, c_n - c \rangle = \langle x^*, l \rangle m_n$. Hence

$$\|c_n - (c + x_n)\| \to 0,$$

and, as f is uniformly continuous on the bounded sets,

$$|f(c_n) - f(c + x_n)| \to 0.$$

Therefore

$$c + x_n \in H, f(c + x_n) \to f(c),$$

and $c \in H$ minimizes f over H. As the problem (H, f) is well-posed by assumption, it follows that $c + x_n \to c$ and so, as a result $c_n \to c$. $\square$

Remark 10.2.15 With the same proof it can be shown that if we suppose only Tykhonov well-posedness in the generalized sense, then every sequence of minima of f from the approximating sets C_n has a subsequence converging to a minimum point of f on C. Even more, it is enough to consider elements $c_n \in C_n$ such that $f(c_n) - \inf_{C_n} f \to 0$, to get the same result.

Remark 10.2.16 It is worth noticing that it is not enough to assume that (C, f) is well-posed for a given fixed set C, to get that for every sequence $\{C_n\}$ converging to C, then $c_n \to c$. Consider for instance, in a separable Hilbert space with basis $\{e_n : n \in \mathbb{R}\}$, the function

$$f(x) = \sum_{n=1}^{\infty} \frac{(x, e_n)^2}{n^2} + \text{Max}\{0, \|x\| - 1\}.$$

Then f has bounded level sets, it is continuous, real valued and strictly quasi convex. Then it has one and only one minimum point on every closed convex set. The pair $(\{0\}, f)$ is obviously Tykhonov well-posed, but considering the sequence $C_n := [\frac{1}{n}e_n, e_{n^2}]$, Mosco converging to $\{0\}$, we see that c_n does not strongly converge to zero (figure it out!).

Now we see a result going in the opposite direction. We start by a useful proposition.

Proposition 10.2.17 *Let X be a reflexive Banach space, let $K \subset X$ be a closed convex set such that $0 \neq y_0 \in K$, let $y_0^* \in X^*$ be such that $\langle y_0^*, y_0 \rangle = \|y_0\|$ and $\|y_0^*\|_* = 1$, let $\{d_n\}, \{b_n\}$ be two real valued sequences such that $d_n \to 0$, $b_n \to 0$. Finally, $\forall n \in \mathbb{N}$, let $y_n^* \in X^*$ be such that $\|y_n^*\|_* = 1$. Set*

$$C_n := \{x \in K : \langle y_0^*, x \rangle + b_n \langle y_n^*, x \rangle \geq d_n\},$$
$$C := \{x \in K : \langle y_0^*, x \rangle \geq 0\}.$$

Then $C_n \xrightarrow{M} C$.

Proof. Let $x \in C$. If $\langle y_0^*, x \rangle > 0$, then $x \in C_n$ eventually. This means that

$$\text{Li } C_n \supset \{x \in K : \langle y_0^*, x \rangle > 0\}.$$

Now, for $x \in C$ and $\langle y_0^*, x \rangle = 0$, $ax + (1-a)y_0 \in K$ for $0 \leq a \leq 1$, and $\langle y_0^*, ax + (1-a)y_0 \rangle > 0$ for $a < 1$ whence, being $\text{Li } C_n$ a closed set,

$$\text{Li } C_n \supset \overline{\{x \in K : \langle y_0^*, x \rangle > 0\}} \supset C.$$

Let us prove now that if $x_k \in C_{n_k}$ and $x_k \rightharpoonup x$, then $x \in C$. Since K is weakly closed, $x \in K$. Moreover, as

$$\langle y_0^*, x_k \rangle + b_{n_k} \langle y_{n_k}^*, x_k \rangle \geq d_{n_k},$$

we get, passing to the limit, that

$$\langle y_0^*, x \rangle \geq 0,$$

whence $x \in C$. $\square$

Theorem 10.2.18 *Let X be a reflexive Banach space, let $f : X \to \mathbb{R}$ be convex, continuous and such that f has a unique minimum point on every closed convex $A \subset X$. If for each C_n, C closed convex set it happens that*

$$C_n \xrightarrow{M} C \implies \underset{C_n}{\text{Min}} f \to \underset{C}{\text{Min}} f,$$

then the problem (A, f) is Tykhonov well-posed for each closed convex set $A \subset X$.

Proof. Suppose there exists a closed convex set $K \subset X$ such that f is not Tykhonov well-posed on K. Without loss of generality, we can suppose $0 \in K$ and $f(0) = 0 = \inf_K f$. Let $0 \neq y_0 \in K$. Let $y_0^* \in X^*$ be such that $\langle y_0^*, y_0 \rangle = \|y_0\|$ and $\|y_0^*\|_* = 1$. Define

$$\emptyset \neq A_n := \left\{ x \in K : \langle y_0^*, x \rangle \geq \frac{1}{n^2} \right\}.$$

Since $0 \notin A_n$, $\inf_{A_n} f := a_n > 0$. As f is not Tykhonov well-posed over K, there exists a minimizing sequence $\{m_n\} \subset K$ not converging to 0. Since f has bounded level sets (see Proposition 10.2.11), it is easy to show that $m_n \rightharpoonup 0$. Then it is possible to build a sequence $\{y_n\}$ from $\{m_n\}$, still minimizing and such that, for a suitable $a > 0$,

$$|\langle y_0^*, y_n \rangle| \leq \frac{1}{n^2}, \quad \|y_n\| = a, \quad f(y_n) < a_n.$$

Now let, $\forall n \in \mathbb{N}, y_n^* \in X^*$ such that $\|y_n^*\|_* = 1$ and $\langle y_n^*, y_n \rangle = a$. Define

$$C_n := \{ x \in K : \langle y_0^*, x \rangle + \frac{1}{n} \langle y_n^*, x \rangle \geq \frac{a}{n} + \langle y_0^*, y_n \rangle \},$$
$$C := \{ x \in K : \langle y_0^*, x \rangle \geq 0 \}.$$

Then $y_n \in C_n$ and, from Proposition 10.2.17, $C_n \xrightarrow{M} C$. Let $c := \text{Min}_C f$, $c_n := \text{Min}_{C_n} f$. Observe that $c = 0$. As $y_n \in C_n$, then $f(c_n) \leq f(y_n) < a_n$, whence $c_n \notin A_n$. It follows that $\langle y_0^*, c_n \rangle < \frac{1}{n^2}$. As $c_n \in C_n$, then $\langle y_0^*, c_n \rangle + \frac{1}{n} \langle y_n^*, c_n \rangle \geq \frac{a}{n} + \langle y_0^*, y_n \rangle$, implying $\langle y_n^*, c_n \rangle \geq a - \frac{2}{n}$. This in turn implies that the sequence $\{c_n\}$ does not (strongly) converge to $0 = c$, against the assumptions, since c_n minimizes f over C_n, $C_n \xrightarrow{M} C$ and 0 minimizes f over C.   □

Exercise 10.2.19 Show that in Proposition 10.2.17, actually $C_n \xrightarrow{AW} C$.

Remark 10.2.20 Theorem 10.2.18 can be easily generalized, with essentially the same proof just by considering Hausdorff convergence in the assumptions rather than Mosco convergence. For the set K in the proof can be assumed to be a bounded set. If it is not, we can repeat the same argument of the proof with K intersected with a suitable sublevel set of f. Then the set C in the proof is bounded, hence AW convergence (see Exercise 10.2.19) implies Hausdorff convergence (see Exercise 8.6.16).

In Theorem 10.2.14, as we have seen, we must assume Tykhonov well-posedness for a whole family of problems. Assuming Tykhonov well-posedness of the limit problem (C, f) does not suffice. Mosco convergence is in some sense too weak to get stability, having only well-posedness at the limit. The next results show instead that such a result can be achieved with the finer AW convergence. To prepare the proof of the two final results of this section, we start with two preliminary propositions.

Proposition 10.2.21 *Let X be a normed space, $f_n, f \in \Gamma(X)$ be such that $f_n \overset{AW}{\to} f$. Then, $\forall a > \inf_X f$,*

$$(f_n)^a \overset{AW}{\to} f^a.$$

Proof. Let $a > \inf_X f$ and fix b such that $a > b > \inf_X f$. Let $\bar{x} \in X$ be such that $f(\bar{x}) < b$. Let $\bar{r} > \max\{|a|, \|\bar{x}\|\}$. There are $N_1 \in \mathbb{N}$ and $z_n \to \bar{x}$ such that $f_n(z_n) < b, \forall n > N_1$. Let $\varepsilon > 0$. We must find $N \in \mathbb{N}$ such that $\forall n > N$, $\forall r > \bar{r}$, the following relations hold:

$$e(f^a \cap rB, (f_n)^a) \le \varepsilon; \tag{10.3}$$

$$e((f^a)_n \cap rB, f^a) \le \varepsilon. \tag{10.4}$$

Fix $r > \bar{r}$ and let $c > 0$ be such that $c + \frac{2cr}{a-b+c} < \varepsilon$. Finally, let $N_2 \in \mathbb{N}$ be such that $h_r(\text{epi } f_n, \text{epi } f) < c, \forall n > N_2$.

Let us show that the choice of $N = \max\{N_1, N_2\}$ does the job. We shall only verify (10.3), since (10.4) follows in the same way. Let $x \in X$ be such that $f(x) \le a$ and $\|x\| \le r$. Let $n > N$. Then, there exists $(x_n, r_n) \in \text{epi } f_n$ such that

$$\|x_n - x\| < c, \quad |r_n - a| < c.$$

Note that $f_n(x_n) \le r_n \le a + c$. If it happens that $f_n(x_n) \le a$ the proof is over. But this does not always happen! However, we now exploit the sequence $\{z_n\}$ built up before. For, it is possible to find on the line segment $[z_n, x_n]$, a point $y_n \in (f^a)_n$ at distance less that ε from x, and this allows us to conclude the proof. To see this, let $\lambda = \frac{a-b}{a-b+c}$. Then

$$f_n(\lambda x_n + (1-\lambda)z_n) \le \frac{a-b}{a-b+c}(a+c) + \frac{c}{a-b+c}b = a,$$

and

$$\|\lambda x_n + (1-\lambda)z_n - x\| \le \lambda\|x_n - x\| + (1-\lambda)\|z_n - x\| \le \lambda c + \frac{c}{a-b+c}2r < \varepsilon.$$

$\square$

The next result deals with the stability of the value of the problem.

Proposition 10.2.22 *Let X be a normed space, and let $f_n, f \in \Gamma(X)$ be such that $f_n \overset{AW}{\to} f$. If there exists $a > \inf f$ such that f^a is bounded, then $\inf f_n \to \inf f$.*

Proof. It is enough to show that $\liminf_{n\to\infty} \inf_X f_n \ge \inf f$. There is nothing to prove if $\inf f = -\infty$ (this never happens if X is a reflexive Banach space). Suppose then $\inf f \in \mathbb{R}$ and, by contradiction, that there are $\varepsilon > 0$, a subsequence from $\{f_n\}$ (always named $\{f_n\}$) and x_n such that $f_n(x_n) < \inf f - 2\varepsilon$ for all n. As f^a is a bounded set, then the sets $(f_n)^a$ are equibounded, since

$(f_n)^a \overset{\text{AW}}{\to} f^a$ and so $(f_n)^a \overset{\text{H}}{\to} f^a$ (see Exercise 8.6.16). Thus there exists $r > |\inf_X f| + 3\varepsilon$ such that $(f_n)^{\inf_X f - 2\varepsilon} \subset rB_X$. As there exists n such that $e(\text{epi } f_n \cap rB_{X \times \mathbb{R}}, \text{epi } f) < \varepsilon$, then there exists $(y_n, \alpha_n) \in \text{epi } f$ such that

$$\|y_n - x_n\| < \varepsilon \text{ and } |\alpha_n - f_n(x_n)| < \varepsilon,$$

but this implies $f(y_n) \leq \inf_X f - \varepsilon$, which is impossible.   □

The situation with Mosco convergence is clarified by the following exercise.

Exercise 10.2.23 Let X be a separable Hilbert space with basis $\{e_n : n \in \mathbb{N}\}$. Let $f_n(x) = \sum_{k \neq n}(x, e_k)^2 + \frac{1}{n}(x, e_n)$. Find f, the Mosco limit of $\{f_n\}$. Is f Tykhonov well-posed? Does $\inf f_n \to \inf f$?

Now we have a stability result for the minimum points.

Theorem 10.2.24 *Let X be a normed space, $f_n, f \in \Gamma(X)$ such that $f_n \overset{\text{AW}}{\to} f$. Moreover, suppose f Tykhonov well-posed in the generalized sense. Then for $\varepsilon_n > 0$ with $\varepsilon_n \to 0$ and x_n such that $f_n(x_n) \leq \inf f_n + \varepsilon_n$ it holds that $\{x_n\}$ has a subsequence converging to a minimum point for f.*

Proof. It is enough to show that $\forall \varepsilon > 0, \exists N, \forall n \geq N$,

$$x_n \in B_{2\varepsilon}[\text{Min } f].$$

By the well-posedness assumption, $\exists a > \inf_X f$ such that

$$f^a \subset B_\varepsilon[\text{Min } f].$$

As $(f_n)^a \to f^a$ in the Hausdorff sense, (see Proposition 10.2.21 and Exercise 8.6.16), $\exists N_1, \forall n \geq N_1$,

$$(f_n)^a \subset B_\varepsilon[f^a] \subset B_{2\varepsilon}[\text{Min } f].$$

As $\inf f_n \to \inf f$, then $f_n(x_n) \to \inf f$, whence there is N_2 such that $\forall n \geq N_2, x_n \in (f_n)^a$. The choice of $N = \max\{N_1, N_2\}$ does the job.   □

As a result, given a Tykhonov well-posed problem, if we are able to approximate in the AW sense a given function by functions whose minima are easier to find, then it is enough to get approximate solutions of the perturbed problems in order to be close to the true solution of the initial problem. In passing, we observe that this is an interesting, yet qualitative result. However, giving qualitative estimates is another very important issue that is not considered in this book.

The above theorem deals with an unconstrained problem (at least explicitly). An analogous result holds in the presence of an explicit constraint set, in the following sense. From Theorem 10.2.24 and Theorem 9.2.5 we get

Theorem 10.2.25 *Let X be a normed space, $f_n, f \in \Gamma(X)$, let $C_n, C \subset X$ be closed convex sets such that*

$$f_n \overset{\mathrm{AW}}{\to} f, \quad C_n \overset{\mathrm{AW}}{\to} C.$$

If there exists $c \in C$ such that f is continuous at c, and the problem (C, f) is Tykhonov well-posed in the generalized sense, then for every $c_n \in C_n$ such that $f_n(c_n) - \inf_{C_n} f_n \to 0$, one has that $\{c_n\}$ has a subsequence converging to a minimum point for f over C.

Proof. Apply the sum theorem to $f_n + I_{C_n}$ $f + I_C$ and use Theorem 10.2.24.
□

Corollary 10.2.26 *Let X be a normed space, $f \in \Gamma(X)$, $C \subset X$ a closed convex set and suppose there is a point $c \in C$ where f is continuous. Moreover, suppose the problem (C, f) is Tykhonov well-posed. Then the problem (C, f) is strongly well-posed.*

Proof. Let $\{x_n\} \subset X$ be such that $d(x_n, C) \to 0$ and $\limsup f_n(x_n) \le \inf f$. If $\{x_n\} \subset C$ eventually, there is nothing to prove. Otherwise, for $x_n \notin C$, consider the sequence of closed convex sets $C_n := \mathrm{co}\{C \cup x_n\}$. It is easy to prove that $C_n \overset{\mathrm{AW}}{\to} C$. Apply Theorem 10.2.25 to get that $\{x_n\}$ has a subsequence converging to a point of C minimizing f over C. Uniqueness of the minimum point provides the result.
□

Finally, let us remark that the results obtained with AW convergence in Proposition 10.2.22, in Theorem 10.2.24 and in Theorem 10.2.25 equally hold for the weaker bounded–proximal convergence. As the most careful reader has noticed, it is the upper part of the convergence that plays the game, which is the same for the two topologies.

10.3 A new well-posedness concept

The final section of this chapter deals with another well-posedness concept, which was introduced in recent years and is important and interesting since in some sense it unifies the ideas of Tykhonov well-posedness and stability. We shall simply call it well-posedness.

The setting is the following: we consider a metric space $\mathcal{A}$, called the *data space*, and another metric space X, called the *domain space*. An extended real valued function f_a, defined on X, is associated to each $a \in \mathcal{A}$. So that each $a \in \mathcal{A}$ represents a minimum problem: $\inf_{x \in X} f_a(x)$.

Definition 10.3.1 *We shall say that the problem $a \in \mathcal{A}$ is well-posed if*

(i)   *there exists a unique $x_0 \in X$ such that $f_a(x_0) \le f_a(x), \forall x \in X$;*

(ii)   for any $\{a_n\} \subset \mathcal{A}$ such that $a_n \to a$, $\inf f_{a_n}$ is finite eventually, and if $\{x_n\} \subset X$ is such that $f_{a_n}(x_n) - \inf f_{a_n} \to 0$, then $x_n \to x_0$.

This notion was firstly introduced by Zolezzi (in a slightly different context) [Zo], with the name of well-posedness by perturbations; later and independently it was given by Ioffe and Zaslavski [IZ], with the additional condition $\inf f_{a_n} \to \inf f_a$, i.e., continuity at a of the value function. I prefer to keep this third condition separated from the definition. We shall see in many examples that continuity of the value is often an automatic consequence of the first two conditions. Simply observe, for instance, that if the distance on the data space $\mathcal{A}$ induces Kuratowski convergence of $\{f_{a_n}\}$ to f_a, then continuity of the value function is a direct consequence of (ii) above (see Theorem 8.6.6).

The meaning of this definition is clear. For a well-posed problem, finding approximate solutions of "nearby" problems drives toward the true solution. Thus, when facing one of these problems, and the solution is hard to find numerically, we can try to approximate the objective function with simpler ones and then apply some algorithm to get an approximate solution of the approximate problems. As long as the approximation becomes more and more accurate, the approximate solutions come closer and closer to the effective solution of the initial problem. An example of this could be the following well-known procedure. Suppose we have to minimize a real valued function defined on a separable Hilbert space with basis $\{e_n : n \in \mathbb{R}\}$. Since the procedure to find the solution can be very complicated by the fact of being in an infinite dimensional setting, one can try to solve the problem on a sequence of finite dimensional spaces X_n "invading" X, e.g., $X_n = \mathrm{sp}\{e_1, \ldots, e_n\}$. Clearly this procedure, called the Riesz–Galerkin method, fits in the above model (with a suitable topology on the data space, of course).

To clarify how the above abstract setting can describe more concrete situations and the fact that it is useful to introduce a data space, which is not necessarily of space of functions, let us describe a problem we shall meet later. Suppose we have two convex, real valued functions f, g defined on $\mathbb{R}^n$, and consider the following problem:

$$\text{minimize} \quad f(x) - \langle p, x \rangle$$
$$\text{such that} \quad g(x) \leq \alpha,$$

where $p \in \mathbb{R}^n, \alpha \in \mathbb{R}$. We are interested in perturbing the linear term of the objective function and the right-hand side of the inequality constraint, while we want to keep fixed the functions f, g. So that in this case a typical element a of the data space is a pair (p, α) and the function f_a is defined in the usual way:

$$f_a(x) = \begin{cases} f(x) - \langle p, x \rangle & \text{if } g(x) \leq \alpha, \\ \infty & \text{otherwise.} \end{cases}$$

In other examples, the data space will simply be a prescribed family of functions to be minimized. Let us now see a first example.

Example 10.3.2 Let $\mathcal{A}$ be the set of convex, lower semicontinuous, extended real valued functions on $\mathbb{R}^n$, endowed with a metric compatible with Kuratowski convergence (of the epigraphs). Let $f \in \mathcal{A}$ be such that f has a unique minimum point. Then f is well-posed. Suppose, without loss of generality, $f(0) = 0$ is the minimum point of f. Take $f_n \xrightarrow{\mathrm{K}} f$. Let us show that $\inf f_n \to \inf f$. It is enough to show that $\liminf \inf f_n \geq \inf f$. Suppose not. Then there is $a < 0$ such that along a subsequence, $\inf f_n < a$. (Without explicitly stating it, we shall pass at various time to subsequences and always use the same label n.) Thus there is z_n such that $f_n(z_n) \leq a$. Suppose $\{z_n\}$ has a limit point z. From Kuratowski convergence, we have $f(z) \leq \liminf f_n(z_n) \leq a$, which is impossible. Suppose then $\|z_n\| \to \infty$. There is $y_n \to 0$ such that $\limsup f_n(y_n) \leq \inf f = 0$. Consider $w_n = \lambda_n y_n + (1 - \lambda_n) z_n$, with λ_n chosen in such a way that $\|w_n\| = 1$. Then $\limsup f_n(w_n) \leq a$ and, for w a limit point of $\{w_n\}$, we have $f(w) \leq a$. This is impossible and we deduce $\inf f_n \to \inf f$. The rest of the proof follows more or less the same pattern, and is left to the reader.

Let us start by seeing how the Furi–Vignoli criterion (see Proposition 10.1.6) can be extended in this setting.

Proposition 10.3.3 *Let (X, d) be a complete metric space, let $a \in \mathcal{A}$ be such that the associated function f_a is lower semicontinuous. Suppose a is well-posed. Then*

$$\lim_{\delta \to 0} \mathrm{diam}\overline{\{\bigcup f_b^{\inf f_b + \delta} : d(a, b) < \delta\}} = 0.$$

Conversely, suppose

(i)   *the value function $\inf(\cdot)$ is finite around a;*

(ii) $\lim_{\delta \to 0} \mathrm{diam}\overline{\{\bigcup f_b^{\inf f_b + \delta} : d(a, b) < \delta\}} = 0.$

Then a is well-posed.

Proof. Suppose a is well-posed and, by contradiction, that there are two sequences $\{a_n\}$ and $\{b_n\}$ such that there are x_n and y_n with $f_{a_n}(x_n) \leq \inf f_{a_n} + \varepsilon_n$, $f_{b_n}(y_n) \leq \inf f_{b_n} + \varepsilon_n$ and $d(x_n, y_n) \geq a$, for some $a > 0$. Then at least one of the two sequences $\{x_n\}$, $\{y_n\}$ does not converge to the minimum point of f_a, contrary to condition (iii) of well-posedness. Now suppose that (i) and (ii) hold and let us see that a is well-posed. Set $A_\delta := \overline{\{\bigcup f_b^{\inf f_b + \delta} : d(a, b) < \delta\}}$. Since $f_a^{\inf f_a + \delta} \subset A_\delta$ for all $\delta > 0$, it is clear for the Furi–Vignoli criterion that f_a is Tykhonov well-posed; in particular it has a solution $\bar{x} \in \bigcap_{\delta > 0} A_\delta$. Now, let $a_n \to a$ and take $\{x_n\}$ as in point (ii) of Definition 10.3.1. Then, for all $\delta > 0$, $\{x_n\} \subset A_\delta$ eventually. Thus, by (i) $\{x_n\}$ is a Cauchy sequence, and so it has a limit $x_0 \in \bigcap_{\delta > 0} A_\delta$. But by assumption (ii) $\bigcap_{\delta > 0} A_\delta$ is a singleton, and thus $x_0 = \bar{x}$ and the proof is complete.   □

Also the condition involving a forcing function, in the characterization of Tykhonov well-posedness, can be rephrased in this context. First of all, we have to change the definition of forcing function a bit, as follows. Let $D \subset \mathbb{R}_+^2$ be such that $(0,0) \in D$. A function $c \colon D \to [0, \infty)$ is said to be forcing if $c(0,0) = 0$, $(t_n, s_n) \in D$ for all n, $s_n \to 0$, and $c(t_n, s_n) \to 0$ imply $t_n \to 0$.

The following result holds.

Proposition 10.3.4 *Let (X, d) be a metric space, let $(\mathcal{A}, \delta)$ be another metric space and suppose $a \in \mathcal{A}$ well-posed. Then there exist a forcing function c and $\bar{x} \in X$ such that*

$$f_b(x) \geq \inf f_b + c\big[\big(d(x, \bar{x}), (\delta(a, b))\big)\big], \qquad (10.5)$$

for all $x \in X$ and $b \in \mathcal{A}$. Conversely, suppose $\inf(\cdot)$ finite around a, and that there exist a forcing function c and a point $\bar{x}$ fulfilling (10.5). Then a is well-posed.

Proof. Suppose a is well-posed, with solution $\bar{x}$. Define for small $s > 0$,

$$c(t, s) = \inf_{\delta(a,b)=s} \inf_{d(x,\bar{x})=t} f_b(x) - \inf f_b.$$

It is obvious that (10.5) is satisfied. It is also clear that $c(0,0) = 0$. Now suppose $c(t_n, s_n) \to 0$ and $s_n \to 0$. Then there are b_n such that $\delta(b_n, a) = s_n$ and x_n such that $d(x_n, \bar{x}) = t_n$ such that $f_{b_n}(x_n) - \inf f_{b_n} \to 0$. By well-posedness then $x_n \to \bar{x}$, and thus $t_n \to 0$, which implies that c is forcing. Conversely, let $\{x_n\}$ be such that $f_a(x_n) \to \inf f_a$. Since $f_a(x_n) \geq c[d(x_n, \bar{x}), 0] + \inf f_a$, it follows that $c[d(x_n, \bar{x}), 0] \to 0$ and thus $x_n \to \bar{x}$, since c is forcing. By lower semicontinuity of f_a, $\bar{x}$ minimizes f_a. An analogous argument shows that if $a_n \to a$ and if $f_{a_n}(x_n) - \inf f_{a_n} \to 0$, then $x_n \to \bar{x}$. $\qquad \square$

We now give some examples showing that in several important classes of problems the weaker (in principle) notion of Tykhonov well-posedness actually implies the stronger one introduced in this section. We shall only outline the proofs.

Example 10.3.5 Let (X, d) be a complete metric space. Let $\mathcal{F}(X)$ be a family of functions in X endowed with the metrizable topology of uniform convergence on the bounded sets in X. Then $\mathcal{F}(X)$ is one of the following four:

(i) $\mathcal{F}(X) := \{f \colon X \to \mathbb{R},\ f$ is lower semicontinuous and $f(x) \geq \psi(x)$ for any $x \in X\}$, where ψ is a bounded from below coercive function in X (i.e., $\psi(x) \to \infty$ if $\rho(x, \theta) \to \infty$);

(ii) $\mathcal{F}(X) := \{f \colon X \to \mathbb{R},\ f$ is continuous and $f(x) \geq \psi(x)$ for any $x \in X\}$, with ψ a bounded from below coercive function in X;

(iii) X is a real Banach space and $\mathcal{F}(X) := \{f \colon X \to \mathbb{R},\ f$ is continuous, quasi-convex and bounded from below on the bounded sets$\}$;

(iv) X is a real Banach space and $\mathcal{F}(X) := \{f \colon X \to \mathbb{R},\ f$ is continuous and convex$\}$.

In these cases, the data space $\mathcal{A}$ coincides with $\mathcal{F}(X)$. Thus, we are given a function $f \in \mathcal{F}(X)$ which is Tykhonov well-posed, a sequence $\{f_n\}$ converging to f, and we must show that the two conditions of Definition 10.3.1 are fulfilled. First, observe that the following relation always holds:

$$\limsup \inf f_n \leq \inf f.$$

Next, the strategy to prove that sequences $\{x_n\}$ as in condition (ii) of Definition 10.3.1 converge to the unique point of the limit function f is the same in all cases. Since the limit function is assumed to be Tykhonov well-posed, one shows that $\{x_n\}$ is a minimizing sequence for f. In the first two cases, we have that

$$\inf \psi \leq \limsup \inf f_n,$$

and thus, as

$$f_n(x_n) \leq \inf f + 1$$

eventually, then $\{x_n\}$ is a bounded sequence. Then the proof is complete, since on bounded sets we have uniform convergence of $\{f_n\}$ to f.

In the other two cases, f has bounded level sets. Thus there is $r > 0$ such that $\{x : f(x) \leq \inf f + 3\} \subset rB$. As $f(x) \geq \inf f + 3$ if $\|x\| = r$, then eventually $f_n(x) \geq \inf f + 2$ if $\|x\| = r$ and also $f_n(x) \leq \inf f_n + 1$ if $\|x\| = r$. By quasi convexity it follows that the level sets of the functions f_n are equibounded; thus the game is again played, as before, on a bounded set, and uniform convergence there suffices.

Remark 10.3.6 The coercivity assumption we made in the first two cases cannot be substituted by a weaker condition, as, for instance, lower equiboundedness. Consider $f(x) = x^2$ and the sequence $f_n(x) = \max\{x^2 - \frac{1}{n}x^4, 0\}$.

Example 10.3.7 Let X be a normed space and let

$$\mathcal{A} = \Gamma(X),$$

endowed with the Attouch–Wets convergence. More generally, we can consider constrained problems, and consider $\mathcal{A}$ as the product space $\Gamma(X) \times C(X)$ (with the product topology engendered by the Attouch–Wets convergence in both spaces). In this case an element of $\mathcal{A}$ is a pair (f, A), where f is the objective function, and A is the constraint set. Theorem 10.2.24 shows that Tykhonov well-posedness implies well-posedness in the unconstrained case; Theorem 10.2.25 shows that Tykhonov well-posedness implies well-posedness at every $a = (f, A)$ fulfilling a condition of the form: there is a point $a \in A$ where f is continuous. Observe that these pairs are a dense subset of $\mathcal{A}$ (this follows for instance from Theorem 9.2.11).

Example 10.3.8 Let us recall the mathematical programming problem. Let $C \subset X$ be a nonempty, closed convex set in the Euclidean space X, and suppose we are given a convex, lower semicontinuous function $k \colon C \to \mathbb{R}$ and another function $g \colon X \to \mathbb{R}^m$ which is continuous and with convex components. Let us consider the problem

$$(\mathcal{P}) \qquad\qquad \inf_{\substack{x \in C \\ g(x) \le 0}} k(x) = \inf_{x \in X} f(x),$$

where

$$f(x) := \begin{cases} k(x) & \text{if } x \in C \text{ and } g(x) \le 0. \\ \infty & \text{otherwise.} \end{cases}$$

The condition $g(x) \le 0$ is, as usual, intended coordinatewise. We can take in this case $\mathcal{A} = \mathbb{R}^m$, and to $a \in \mathcal{A}$ is associated the function

$$f_a(x) := \begin{cases} k(x) & \text{if } x \in C \text{ and } g(x) \le a, \\ \infty & \text{otherwise.} \end{cases}$$

It is natural to consider the Euclidean metric in $\mathcal{A}$. The following proposition holds.

Proposition 10.3.9 *Suppose there is at least one $\bar{x}$ fulfilling a constraint qualification condition of the form $g_i(\bar{x}) < 0, \forall i$. Suppose also there are a vector $a \in \mathbb{R}^m$ and $b \in \mathbb{R}$ such that $a_i > 0$, $i = 1, \ldots, m$ and*

$$A := \{x \in X : g(x) \le a, k(x) \le b\}$$

is nonempty and bounded. Then if the problem $(\mathcal{P})$ has at most one solution, it is actually well-posed.

Proof. Observe that the constraint qualification condition guarantees that for b sufficiently close to a, the constraint set $g(x) \le b$ is nonempty. This, together with the assumption that the set A above is bounded, guarantees that the value $\inf f_b$ is finite (and attained) around a. The rest is absolutely easy to check and is left to the reader. $\qquad\square$

Exercise 10.3.10 Consider $\mathcal{A} = \Gamma(X)$ endowed with the Attouch–Wets convergence. Let $f \in \mathcal{A}$ and suppose f^* is Fréchet differentiable at the origin. Prove that f is well-posed.

10.4 A digression: projecting a point on a closed convex set

In this section we see some facts concerning *the best approximation problem*, i.e., the problem of projecting, over a closed convex subset $C \subset X$, a point $x \in X$ outside it. This means minimizing the function $f(x) = \|x - c\|$ over C. It is an easy consequence of the Weierstrass theorem that if the underlying space is finite dimensional, then the problem does have solution, without even assuming C to be convex. If X is reflexive infinite dimensional, an application of the Weierstrass theorem (using the weak topology on X) again provides the existence of at least one solution (of course in this case C is assumed to be convex). In general, the problem could also have more than one solution, in the convex case too. If we consider $X = \mathbb{R}^2$, endowed with the box norm, (i.e., $\|(x, y)\| = \max\{|x|, |y|\}$, it is easy to see that the problem of projecting the vector $(0, 2)$ on the unit ball has more than one solution. When X is a Hilbert space, the projection is unique and denoting it by $p_C(x)$, we have that $y = p_C(x)$ if and only if $y \in C$ and

$$\langle x - y, c - y \rangle \leq 0, \tag{10.6}$$

for all $c \in C$ (see Exercise 4.1.4). In this section we want to generalize the above formula and to make some consideration on what happens when moving either the point x or the set C.

First, let us remember that the subdifferential of the function $\| \cdot \|$ outside the origin is the duality map

$$\delta(x) = \{x^* \in X^* : \|x^*\|_* = 1 \text{ and } \langle x^*, x \rangle = \|x\|\}$$

(see Example 3.2.7).

We now see how to extend the formula (10.6) whose geometrical meaning in Hilbert space is very clear.

Proposition 10.4.1 *Let X be a reflexive space, let C be a closed convex set, and $\bar{x} \notin C$. Then $y \in P_C(\bar{x})$ if and only if $y \in C$ and there is $x^* \in \delta(\bar{x} - y)$ such that*

$$\langle x^*, c - y \rangle \leq 0, \tag{10.7}$$

$\forall c \in C$.

Proof. First, observe that $P_C(\bar{x}) \neq \emptyset$. Then $y \in P_C(\bar{x})$ if and only if y minimizes $f(x) = \|x - \bar{x}\| + I_C(x)$, if and only if $0 \in \partial(\|y - \bar{x}\| + I_C(y)) = \partial(\|y - \bar{x}\|) + N_C(y)$. Thus $y \in P_C(\bar{x})$ if and only if there exists x^* such that $\langle x^*, c - y \rangle \leq 0 \; \forall c \in C$ and $-x^* \in \delta(y - \bar{x})$, i.e., $x^* \in \delta(\bar{x} - y)$. □

Exercise 10.4.2 Let $X = \mathbb{R}^2$, equipped with the following norm: $\|(x, y)\| = \max\{|x|, |y|\}$. Project $(0, 2)$ on the unit ball, and observe that $(1, 1)$ is one projection. Prove that $\delta((0, 2) - (1, 1)) = \{(x, y) : y = x + 1, -1 \leq x \leq 0\}$. However, only $x^* = (0, 1)$ satisfies (10.7).

Clearly, the structure of the set $P_C(x)$ is related to the geometry of the space. We now want to investigate this fact in more detail, and we start by providing a useful definition.

Definition 10.4.3 A Banach space X, normed by $\| \cdot \|$, is said to be *strongly smooth* if the function $\| \cdot \|$ is Fréchet differentiable outside the origin.

It follows that in a strongly smooth Banach space $(\| \cdot \|)'(x) = \delta(x)$. And as a corollary of Proposition 10.4.1, we can state:

Corollary 10.4.4 *Let X be a reflexive Banach space such that $\| \cdot \|$ is Gâteaux differentiable outside the origin, let C be a closed convex set and let $x \notin C$. Then $y = p_C(x)$ if and only if $y \in C$ and*

$$\langle \delta(x - y), c - y \rangle \leq 0,$$

$\forall c \in C.$

Proof. We give a simple, alternative proof of the statement. Take any $c \in C$, let $y \in C$ and let $g \colon [0, 1] \to \mathbb{R}$ be so defined:

$$g(s) = \|x - y + s(y - c)\|.$$

Clearly, $y = p_C(x)$ if and only if 0 minimizes g. And 0 minimizes g if and only if $g'(0) \geq 0$. Moreover,

$$0 \leq g'(0) = \langle (\| \cdot \|)'(x - y), y - c \rangle = \langle \delta(x - y), y - c \rangle.$$

$\square$

Now we want to analyze the Tykhonov well-posedness of the best approximation problem. As already remarked, this is a minimization problem, and thus the well-posedness machinery can be applied. The interesting result is that not only are the existence and uniqueness of the projection point related to properties of the Banach space X, but also Tykhonov well-posedness. And the properties on X characterizing Tykhonov well-posedness also have a characterization in terms of the dual space X^*.

Definition 10.4.5 A Banach space X is said to be an *E-space* if
(i) X is reflexive;
(ii) X is *strictly convex*:

$$x \neq y, \|x\| = \|y\| = 1 \text{ implies } \|ax + (1 - a)y\| < 1 \; \forall a \in (0, 1);$$

(iii) $x_n \rightharpoonup x$ and $\|x_n\| \to \|x\|$ imply $x_n \to x$ (the Kadeč–Klee property).

X is said to be strictly convex when it fulfills property (ii), as this implies that the boundary of the unit ball, and so of all balls, does not contain line segments; the Kadeč–Klee property is instead equivalent to

$$\|x_n\| = \|x\| = 1, x_n \rightharpoonup x \Longrightarrow x_n \to x.$$

Let us see now the following fundamental theorem:

Theorem 10.4.6 *Let X be a Banach space. Then the following are equivalent:*

(i) X is an E-space;
(ii) $\forall x_0 \in X, \forall C \subset X$, C a closed convex set, the problem of minimizing the function $c \mapsto \|c - x_0\|$ over C is Tykhonov well-posed;
(iii) $\forall x^* \in X^*$ such that $\|x^*\|_* = 1$, the problem of minimizing $\|x\|$ over the set $C := \{c \in X : \langle x^*, c \rangle = 1\}$ is Tykhonov well-posed;
(iv) $\forall 0 \neq x^* \in X^*$, the problem of minimizing

$$I_B(\cdot) - \langle x^*, \cdot \rangle$$

   is Tykhonov well-posed;
(v) X^* is strongly smooth.

Proof. Step 1. Let us start by proving that (i) implies (ii). The existence of the projection point is an easy consequence of the reflexivity assumption. Let us show uniqueness. Suppose c_1, c_2 minimize $\|\cdot - x_0\|$ on C. Then

$$x = \frac{c_1 - x_0}{\|c_1 - x_0\|}, \quad y = \frac{c_2 - x_0}{\|c_1 - x_0\|}$$

are norm one elements. If $x \neq y$, then

$$1 > \left\| \frac{1}{2}(x + y) \right\| = \frac{\left\| \frac{c_1 + c_2}{2} - x_0 \right\|}{\|c_1 - x_0\|},$$

but this is impossible. Thus $x = y$ and $c_1 = c_2$. To conclude, let us prove the convergence of the minimizing sequences. Let $\{x_n\} \subset C$ be such that $\|x_n - x_0\| \to \|\bar{x} - x_0\|$, where $\bar{x}$ denotes the projection of x_0 over C. It is easy to verify that $x_n \rightharpoonup \bar{x}$, whence $x_n - x_0 \rightharpoonup \bar{x} - x_0$. From the Kadeč–Klee property we deduce $x_n - x_0 \to \bar{x} - x_0$, hence $x_n \to \bar{x}$.

Step 2. (ii) implies (iii). This is obvious.

Step 3. Let us now prove that (iii) implies (i). First, let us observe that for every $x^* \in X^*$ such that $\|x^*\|_* = 1$, one has that

$$\inf\{\|x\| : \langle x^*, x \rangle = 1\} = 1.$$

To prove this, it is enough to produce a sequence $\{y_n\}$ such that $\langle x^*, y_n \rangle = 1$ for all n and $\|y_n\| \to 1$. So, let $\{x_n\} \subset B$ be such that $\langle x^*, x_n \rangle \to 1$ and let $y \in X$ be such that $\langle x^*, y \rangle = 2$. Let $a_n \in (0, 1)$ such that $\langle x^*, a_n x_n + (1 - a_n)y \rangle = 1$, for all n. As $a_n \to 1$ we get

$$\limsup \|a_n x_n + (1 - a_n)y\| \leq \limsup(a_n \|x_n\| + (1 - a_n)\|y\|) = 1.$$

Let us now show that X must be a reflexive space. As the problem of minimizing $\|\cdot\|$ over $\{\langle x^*, \cdot \rangle = 1\}$ is well-posed, then it has a solution x. Hence for all $x^* \in X^*$ such that $\|x^*\|_* = 1$, there exists $x \in X$ such that $\|x\| = 1$ and $\langle x^*, x \rangle = 1$. By a theorem of James this implies that X is reflexive. Let us suppose now that there are $x \neq y$ such that $\|x\| = \|y\| = \|\frac{x+y}{2}\| = 1$.

Let $z^* \in X^*$ be such that $\|z^*\|_* = 1$ and $\langle z^*, \frac{x+y}{2} \rangle = 1$. As $\langle z^*, x \rangle \leq 1$ and $\langle z^*, y \rangle \leq 1$, then $\langle z^*, x \rangle = 1$ and $\langle z^*, y \rangle = 1$. Hence x, y are two distinct solutions of the problem of minimizing $\|x\|$ over the set $\{\langle z^*, \cdot \rangle = 1\}$, and this is contrary to the assumptions. To conclude, let us show that the Kadeč–Klee property must hold. Let $x_n \rightharpoonup x$ and $\|x_n\| = \|x\| = 1$. Let $x^* \in X^*$ be such that $\langle x^*, x \rangle = 1$ and $\|x^*\|_* = 1$. Let $y_n = x_n + (1 - \langle x^*, x_n \rangle)x$. Then $\langle x^*, y_n \rangle = 1$ and $\limsup \|y_n\| \leq \limsup(1 + 1 - \langle x^*, x_n \rangle) = 1$. As the problem of minimizing $\|\cdot\|$ over $\{\langle x^*, \cdot \rangle = 1\}$ is Tykhonov well-posed, with solution x, and $\{y_n\}$ is a minimizing sequence, then $y_n \to x$, and so $x_n \to x$.

Step 4. We now prove that (iv) and (v) are equivalent. Setting $f(x) = I_B(x)$, we have that $f^*(x^*) = \|x^*\|_*$. Then, from (v) we have that f^* is Fréchet differentiable at $0^* \neq x^*$ and thus, by the Asplund–Rockafellar theorem (see Theorem 10.1.11), $f(\cdot) - \langle x^*, \cdot \rangle$ is Tykhonov well-posed. And conversely.

Step 5. We now show that (i) implies (iv). First, observe that it is enough to show the claim only if $\|x^*\|_* = 1$. And also we have that

$$\inf I_B(\cdot) - \langle x^*, \cdot \rangle = -1.$$

Now, reflexivity implies that the problem has a solution, say x. Suppose y is another solution. They satisfy $\langle x^*, x \rangle = \langle x^*, y \rangle = 1$, thus $\langle x^*, \frac{1}{2}(x+y) \rangle = 1$, and this in particular implies $\|\frac{1}{2}(x+y)\| = 1$. It follows that $x = y$, by strict convexity. Finally, let $\{x_n\}$ be a minimizing sequence. Clearly, it weakly converges to x. Moreover, as $\langle x^*, x_n \rangle \to 1$, then $\|x_n\| \to 1 = \|x\|$, and so, by the Kadeč–Klee property, $x_n \to x$.

Step 6. To conclude the proof of the theorem, we prove that (iv) implies (iii). First, observe that the minimum problems in (iii) and (iv). have always the same (possibly empty) set. For, any solution x of (iv) satisfies $\langle x^*, x \rangle = 1$ and $\|x\| = 1$. Thus $x \in C$ and since every element of C must have norm of at least one, x actually solves the problem in (iii), and vice-versa. Thus, by (iv) we have existence and uniqueness of the solution of the problem in (iii). Now, let $\{x_n\}$ be a minimizing sequence for the problem in (iii). Thus $\langle x^*, x_n \rangle = 1$ and also $\|x_n\| = 1$. Let $y_n = \frac{x_n}{\|x_n\|}$. Then $I_B(y_n) - \langle x^*, y_n \rangle \to -1$ and thus $\{y_n\}$ is a minimizing sequence for the problem in (iv). Thus, by Tykhonov well-posedness, $y_n \to x$. This implies $x_n \to x$ and the proof is complete. $\square$

Corollary 10.4.7 *Let X be a reflexive and strictly convex Banach space. If, for all C_n, C*

$$C_n \xrightarrow{M} C \implies p_{C_n}(0) \to p_C(0),$$

then X has the Kadeč–Klee property.

Proof. From Theorem 10.2.18 we know that the best approximation problem is Tykhonov well-posed for every closed convex set C. We conclude, by appealing to Theorem 10.4.6. $\square$

We now consider the stability of the projection problem. First, we keep fixed the closed convex set C and we move the point x to be projected on C. Later, we keep x fixed and we perturb C. A first result can be given when X is a Hilbert space.

Proposition 10.4.8 *Let X be a Hilbert space, let C be a closed convex subset of X. Then the map $x \mapsto p_C(x)$ is 1-Lipschitz.*

Proof. From

$$\langle x - p_C(x), p_C(y) - p_C(x) \rangle \leq 0$$

and

$$\langle y - p_C(y), p_C(x) - p_C(y) \rangle \leq 0$$

we get

$$\langle x - p_C(x) + p_C(y) - y, p_C(y) - p_C(x) \rangle \leq 0,$$

and this in turn implies

$$\|p_C(x) - p_C(y)\|^2 \leq \langle x - y, p_C(x) - p_C(y) \rangle,$$

from which the result easily follows.   □

On the other hand, there exist examples in which X is reflexive and strictly convex (thus $p_C(x)$ is well defined for every closed convex set C), C is a linear subspace and $x \mapsto p_C(x)$ is not continuous. However, here the Kadeč–Klee property plays a role.

Proposition 10.4.9 *Let X be an E-space, let C be a closed convex subset of X. Then the map $x \mapsto p_C(x)$ is continuous.*

Proof. Let $x \in X$ and let $\{x_n\}$ be a sequence such that $x_n \to x$. Then

$$\|x - p_C(x)\| \leq \|x - p_C(x_n)\| \leq \|x - x_n\| + \|x_n - p_C(x_n)\|$$
$$\leq \|x - x_n\| + \|x_n - p_C(x)\| \leq 2\|x - x_n\| + \|x - p_C(x)\|.$$

It follows that

$$\|x - p_C(x_n)\| \to \|x - p_C(x)\|.$$

Thus $\{p_C(x_n)\}$ is a minimizing sequence for the Tykhonov well-posed problem of projecting x over C, from which the conclusion follows.   □

We now turn our attention to the second problem, i.e., we perturb the set C where the given point x is to be projected. A first result is the following proposition, whose proof is given here for easy reference. The result however follows from a previous one.

Proposition 10.4.10 *Let X be a reflexive Banach space, let $x \in X$ and let $\{C_n\} \subset C(X)$ be such that $C_n \xrightarrow{M} C$. Finally, let $y_n \in p_{C_n}(x)$. Then $\{y_n\}$ is bounded and any weak limit y of $\{y_n\}$ is a projection of x over C.*

Proof. Fix an arbitrary $c \in C$. Then there exists $c_n \in C_n$ such that $c_n \to c$. Thus

$$\|y_n - x\| \le \|c_n - x\| \le \|c - x\| + 1$$

eventually, and this shows at first that $\{y_n\}$ is bounded. Now, let $\{y_j\}$ be a subsequence of $\{y_n\}$ with $y_j \in C_j$ for all j and $y_j \rightharpoonup y$. Then $y \in C$ by Mosco convergence. Moreover,

$$\|y - x\| \le \liminf \|y_j - x\| \le \liminf \|c_j - x\| \le \|c - x\|,$$

and this completes the proof. ☐

A first result on the connection between Mosco convergence of sets and convergence of projections is:

Theorem 10.4.11 *Let X be an E-space and let $C_n, C \subset X$ be closed convex subsets of the Banach space X. If X is an E-space, then $C_n \overset{\mathrm{M}}{\to} C$ implies $p_{C_n}(x) \to p_C(x), \forall x \in X$. If X is strongly smooth, then $p_{C_n}(x) \to p_C(x), \forall x \in X$ implies $C_n \overset{\mathrm{M}}{\to} C$.*

Proof. Suppose $C_n \overset{\mathrm{M}}{\to} C$. From Theorem 10.4.6 we know that setting $f_x(y) = \|x - y\|$, the problem of minimizing f over any closed convex set K is Tykhonov well-posed. Then, from Theorem 10.2.14, we get that $p_{C_n}(x) \to p_C(x)$. Conversely, suppose X is strongly smooth, let $\{C_n\} \subset C(X)$, $C \in C(X)$ and suppose that, for all $x \in X$, $p_{C_n}(x) \to p_C(x)$. Let $c \in C$. Since $p_{C_n}(c) - c \to p_C(c) - c = 0$, then $p_{C_n}(c) \to c$ and this shows that $C \subset \mathrm{Li}\, C_n$. Now, suppose $x_k \in C_{n_k}$, where $\{n_k\}$ is a subsequence of the integers, and $x_k \rightharpoonup x$. We must prove that $x \in C$. From Proposition 10.4.4 we have that

$$\langle \delta(x - p_{C_{n_k}}(x)), x_k - p_{C_{n_k}}(x) \rangle \le 0.$$

On the other hand, as $\| \cdot \|$ is Fréchet differentiable outside the origin, from Corollary 3.5.8 we know that it is actually C^1 on this open set, so that δ is continuous. Thus we can pass to the limit in the above relation, to get

$$0 \ge \langle \delta(x - p_C(x)), x - p_C(x) \rangle = \|x - p_C(x)\|.$$

Thus $x \in C$ and this completes the proof. ☐

Exercise 10.4.12 Let X be an E-space, and $\bar{x} \in X$. Let $\mathcal{A} = C(X)$, endowed with a distance compatible with the Mosco topology. For $a = C$, let $f_a(x) = \|x - \bar{x}\| + I_C(x)$. Prove that a is well-posed for all a.

Let X be a Hilbert space and let $\mathcal{A} = X \times C(X)$, with $C(X)$ endowed with a distance compatible with the Mosco topology. For $a = (\bar{x}, C)$, let $f_a(x) = \|x - \bar{x}\| + I_C(x)$. Prove that a is well-posed for all a.

Now we extend the result given in Proposition 8.3.5 on the connections between Mosco and Wijsman convergence on the closed convex subsets of a Banach space X. To prove it, we need an auxiliary result which is interesting in itself. It deals with differentiability of the distance function (compare it with Proposition 4.1.5 and the following Theorem 10.4.15).

Proposition 10.4.13 *Let X be a reflexive, strictly convex and strongly smooth Banach space. Let C be a closed convex set and consider the function $f(x) := d(x, C)$. Then f is Fréchet differentiable at every $x \notin C$ and $f'(x) = \delta(x - p_C(x))$.*

Proof. First, observe that $p_C(z)$ is a singleton for every $C \in C(X)$ and $z \in X$. Moreover, as $\partial(\|x\|) = \delta(x)$, we have that

$$\|y - p_C(y)\| \geq \|x - p_C(x)\| + \langle \delta(x - p_C(x)), y - p_C(y) - (x - p_C(x)) \rangle.$$

Since

$$\langle \delta(x - p_C(x)), p_C(x) - p_C(y) \rangle \geq 0,$$

we then get

$$d(y, C) \geq d(x, C) + \langle \delta(x - p_C(x)), y - x \rangle. \tag{10.8}$$

Moreover, from Fréchet differentiability of the norm,

$$d(y, C) = \|y - p_C(y)\| \leq \|y - p_C(x)\| \tag{10.9}$$
$$= d(C, x) + \langle \delta(x - p_C(x)), y - x \rangle + \varepsilon_y \|y - x\|,$$

where $\varepsilon_y \to 0$ when $y \to x$. Combining (10.8) and (10.9) we get the claim. $\square$

Theorem 10.4.14 *Let X be a separable, strongly smooth E-space and let $C_n, C \subset C(X)$. Then the following are equivalent:*

(i) $C_n \overset{M}{\to} C$;

(ii) $C_n \overset{W}{\to} C$;

(iii) $p_{C_n}(x) \to p_C(x)$, *for all $x \in X$.*

Proof. We already know that (i) implies (ii) in any Banach space. We also know that (i) is equivalent to (iii) if X is an E-space. It is then enough to prove that (ii) implies (i), and to do this we only need to see that if $c_k \in C_{n_k}$ for all k and $c_k \rightharpoonup x$, then $x \in C$ (remember that the lower part of the convergences are always the same). We have that for all $y \in X$,

$$d(y, C_{n_k}) \geq d(x, C_{n_k}) + \langle \delta(x - p_{C_{n_k}}(x)), y - x \rangle. \tag{10.10}$$

The sequence $\{\delta(x - p_{C_{n_k}}(x)\}$ is norm one. Let z be any of its limit points. Passing to the limit in (10.10) yields

$$d(y, C) \geq d(x, C) + \langle z, y - x \rangle,$$

for all $y \in X$. Thus $z = \delta(x - p_C(X))$ is norm one and so, since X^* is an E-space (by assumption X is strongly smooth), finally we have

$$\delta(x - p_{C_{n_k}}(x)) \to \delta(x - p_C(x)).$$

Setting $y = c_k$ in (10.10), we have

$$0 \geq d(x, C_{n_k}) + \langle \delta(x - p_{C_{n_k}}(x)), c_k - x \rangle.$$

Passing to the limit, we finally get $d(x, C) = 0$ and this ends the proof. $\square$

To conclude this chapter, we provide a formula for the subdifferential of the function $d(\,\cdot\,,C)$, where C is a closed convex set of a general Banach space. We have already seen a formula valid on reflexive spaces. In this case, the existence of the projection on C of any point outside C simplifies the calculations. Unfortunately, in a nonreflexive Banach space the projection of a point outside C does not always exist. Before establishing the result, let us collect the main properties of this nice function. It is convex, 1-Lipschitz, it can be written as an inf-convolution, and if the norm in X is sufficiently smooth, it is Fréchet differentiable outside C (this last result is Proposition 10.4.13).

Theorem 10.4.15 *Let X be a Banach space, let $C \subset X$ be a nonempty closed set. Then*

$$\partial d(\,\cdot\,,C)(x) = \begin{cases} N_{\hat{C}}(x) \cap \partial B_{X^*} & \text{if } x \notin C, \\ N_C(x) \cap B_{X^*} & \text{if } x \in C, \end{cases}$$

where $\hat{C} = \{z \in X : d(z,C) \leq d(x,C)\}$.

Proof. First, let us suppose $x \notin C$. Observe that $\hat{C}$ is nothing other than the level set at height $0 < a = d(x,C)$ of the function $d(\,\cdot\,,C)$. From Theorem 4.3.11 we know that $N_{\hat{C}}(x) = \text{cone}\,\partial d(\,\cdot\,,C)(x)$. Thus the result is established if we prove that $x^* \in \partial d(\,\cdot\,,C)(x)$, $x \notin C$, imply $\|x^*\|_* = 1$. Since $d(\,\cdot\,,C)$ is 1-Lipschitz, we immediately have $\|x^*\|_* \leq 1$. Now, fix $\varepsilon > 0$ and take $c \in C$ such that $d(x,C) + \varepsilon \geq \|x - c\|$. Since $x^* \in \partial d(\,\cdot\,,C)(x)$, the following inequality holds.

$$0 \geq d(x,C) + \langle x^*, c - x \rangle \geq \|c - x\| - \varepsilon + \langle x^*, c - x \rangle.$$

Thus

$$\|x^*\|_* \geq 1 - \frac{\varepsilon}{a},$$

and this provides the required inequality, since $\varepsilon > 0$ is arbitrary. Now, let $x \in C$. It is not difficult to see that $\partial d(\,\cdot\,,C)(x) \subset N_C(x) \cap B_{X^*}$. Next, suppose $x^* \in N_C(x)$ and $\|x^*\|_* \leq 1$. The relation to be proved,

$$d(z,C) \geq \langle x^*, z - x \rangle,$$

is clearly nontrivial only if $\langle x^*, z - x \rangle > 0$. Set

$$H = \{w \in X : \langle x^*, w \rangle = \langle x^*, x \rangle\}, \quad H_- = \{w \in X : \langle x^*, w \rangle \leq \langle x^*, x \rangle\}.$$

Observe that $C \subset H_-$, and thus $d(z,C) \geq d(z,H)$. Now we show $d(z,H) \geq \langle x^*, z-x \rangle$ and this will end the proof. Every $h \in H$ can be written as $h = x+v$, with v such that $\langle x^*, v \rangle = 0$. Thus

$$d(z,H) = \inf_{h \in H} \|z - h\| = \inf_{v : \langle x^*, v \rangle = 0} \|z - (x + v)\|.$$

From this we conclude, since $\|x^*\|_* \leq 1$ and so

$$\|z - x - v\| \geq \langle x^*, z - x \rangle,$$

for all v such that $\langle x^*, v \rangle = 0$.   □

Generic well-posedness

You may say I'm a dreamer,
but I'm not the only one.
(J. Lennon, "Imagine")

Given a generic function $f: X \to \mathbb{R}$ to be minimized, when discussing the Weierstrass existence theorem we have argued that, without a topology on X rich enough in both closed and compact sets at the same time, it is impossible to give a general result of existence of minima. We also saw that in such a case it is important to establish that, in a given class of problems, "many" of them have existence and further properties, like some form of well-posedness. In this chapter we want to look at this topic, which has been widely studied in recent years, with results that are still in progress. The subject has very many aspects, thus we have to make a choice. For instance, we have to specify what we mean by "many" problems in a class. A first idea could be that in a given class a dense subset contains many elements. A set could also be considered "big" in the Baire category sense, or when a concept of measure is available, if its complement has null measure. It is enough to consider the idea of Baire category in order to understand that one can think of many different types of results, since different topologies, even if comparable, give rise to noncomparable results. Indeed, a set (in a Baire space) is declared big (of second category) if it contains a dense G_δ set, and small (of first category) if its complement is big. We remember that G_δ means a countable intersection of open sets (which, by definition, is nonempty in a Baire space), and thus, taking a finer topology we get a less strong result. On the other hand, denseness goes in exactly the opposite way. Thus we are forced to select results, as a complete overview will probably require a whole book. We shall focus mainly on sets of problems, usually described by functions to be minimized, and we consider the topology of uniform convergence on bounded sets, or similar ones, like the Attouch–Wets. Occasionally, we shall mention other topologies or problems differently described, for instance constrained problems described by a pair (set, function). More importantly, we shall appeal to a relatively recent concept of smallness, called σ-porosity, which has a very interesting feature. In any Baire space the complement of a σ-porous set is of second category, and in any Euclidean space a σ-porous set is of null measure. This is important, as it is well known that there are small sets in the Baire category sense which are of full measure. Actually, it can be shown that there are sets which are of first category and of null measure at the same time, yet they are not σ-porous. Thus, we shall dedicate the first section of this chapter to the illustration of this concept.

11.1 Porosity

We now introduce two notions of porosity (and related σ-porosity), which are both used in the subsequent results. The second one requires a stronger property, but it is probably easier to understand, and it is used in all results but one. Let us, in any case, underline that as we shall see, even σ-porosity in the weaker sense enjoys the properties we mentioned before. The reference paper for this concept is [Zaj].

Here is the first definition.

Definition 11.1.1 Let M be a metric space and let $A \subset M$. Let $x \in M$, $R > 0$, and denote by $\sigma(x, A, R)$ the supremum of all $r > 0$ such that there exists $z \in M$ such that $B(z; r) \subset B(x; R) \setminus A$. The number

$$\limsup_{R \to 0} \frac{\sigma(x, A, R)}{R}$$

is called the *porosity* of A at x. A set A is said to be *porous* if the porosity at x is positive for every $x \in A$. A set A is called σ-*porous* if it is a countable union of porous sets.

Example 11.1.2 We shall consider some subsets of the real line in order to study their porosity. First, let us observe that, of course, every isolated point of A is a porosity point for A. Now, let us consider at first $A = \{0 \cup \bigcup_{n \geq 1} \frac{1}{n}\}$. Is it a porous set? Of course, we must check porosity only at 0. But clearly the porosity of A at 0 is positive since A does not have elements on the left with respect to 0. Consider now

$$A = \left\{ 0 \cup \bigcup_{n \geq 1} \frac{1}{n} \cup \bigcup_{n \geq 1} \frac{-1}{n} \right\}.$$

Again, we must check porosity only at 0. Take a sequence $\{R_k\}$ such that $R_k \to 0$ and fixing k, let n be such that $\frac{1}{n+1} \leq R_k < \frac{1}{n}$. Then

$$\sigma(0, A, R_k) \leq \frac{1}{n} - \frac{1}{n+1},$$

which implies

$$\frac{\sigma(0, A, R_k)}{R_k} \leq \frac{1}{n},$$

and this implies that A is not porous at 0. Finally, consider

$$A = \left\{ 0 \cup \bigcup_{n \geq 1} -\frac{1}{e^n} \cup \bigcup_{n \geq 1} \frac{1}{e^n} \right\}.$$

By choosing $R_n = \frac{1}{e^n}$, it is easily seen that

$$\frac{\sigma(0, A, R_k)}{R_k} = \frac{e - 1}{2e},$$

and this shows that A is a porous set.

The previous examples highlight the importance of making accurate estimations in evaluating the porosity of a set at a given point. They also show that a porous set can have a nonporous closure, and that the union of two porous sets need not be porous.

Now we want to see that a σ-porous set is really small. Let us recall the following definition.

Definition 11.1.3 A set A in the metric space M is said to be *nowhere dense* if

$$\forall x \in M, \forall R > 0 \; \exists y \in M, r > 0, \; B(y; r) \subset B(x; R) \setminus A.$$

A set A is called *meager*, or *a first category set* if it is a countable union of nowhere dense sets.

In other words, the set A is nowhere dense if every ball in the space contains a ball not meeting A.

We now prove that a porous set is nowhere dense; this immediately implies that a σ-porous set is of first category.

Proposition 11.1.4 *Let A be a porous set. Then A is nowhere dense.*

Proof. Fix $x \in M$ and $R > 0$. Suppose, without loss of generality, $B(x; R) \cap A \neq \emptyset$. Take $z \in B(x; R) \cap A$ and $\bar{R} > 0$ such that $B(z; \bar{R}) \subset B(x; R)$. Since A is porous at z, there are $R_n \to 0$ and $q > 0$ such that

$$\frac{\sigma(z, A, R_n)}{R_n} > q.$$

Fix n so large that $R_n < \bar{R}$ and $\sigma(z, A, R_n) > qR_n$. By porosity, there is $r(> qR_n)$ such that

$$B(y; r) \subset B(z; R_n) \setminus A \subset B(z; \bar{R}) \setminus A \subset B(x; R) \setminus A.$$

$\square$

A porous set A is small also from the point of view of Lebesgue measure. To see this, let us suppose that A is a subset of the metric space $\mathbb{R}^m$, and let us start by considering the following definition.

Definition 11.1.5 Let $x \in \mathbb{R}^m$. Then x is said to be an *outer density point* for A if
$$\lim_{R \to 0} \frac{m^*[B(x;R)] \cap A}{m[B(x;R)]} = 1,$$
where m^* denotes the outer measure of a set.

Recall the *density theorem* by Lebesgue: for an arbitrary set A, the set of all points of A which are *not* density points is of null measure. Thus the following proposition will imply that a σ-porous set is of null measure.

Proposition 11.1.6 *Let A be a set which is porous at a given point x. Then x is not an outer density point for A.*

Proof. Remember that there is $c > 0$ such that $m[B(x;R)] = cR^m$. Since A is porous at x, there are $R_n \to 0$ and $q > 0$ such that
$$\frac{\sigma(x, A, R_n)}{R_n} > q.$$

Then eventually there exists z_n such that $B(z; qR_n) \subset B(x; R_n) \setminus A$. It follows that
$$\lim_{R_n \to 0} \frac{m^*[B(x;R_n)] \cap A}{m[B(x;R_n)]} \leq \lim_{R_n \to 0} \frac{cR_n^m - c(qR_n)^m}{cR_n^m} = 1 - q^m.$$

$\square$

Definition 11.1.7 Let (M, d) be a metric space and $A \subset M$. The set A is called strongly porous in M if there are $\lambda \in (0,1)$ and $r_0 > 0$ such that for any $x \in M$ and $r \in (0, r_0)$ there is $y \in M$ such that $B(y; \lambda r) \subset B(x; r) \setminus A$. A is called *strongly σ-porous in M* if it is a countable union of porous sets in M.

Observe that no set in Example 11.1.2 is strongly porous. Clearly, a strongly (σ-)porous set is also a (σ-)porous set. Thus the properties of the (σ-)porous sets are fulfilled, obviously, by the strongly (σ-)porous sets. Moreover, the following proposition holds.

Proposition 11.1.8 *The set A is strongly porous in M if there are $\lambda \in (0,1)$ and $r_0 > 0$ such that for any $a \in A$ and $r \in (0, r_0)$, there is $y \in M$ such that $B(y; \lambda r) \subset B(x; r) \setminus A$.*

Proof. In other words, the claim is that the required property needs to be checked *only* at the points of A. Thus, suppose we have λ fulfilling the property for all $a \in A$, and let us find $\bar{\lambda}$ and $\bar{r_0}$ fulfilling the property for all $x \in M$. We shall show that the choice of $\bar{r_0} = r_0$ and $\bar{\lambda} = \frac{\lambda}{2}$ works. Take $x \in M$ and suppose, without loss of generality, $B(x; \frac{1}{2}r) \cap A \neq \emptyset$. Let $y \in B(x; \frac{1}{2}r) \cap A$. Then $B(y; \frac{r}{2}) \subset B(x; r)$. By assumption, there is $z \in M$ such that

$$B\left(z; \frac{\lambda}{2}r\right) \subset B\left(y; \frac{r}{2}\right) \setminus A \subset B(x; r) \setminus A.$$

□

Corollary 11.1.9 *A is strongly porous if and only if $\bar{A}$ is strongly porous.*

11.2 Some observations on concave/convex functions

In this section we see some properties of concave/convex functions, having in mind the fact that in some problems of convex programming the value function is of this type. Since we are interested in well-posedness of convex programs, and this involves the study of the associated value function, we shall concentrate on results useful for this scope.

So, let U, V be Banach spaces and $h \colon U \times V \to [-\infty, \infty]$ be a given function. We suppose that

- $h(\cdot, v)$ is concave for all $v \in V$;
- $h(u, \cdot)$ is convex for all $u \in U$.

Set

$$\text{dom } h = \{(u, v) : |h(u, v)| < \infty\}.$$

Dom h is called the effective domain of h. First, let us observe that dom h need not be convex. The set $\{x : |f(x)| < \infty\}$ for a convex function taking also value $-\infty$ need not be convex. However, it is possible to prove the following:

Proposition 11.2.1 *Suppose* int dom $h \neq \emptyset$. *Then there are open sets $A \subset X$, $B \subset Y$ such that* int dom $h = A \times B$.

Proof. It is enough to prove that if $(a, b) \in$ int dom h and $(c, d) \in$ int dom h, then also $(a, d) \in$ int dom h. There is $\varepsilon > 0$ such that

$$\{(u, v), (w, z) : \|a - u\| < \varepsilon, \|b - v\| < \varepsilon, \|c - w\| < \varepsilon, \|d - z\| < \varepsilon\} \subset \text{int dom } h.$$

We claim that

$$\{(s, t) : \|c - s\| < \varepsilon, \|b - t\| < \varepsilon\} \subset \text{dom } h.$$

Suppose not. Then there is an element $(s, t) : \|c - s\| < \varepsilon, \|b - t\| < \varepsilon$ and $|h(s, t)| = \infty$. Suppose $h(s, t) = -\infty$. Consider the convex function $h(s, \cdot)$. Then $h(s, t) = -\infty$, $h(s, d) \in \mathbb{R}$. Thus it must be that $h(s, \lambda t + (1 - \lambda)d) = -\infty$ for all $\lambda \in [0, 1)$, a contradiction. The case $h(s, t) = -\infty$ can be seen in the same way. □

Proposition 11.2.2 *Suppose h is lower and upper bounded around a point $(\bar{u}, \bar{v})$ in int dom h. Then h is locally Lipschitz around $(\bar{u}, \bar{v})$.*

Proof. Suppose h is lower and upper bounded on a ball B centered at $(\bar{u}, \bar{v})$. As $h(u, \cdot)$ is a convex function lower and upper bounded on a neighborhood of $\bar{v}$, and for all u in a neighborhood of $\bar{u}$, then there is a constant $k > 0$ such that

$$|h(u, v) - h(u, w)| \leq k\|v - w\|,$$

for v, w in a suitable neighborhood of $\bar{v}$, for all $u \in B$. The constant k can be chosen independently from u, since it can be chosen only in dependence of the upper and lower bounds of the function h in B (see Lemma 2.1.8). In exactly the same way, we see that there is a constant, which we continue to call k, such that

$$|h(u, v) - h(t, v)| \leq k\|t - u\|,$$

for t, u in a suitable neighborhood of $\bar{u}$ and for all v in some suitable neighborhood of $\bar{v}$. Thus, for t, u in a suitable neighborhood of $\bar{u}$ and v, w in a suitable neighborhood of $\bar{v}$, we have

$$|h(u, v) - h(t, w)| \leq k(\|u - t\| + \|v - w\|).$$

□

Definition 11.2.3 The *subdifferential* of a concave/convex function h is defined to be

$$\partial h(x, y) = \{(p, q) : p \in \partial(-h)(\cdot, y)(x), q \in \partial h(x, \cdot)(y)\}.$$

Exercise 11.2.4 Prove that ∂h is a maximal monotone operator from $X \times Y$ into $X^* \times Y^*$.

The next result concerns the points of Fréchet differentiability of a concave/convex function h as above, and it is a generalization of the same result of Preiss–Zajíček [PZ] for convex functions.

Theorem 11.2.5 *Let X and Y be Banach spaces with separable duals. Let $A \subset X$ and $B \subset Y$ be open convex sets, and let h be a continuous concave/convex function on $A \times B$. Then the collection of $(x, y) \in A \times B$ such that either $h(\cdot, y)$ is not Fréchet differentiable at x or $h(x, \cdot)$ is not Fréchet differentiable at y is σ-porous.*

Proof. Set

$$\mathcal{A} = \{(x, y) \in A \times B : h(\cdot, y) \text{ is not Fréchet differentiable at } x\},$$
$$\mathcal{B} = \{(x, y) \in A \times B : h(x, \cdot) \text{ is not Fréchet differentiable at } y\}.$$

We must show that both $\mathcal{A}$ and $\mathcal{B}$ are σ-porous. By symmetry, it is enough to show that $\mathcal{B}$ is σ-porous.

Since for every $x \in A$ $h(x, \cdot) \colon B \to \mathbb{R}$ is continuous, $\partial h(x, \cdot)(y) \neq \emptyset$ at every $y \in B$. Now fix any $(x, y) \in B$ and choose $Y^* \ni q_{xy} \in \partial h(x, \cdot)(y)$.

Since $h(x, \cdot)$ is not Fréchet differentiable at y, then

$$\limsup_{\|z\| \to 0} \left\| \frac{h(x, y + z) - h(x, y) - \langle q_{xy}, z \rangle}{\|z\|} \right\| > 0. \tag{11.1}$$

For any $n \in \mathbb{N}$ we set

$$\mathcal{B}_n = \left\{ (x, y) \in \mathcal{B}, \ \limsup_{\|z\| \to 0} \frac{(h(x, y + z) - h(x, y) - \langle q_{xy}, z \rangle}{\|z\|} > \frac{1}{n} \right\}.$$

Clearly $\mathcal{B} = \bigcup_n \mathcal{B}_n$, so it is enough to verify that each $\mathcal{B}_n$ is σ-porous. Since X^* is separable, we can find sets $\mathcal{B}_{nm}$ such that $\mathcal{B}_n = \bigcup_m \mathcal{B}_{nm}$ and

$$\|q_{xy} - q_{uv}\| \le \frac{1}{6n}$$

whenever $(x, y), (u, v) \in \mathcal{B}_{nm}$. For instance, take a dense sequence p_m in X^* and set

$$\mathcal{B}_{nm} = \left\{ (x, y) \in \mathcal{B}_n : \|q_{xy} - p_m\| \le \frac{1}{12n} \right\}.$$

We shall show that each $\mathcal{B}_{nm}$ is porous. Fix such a set, and let $(x, y) \in \mathcal{B}_{nm}$. As h is a concave/convex function which is continuous at (x, y), it is locally Lipschitz in a neighborhood of (x, y), that is there are $R > 0$ and $K > 0$ such that $|h(u, v) - h(u', v')| \le K(\|u - u'\| + \|v - v'\|)$ if $\|u - x\| \le R$, $\|u' - x\| \le R$, $\|v - y\| \le R$, $\|v' - y\| \le R$. So that $\|q_{xy}\| \le K$. It follows from (11.1) that we can find $\eta > 0$ and a sequence z_k with $r_k = \|z_k\| \to 0$ such that

$$h(x, y + z_k) - h(x, y) - \langle q_{xy}, z_k \rangle > \left(\frac{1}{n} + 2\eta \right) r_k,$$

for all k. Set $\delta_k = \frac{r_k \eta}{K}$. Now fix k. Then for any u such that $\|u - x\| < \delta_k$, we have

$$h(u, y + z_k) - h(u, y) - \langle q_{xy}, z_k \rangle > \frac{r_k}{n}. \tag{11.2}$$

Set $\lambda = \frac{1}{3Kn}$. We shall show that

$$B(x; \lambda \delta_k) \times B(y + z_k; \lambda r_k) \cap \mathcal{B}_{nm} = \emptyset \tag{11.3}$$

and this will end the proof. Assume the contrary: there exists $(u, v) \in \mathcal{B}_{nm}$ such that $\|x - u\| \le \lambda \delta_k$ and $\|v - (y + z_k)\| \le \lambda r_k$. This means in particular that

$$\|u - x\| \le \delta_k \text{ and } \|y - v\| < 1 + \lambda r_k.$$

Now observe that

$$|h(u, y + z_k) - h(u, v)| \le K\|(y + z_k) - v\| \le K\lambda r_k \le \frac{r_k}{3n}. \tag{11.4}$$

On the other hand, we have

$$h(u, y + z_k) - h(u, v) = h(u, y + z_k) - h(u, y) + h(u, y) - h(u, v) \text{ (by (11.2))}$$
$$> \langle q_{xy}, z_k \rangle + \frac{r_k}{n} + \langle q_{uv}, y - v \rangle = \langle q_{xy}, (y + z_k) - v \rangle$$
$$+ \frac{r_k}{n} + \langle q_{uv} - q_{xy}, y - v \rangle$$
$$\geq \frac{r_k}{n} - K\lambda r_k - \frac{1}{6n}(1 + \lambda r_k) = \frac{r_k}{3n}.$$

This last inequality contradicts (11.4), and thus (11.3) is proved.   □

11.3 Genericity results

In this section we want to show that in some classes of unconstrained and constrained (convex) minimum problems, most of the problems are well-posed, in one sense or another. Here "most" is intended in the Baire category sense. Variational principles play an important role in this context, as we shall see.

Let us start by seeing how the Ekeland variational principle can be used to get this type of results. For the convenience of the reader, we recall it here in the somewhat stronger version given in Exercise 10.1.10.

Proposition 11.3.1 *Let (X, ρ) be a complete metric space and let $f \colon X \to (-\infty, \infty]$ be a lower semicontinuous, lower bounded function. Let $\varepsilon > 0$, $r > 0$ and $\bar{x} \in X$ be such that $f(\bar{x}) < \inf_X f + r\varepsilon$. Then there exists $\hat{x} \in X$ enjoying the following properties:*

(i) $\rho(\hat{x}, \bar{x}) < r$;
(ii) $f(\hat{x}) < f(\bar{x}) - \varepsilon \rho(\bar{x}, \hat{x})$;
(iii) *the function $f(\cdot) + \varepsilon \rho(\hat{x}, \cdot)$ is Tykhonov well-posed.*

Condition (iii) above essentially states a density result for Tykhonov well-posed problems. Let us see this in an example. Consider the space $\mathcal{F}$ of the real valued, lower semicontinuous positive functions on the complete metric space (X, ρ), which is assumed to be unbounded. We endow $\mathcal{F}$ with a distance compatible with uniform convergence on bounded sets. For instance, fix a certain element $\theta \in X$, and set for any two $f, g \in \mathcal{F}$ and $n \in \mathbb{N}$,

$$\|f - g\|_n = \sup_{\rho(x, \theta) \leq n} |f(x) - g(x)|.$$

If $\|f - g\|_n = \infty$ for some n, then we set $d(f, g) = 1$. Otherwise,

$$d(f, g) = \sum_{n=1}^{\infty} 2^{-n} \frac{\|f - g\|_n}{1 + \|f - g\|_n}. \tag{11.5}$$

In such a way $(\mathcal{F}, d)$ is a complete metric space.

We can now state:

Proposition 11.3.2 *In $(\mathcal{F}, d)$ the set of the functions which are Tykhonov well-posed is dense.*

Proof. Fix $\sigma > 0$. Take j so large that setting $g(x) = f(x) + \frac{1}{j}\rho(x, \theta)$, then $d(f, g) < \frac{\sigma}{2}$. Now, observe that $\lim_{\rho(x,\theta)\to\infty} g(x) = \infty$, and thus there exists M such that $g^1 \subset B(\theta; M)$. Let $s = \sum \frac{1}{2^n}(n + M)$. Apply the principle with $\varepsilon = \frac{\sigma}{2s}$ (r arbitrary) to find $\hat{x}$ such that $\rho(\hat{x}, \theta) \leq M$ and $\hat{x}$ is the unique minimizer of

$$h(\,\cdot\,) = g(\,\cdot\,) + \varepsilon\rho(\,\cdot\,, \hat{x}).$$

Since $|h(x) - g(x)|_n \leq \varepsilon(n + M)$, it follows that $d(h, g) \leq \varepsilon s = \frac{\sigma}{2}$. Then $d(f, h) < \sigma$, and the proof is complete. $\qquad\square$

This is just an example of how to use the Ekeland principle to get such results. It is not difficult to imagine that the same line of reasoning can be made for other classes $\mathcal{F}$ of functions (including spaces of convex functions, since the perturbation term $\varepsilon\|x - \hat{x}\|$ keeps convexity), endowed with different hypertopologies, such as the Attouch–Wets, for instance.

But we want to get more than a density result. At least, we want to have that the Tykhonov well-posed problems are a big set in the Baire category sense. To do this, a very useful tool is the Furi–Vignoli criterion for Tykhonov well-posedness. We recall it here.

Proposition 11.3.3 *Let X be a complete metric space and let $f: X \to (-\infty, \infty]$ be a lower semicontinuous function. The following are equivalent:*

(i)   *f is Tykhonov well-posed;*

(ii) $\inf_{a > \inf f} \operatorname{diam} f^a = 0.$

Now, suppose we have a Baire space $(\mathcal{F}, d)$ of functions. Observe that setting

$$V_j = \left\{ f \in \mathcal{F} : \inf_{a > \inf f} \operatorname{diam} f^a < \frac{1}{j} \right\},$$

if we can prove that the sets V_j are open, then, by the Furi–Vignoli criterion, the Tykhonov well-posed problems are a G_δ set. In turn, openness will be a consequence of continuity (actually, lower continuity, but the upper part is usually for free) of the function $f \mapsto \operatorname{diam} f^a$. Applying then a density argument via the Ekeland variational principle, we are able to conclude that the Tykhonov well-posed problems are a second category set. Let us see some examples.

Theorem 11.3.4 *Let X be a Banach space and consider the set $\Gamma(X)$, equipped with the Attouch–Wets topology. Then most of the problems in $\Gamma(X)$ are Tykhonov well-posed.*

Proof. First, observe that $(\Gamma(X), AW)$ is topologically complete (see Theorem 8.4.10). Secondly, let us see that for the proof we can follow the idea described above. Proposition 10.2.21 guarantees that $f_n^a \xrightarrow{AW} f^a$, whenever $a > \inf f$, and Exercise 8.6.17 shows that the diam function is continuous with this convergence. Thus the sets V_n are open in $(\Gamma(X), AW)$. Now appeal to (a variant of) Proposition 11.3.2 to conclude that each V_n is a dense subset in $(\Gamma(X), AW)$.   □

It is interesting to observe that the same line of reasoning as before does not work for the weaker Mosco topology. This is because the diam function is not lower semicontinuous with Mosco convergence. Not only this, actually the result completely fails with the Mosco topology. In fact, one can prove:

Theorem 11.3.5 *If X is an infinite dimensional reflexive Banach space, then the family U of all functions which are unbounded from below is a dense G_δ set in $(\Gamma(X), M)$.*

Proof. Clearly $U = \bigcap_n U_n$, where

$$U_n := \{f : \exists x,\ f(x) < -n\} = \Gamma(X) \cap (X \times (-\infty, -n)^-),$$

which are open because of the definition of the Mosco topology. Thus U is clearly a G_δ set. Then it remains to show that U is dense in $(\Gamma(X), M)$. We prove it in the particular case of X being a (separable) Hilbert space, the general case can be seen in [BL]. So, let us approximate a given f by a sequence of functions which are unbounded from below. First of all, let us observe that we can suppose f is real valued, since the family of such functions is dense in $(\Gamma(X), M)$. Now, let $\{e_n : n \in \mathbb{N}\}$ be an orthonormal basis on X. Set

$$f_n(x) = f\left(\sum_{i=1}^{n} \langle x, e_i \rangle\right) - \frac{1}{n}\langle x, e_{n+1} \rangle.$$

Since $f_n(ke_{n+1}) = f(0) - \frac{k}{n}$, we see that no f_n is bounded from below. It is now routine to show that $f_n \xrightarrow{M} f$. For, $\lim f_n(x) = f(x)$ and if $x_n \rightharpoonup x$, then, setting $z_n = \sum_{i=1}^{n} \langle x, e_i \rangle e_i$, we have that $z_n \rightharpoonup x$, and thus

$$\liminf f_n(x_n) = \liminf f(z_n) \geq f(x).$$

    □

In the literature it is possible to find various results similar to that of Theorem 11.3.4, but we do not want to insist on this, mainly because we think the porosity results are more powerful and challenging, so that we shall go into more detail later on this kind of result.

The following result deals with problems with constraints.

Theorem 11.3.6 *Let X be a Banach space. Then in $(\Gamma(X) \times C(X), AW \times AW)$, the family of the pairs (A, f) such that f is continuous and (A, f) strongly well-posed, contains a dense and G_δ set.*

Proof. The first step consists in showing that inside $(\Gamma(X), AW)$, the family $\Gamma_c(X)$ of the functions which are everywhere continuous contain a G_δ dense set. This is seen in Exercise 11.3.7. Thus $\Gamma_c(X) \times C(X)$ contains a dense G_δ subset of $\Gamma(X) \times C(X)$. It is then sufficient to prove that

$$\{(f, A) \in \Gamma_c(X) \times C(X) : f + I_A \text{ is strongly well-posed}\}$$

is a dense G_δ set in $\Gamma_c(X) \times C(X)$. Actually, it is enough to see that

$$\{(f, A) \in \Gamma_c(X) \times C(X) : f + I_A \text{ is Tykhonov well-posed}\}$$

is a dense G_δ set in $\Gamma_c(X) \times C(X)$, thanks to Theorem 10.2.25. Now we can turn back to the arguments seen before. Setting

$$V_j = \left\{ f \in \mathcal{F} : \exists a > \inf f + I_A, \operatorname{diam}(f + I_A)^a < \frac{1}{j} \right\},$$

let us see that these sets are open. This follows from the continuity of the function $d \colon \Gamma_c(X) \times C(X) \to [0, \infty)$ such that $d(f, A) = \operatorname{diam}(f + I_A)^a$, which can be established with the help of the Theorem 9.2.5. To conclude the proof, density of Tykhonov well-posed problems follows once again from the Ekeland variational principle. $\qquad\square$

Exercise 11.3.7 Prove that in $(\Gamma(X), AW)$, the family $\Gamma_c(X)$ of the functions which are everywhere continuous contains a G_δ dense set.

Hint. Density follows from the fact that every function f in $(\Gamma(X), AW)$ can be approximated by its n-Lipschitz regularizations (see Theorem 9.2.11). Then, set

$$W_n = \{f \in \Gamma(X) : f \text{ is bounded above on } aB \text{ for some } a > n\},$$

and show that each W_n is open. (Take $f \in W_n$. Then there are a, b such that $f(x) \le a$ if $\|x\| \le n + 4b$. The ball in $X \times \mathbb{R}$ centered at $(0, a + n + 2b)$ and with radius $n + 2b$ is contained in epi f. Then there is an open AW-neighborhood $\mathcal{A}$ of f such that, for all $g \in \mathcal{A}$ the ball with same center and radius $n + b$ is contained in epi g. This implies that g is bounded above on $(n + b)B$.)

We conclude this section by establishing a general variational principle, due to Ioffe and Zaslavski, and by showing how to use it in a simple example. This is done mainly in order to compare this type of approach with that described in the next section, in particular with the porosity principle and its applications. For other applications using this principle, we refer the interested reader to [IZ].

The background necessary to establish the principle is the same one described to give the definition of well-posedness. We are given a domain space (X, ρ) and a data space $(\mathcal{A}, d)$. And to each $a \in \mathcal{A}$ a lower semicontinuous extended real valued function $f_a \colon X \to \mathbb{R} \cup \{+\infty\}$ is associated. We consider the problem of minimizing f_a on X, and we denote by $\inf f_a$ the infimum of f_a on X.

Theorem 11.3.8 *Let $\mathcal{A}$ be as above, and suppose $(\mathcal{A}, d)$ is a Baire space. Suppose there is a dense subset $\mathcal{B}$ of $\mathcal{A}$ such that the following condition $(\mathcal{P})$ holds: for each $a \in \mathcal{B}$ and each $r, k > 0$ there exist $\bar{a} \in \mathcal{A}$ and $\eta, \hat{\eta} > 0$ with the properties*

(i) $B(\bar{a}; \hat{\eta}) \subset B(a; r)$;

(ii) $b \in B(\bar{a}; \hat{\eta}) \quad \Rightarrow \quad \inf f_b > -\infty$ and $\mathrm{diam}(\bigcup_{b \in B(\bar{a}; \hat{\eta})} f_b^{\inf f_b + \eta}) < \frac{1}{k}$.

Then the set $\{a \in \mathcal{A} : a$ is well-posed $\}$ is a G_δ-dense set in $\mathcal{A}$.

Proof. Let $a \in \mathcal{B}$, and use the property with $r = \frac{1}{n}$, $k = n$, to find $\bar{a}_n \in \mathcal{A}$, $\hat{\eta}_n, \eta_n > 0$ such that, for all $b \in B(\bar{a}_n : \hat{\eta}_n)$ it holds that $\inf f_b$ is finite and $\mathrm{diam}(\bigcup_{b \in B(\bar{a}_n; \hat{\eta}_n)} f_b^{\inf f_b + \eta_n}) < \frac{1}{n}$. Define

$$\mathcal{A}_n = \bigcup_{\substack{a \in \mathcal{B} \\ m \geq n}} B(\bar{a}_m; \hat{\eta}_m),$$

and set

$$\bar{\mathcal{A}} = \bigcap_n \mathcal{A}_n.$$

Clearly, $\bar{\mathcal{A}}$ is a dense G_δ set, since the sets $\mathcal{A}_n$ are open and dense for all n. Moreover, in view of Proposition 10.3.3, every $a \in \bar{\mathcal{A}}$ is well-posed. This concludes the proof. $\qquad \square$

Observe that $\eta, \hat{\eta}$ can be always chosen to be the same. However in the definition it is worth distinguishing them.

To see how the previous principle can be used, we provide an example, that we develop in the section dedicated to porosity. This will allow us to show similarities and differences between the two approaches.

Let X be a normed space and let $\mathcal{F}$ be a family of real valued convex functions on X. We put on $\mathcal{F}$ the usual distance d, inducing uniform convergence on bounded sets, defined in (11.5). In the sequel, we shall need the following estimates, which are very easy to prove:

$$f(x) = g(x) \text{ for } \|x\| \leq a \quad \Longrightarrow \quad d(f, g) \leq 2^{-[a]}, \tag{11.6}$$

where $[a]$ denotes the integer part of a;

$$\|f - g\|_n \leq \frac{2^n d(f, g)}{1 - 2^n d(f, g)}, \tag{11.7}$$

provided $2^n d(f, g) < 1$.

To begin with, observe that $\mathcal{F}$ is a closed subspace of the space of the continuous functions defined on X, which is clearly a complete metric space, when endowed with the above distance. Thus $\mathcal{F}$, too, is a complete metric space, and so a Baire space. To study well-posedness, we set $\mathcal{A} = \mathcal{F}$ as the data space, and we shall write $f \in F$ rather $a \in \mathcal{A}$, and so on.

The following result holds.

Proposition 11.3.9 *In $\mathcal{F}$ the well-posed problems are a dense G_δ.*

Proof. We want to apply the Ioffe–Zaslavski principle. Let $\mathcal{B}$ be the set of functions which are lower bounded. Clearly, $\mathcal{B}$ is dense in $\mathcal{F}$. Now, fix r and k and take n so large and $\delta > 0$ so small that

$$\|f - g\|_n \le \delta \Rightarrow d(f,g) < \frac{r}{4}. \tag{11.8}$$

Let $f \in \mathcal{B}$ and take $\bar{x}$ such that

$$f(\bar{x}) \le \inf f + \frac{\delta}{2}. \tag{11.9}$$

Let m be such that $m \ge \|\bar{x}\|$. Now, set

$$\bar{f}(x) = \max\{f(x), f(\bar{x}) + \frac{\delta}{2(m+n)}\|x - \bar{x}\|\}.$$

Observe that if $\|x\| \le n$, either $\bar{f}(x) = f(x)$ or (using (11.9)),

$$0 < \bar{f}(x) - f(x) = f(\bar{x}) + \frac{\delta}{2(m+n)}\|x - \bar{x}\| \le \delta,$$

implying, by (11.8), $d(\bar{f}, f) \le \frac{r}{4}$. Now, let $\bar{\eta} > 0$ be so small that $d(\bar{f}, g) \le \bar{\eta}$ implies $d(f,g) < r$ and

$$\|\bar{f} - g\|_{m+1} < \frac{\delta}{16(m+n)k}.$$

We then have for x such that $\|\bar{x} - x\| = \frac{1}{2k}$,

$$g(x) > f(x) - \frac{\delta}{16(m+n)k} \ge f(\bar{x}) + \frac{\delta}{4(m+n)k} - \frac{\delta}{16(m+n)k}$$

$$\ge g(\bar{x}) + \frac{\delta}{8(m+n)k}.$$

The choice of $\eta = \frac{\delta}{8(m+n)k}$ shows that condition $(\mathcal{P})$ holds, and this completes the proof. $\square$

We finally note that there are in the literature other variational principles, which we only mention here: the Borwein–Preiss principle (see [BP]), in line with the Ekeland principle, but aimed at furnishing smooth perturbations of the initial function; the Deville–Godefroy–Zizler principle (see [DGZ]), which is the first one to explicitly talk about Tykhonov well-posedness; a principle by Ioffe–Revalski and myself, aimed at dealing with problems with functional constraints, which can be found in [ILR], where several other applications are also provided.

11.4 Porosity results

In this section we deal with porosity, rather than genericity, results. We shall establish the principle in a moment. The applications, except for Convex Programming II, are taken from [ILR2].

The background of the porosity principle is the same as that of the Ioffe–Zaslavski principle from the previous section, and refers to the setting established to give the definition of well-posedness: we are given a domain space (X, ρ) and a data space $(\mathcal{A}, d)$. And to each $a \in \mathcal{A}$ a lower semicontinuous extended real valued function $f_a \colon X \to \mathbb{R} \cup \{+\infty\}$ is associated. We consider the problem of minimizing f_a on X, and we denote by $\inf f_a$ the infimum of f_a on X.

Now, here is the new principle.

Theorem 11.4.1 *Let $\mathcal{A}$ be as above and let $\mathcal{B} \subset \mathcal{A}$. Suppose the following condition $(\mathcal{P})$ holds:*
for any $k \in \mathbb{N}$, there are $\lambda \in (0,1)$ and $r_0 > 0$ such that for each $a \in \mathcal{B}$ and each $r \in (0, r_0)$ there exist $\bar{a} \in \mathcal{A}$ and $\eta > 0$ with the properties:

(i) $B(\bar{a}; \lambda r) \subset B(a; r)$;
(ii) $b \in B(\bar{a}; \lambda r)$ *implies* $\inf f_b > -\infty$ *and* $\operatorname{diam}(\bigcup_{b \in B(\bar{a}; \lambda r)} f_b^{\inf f_b + \eta}) < \frac{1}{k}$.

Then the set $\{a \in \mathcal{B} : a \text{ is not well-posed}\}$ is strongly σ-porous in $\mathcal{A}$.

Proof. Set

$$\mathcal{A}_k = \bigcup_{\substack{a \in \mathcal{B} \\ r \leq r_0}} B(\bar{a}; \lambda r),$$

and

$$\bar{\mathcal{A}} = \bigcap_{k=1}^{\infty} \mathcal{A}_k.$$

We shall show that

(i) $\mathcal{B} \setminus \mathcal{A}_k$ is a strongly porous set in $\mathcal{A}$;
(ii)   the problem of minimizing f_a is well-posed for every $a \in \bar{\mathcal{A}}$.

Fix $k \in \mathbb{N}$, and corresponding λ, and r_0 satisfying $(\mathcal{P})$. By Proposition 11.1.8, it is sufficient to check only points of $\mathcal{B}$ to prove porosity of $\mathcal{B} \setminus \mathcal{A}_k$ in $\mathcal{A}$. Take $a \in \mathcal{B}$ and $r \in (0, r_0)$. Then for $\bar{a}$ and λ we have $B(\bar{a}; \lambda r) \subset B(a; r)$ by (i) and $B(\bar{a}; \lambda r) \subset \mathcal{A}_k$ by definition. This proves that $\mathcal{B} \setminus \mathcal{A}_k$ is a porous set in $\mathcal{A}$. To conclude the proof, we use Proposition 10.3.3. a is well-posed if we show that $\operatorname{diam}\{\bigcup f_b^\delta : d(a, b) < \delta\} \to 0$, as $\delta \to 0$. So, given $\tilde{a} \in \bar{\mathcal{A}}$, it suffices to show that for every $\gamma > 0$ there is $\delta > 0$ such that $\operatorname{diam}\{\bigcup f_b^\delta : d(\tilde{a}, b) < \delta\} < \gamma$. Take $k > \frac{1}{\gamma}$. As $\tilde{a} \in \mathcal{A}_k$, then $\tilde{a} \in B(\bar{a}; \lambda r)$, for some $\bar{a} \in \mathcal{A}$, λ and $r > 0$ as above. Keeping in mind condition (ii), to finish, it is sufficient to take $\delta = \min\{\lambda r - d(\tilde{a}, \bar{a}), \eta\}$. $\square$

An immediate corollary to the above theorem provides a result suitable for the applications.

Corollary 11.4.2 *Suppose we write* $\mathcal{A} = \bigcup_{m=0}^{\infty} \mathcal{A}_m$, *where* $\mathcal{A}_0$ *is a strongly σ-porous set in* $\mathcal{A}$, *and that for each* $m \geq 1$, *condition* $(\mathcal{P})$ *holds for* $\mathcal{B} = \mathcal{A}_m$. *Then the set of well-posed problems inside* $\mathcal{A}$ *has a strongly σ-porous complement in* $\mathcal{A}$.

We shall apply this principle to unconstrained convex problems, and to mathematical programming. We shall use the usual distance d defined in (11.5).

11.4.1 Unconstrained convex problems

In this subsection, $\mathcal{F}$ is the space of real valued, convex, continuous functions defined on X, so that we are dealing with the same class as in Proposition 11.3.9, but here we shall consider porosity rather than genericity.

The following proposition shows that the set of functions with unbounded level sets is strongly porous.

Proposition 11.4.3 *Let* $\mathcal{A}_0$ *be the subset of* $\mathcal{F}$ *formed by those* $f \in \mathcal{F}$ *such that either* $\inf f = -\infty$ *or* $f^{\inf f + r}$ *is unbounded for all* $r > 0$. *Then* $\mathcal{A}_0$ *is strongly porous in* $(\mathcal{F}, d)$.

Proof. Let $r_0 = 1$ and $\lambda = \frac{1}{32}$. Fix $r \leq r_0$ and let n be such that $\frac{1}{2^{n-1}} \leq r \leq \frac{1}{2^{n-1}}$. Given $f \in \mathcal{F}$, for every $n = 1, 2 \ldots$, we set

$$w_n = \inf_{\|x\| \leq n+1} f(x)$$

and we define the function f_n as follows:

$$f_n(x) = \max\{f(x), w_n + 2(\|x\| - (n+1))\}, \ x \in X.$$

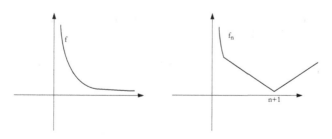

Figure 11.1.

Since $f_n(x) = f(x)$ if $\|x\| \leq n+1$, we get from (11.6)

$$d(f, f_n) \leq \frac{1}{n^{n+1}}. \tag{11.10}$$

Moreover,

$$\inf f_n = \inf_{\|x\| \leq n+1} f_n(x) = w_n,$$
$$\text{and } (\|x\| \geq n+2 \implies f_n(x) \geq \inf f_n + 2). \tag{11.11}$$

Suppose now $d(f_n, g) \leq \frac{1}{2^{n+3}}$. Then by (11.7)

$$\|x\| \leq n+1 \Rightarrow |f_n(x) - g(x)| \leq \|f_n - g\|_{n+1} \leq \frac{2^{n+1}d(f_n, g)}{1 - 2^{n+1}d(f_n, g)} \leq \frac{1}{3},$$

while

$$\|x\| = n+2 \Rightarrow |f_n(x) - g(x)| \leq \|f_n - g\|_{n+2} \leq \frac{2^{n+2}d(f_n, g)}{1 - 2^{n+2}d(f_n, g)} \leq 1.$$

Therefore for z with $\|z\| = n+2$ we have (using also (11.11)) that

$$g(z) \geq f_n(z) - 1 \geq \inf f_n + 1 = \inf_{\|x\| \leq n+1} f_n + 1 \geq \inf_{\|x\| \leq n+1} g + \frac{2}{3}.$$

Since g is convex it follows that

$$\inf g = \inf_{\|x\| \leq n+1} g(x),$$

and thus $g \notin \mathcal{A}_0$ as the level set of g corresponding to $a = \inf g + \frac{1}{3}$ must lie completely in the ball of radius $n+2$. Therefore we conclude that

$$B\left(f_n; \frac{r}{32}\right) \subset B\left(f_n; \frac{1}{2^{n+3}}\right) \subset B\left(f; \frac{1}{2^{n-1}}\right) \setminus \mathcal{A}_0 \subset B(f; r) \setminus \mathcal{A}_0.$$

This ends the proof.   □

Now we are ready to prove the following porosity result. As before, we identify each $f \in \mathcal{F}$ with the corresponding minimization problem on X.

Theorem 11.4.4 *Let X be a Banach space. Then the set of the well-posed problems in $\mathcal{F}$ has a strongly σ-porous complement in $(\mathcal{F}, d)$.*

Proof. Set

$$\mathcal{A} = \mathcal{F};$$
$$\mathcal{A}_0 = \{f \in \mathcal{A} : \text{ either } \inf f = -\infty \text{ or } f^{\inf f + r} \text{ is unbounded } \forall r > 0\};$$
$$\mathcal{A}_m = \{f \in \mathcal{A} \setminus \mathcal{A}_0 : f^{\inf f + r} \bigcap B(0; m) \neq \emptyset \ \forall r > 0\}.$$

Then $\bigcup_{m=0}^{\infty} \mathcal{A}_m = \mathcal{A}$ and $\mathcal{A}_0$ is strongly porous by Proposition 11.4.3. By Corollary 11.4.2 the theorem will be proved if we show that every $\mathcal{A}_m$, $m = 1, 2 \ldots$ satisfies condition $(\mathcal{P})$ of Theorem 11.4.1.

To this end, we have to find, for given $m \in \mathbb{N}$ and $k \in \mathbb{N}$, positive r_0 and $\lambda \in (0, 1)$ such that for any $f \in \mathcal{A}_m$ and any $0 < r < r_0$ there are $\bar{f} \in \mathcal{A}$, $\bar{x} \in X$ and $\eta > 0$ with the properties that

(i) $d(f, \bar{f}) \leq (1 - \lambda)r$;

(ii) $d(g, \bar{f}) < \lambda r$ implies $\inf g > -\infty$ and $g^{\inf g + \eta} \subset B\left(\bar{x}; \frac{1}{2k}\right)$.

So, let us fix $m \geq 1$, $k \geq 1$, and $0 < r < 1$. Set

$$s = \sum_{n=1}^{\infty} 2^{-n}(m + n).$$

We shall show that property $(\mathcal{P})$ holds with the choice of

$$\lambda = \frac{1}{2^{m+8}ks}, \qquad r_0 = 1.$$

To this end, fix $f \in \mathcal{A}_m$. Observe that f is bounded from below. Now, take $\bar{x}$ such that

$$\|\bar{x}\| \leq m \quad \text{and} \quad f(\bar{x}) \leq \inf f + \frac{r}{8}. \tag{11.12}$$

We next define

$$\bar{f}(x) = \max\{f(x), f(\bar{x}) + \frac{r}{8s}\|x - \bar{x}\|\}, \quad x \in X. \tag{11.13}$$

It is routine to verify that $d(f, \bar{f}) \leq \frac{r}{4}$. Suppose now $d(g, \bar{f}) < \lambda r$. From (11.7)

$$\|g - \bar{f}\|_{m+1} < \frac{r}{2^6 ks}. \tag{11.14}$$

Then, for any x with $\|x - \bar{x}\| = \frac{1}{2k}$ we have

$$g(x) > \bar{f}(x) - \frac{r}{2^6 ks} \geq \bar{f}(\bar{x}) - \frac{r}{2^6 ks} + \frac{r}{2^4 ks}$$
$$\geq g(\bar{x}) + \frac{r}{2^4 ks} - \frac{r}{2^5 ks} = g(\bar{x}) + \frac{r}{2^5 ks}.$$

It follows that (ii) is satisfied for $\eta = \frac{r}{2^5 ks}$, and the proof is complete.   □

Let us observe similarities and differences between Proposition 11.3.9 and Theorem 11.4.4, that deal with the same class of functions, but provide in the first a genericity result, and in the second a (stronger) porosity result. When proving the two statements, we need first of all to get rid of a set of "bad" functions: the lower unbounded ones in the genericity result, and those which either are lower unbounded or have unbounded level sets in the porosity result. The first difference between the two approaches is that in the first case we

must show only denseness of the remaining functions; and in the second case
we must show that the bad functions are a much smaller (i.e., porous) set.
Then we must deal with the remaining functions. In both cases, given a certain
parameter r, we take one of such functions f and construct a perturbation $\bar{f}$ of
it, *which is the same in both cases*. In a small ball centered at $\bar{f}$ and of radius
$\bar{\eta}$ we finally prove that all functions enjoy *the same property*, related to the
behavior of the level sets, and this us allows us to conclude. The key point that
makes the difference between the two approaches is that, when constructing
the function $\bar{f}$, we need to find a particular point $\bar{x}$, almost minimizing f, in
a way that depends on r. The perturbation function f depends on the norm
of $\bar{x}$. Changing r, we need to change $\bar{x}$. In the first case, we do *not* have
control on its norm, so we cannot prove that the ratio between η and r is
constant. In the second case, we have control of the norm of $\bar{x}$, independently
from r, and this allows showing that $\bar{\eta}$ can be chosen linearly with respect
to r. This apparently minor fact makes the whole difference in the results,
which is not so minor! To conclude this comment, let me point out that in
the original proof of Theorem 11.4.4 the perturbation $\bar{f}$ is constructed in a
slightly different way. Here we used our construction for the purpose of having
a better comparison between Proposition 11.3.9 and Theorem 11.4.4.

The same ideas, *mutatis mutandis*, used in Theorem 11.4.4 apply, for instance, to the space of the continuous functions on a metric space, uniformly
bounded from below by a given coercive function. More interestingly in our
context, we can prove the same theorem for convex programming problems.
Here having constraints makes things a little more complicated technically,
but the underlying ideas are absolutely the same.

11.4.2 Convex programming I

In this subsection we consider the following problem:

$$\begin{aligned}
\text{minimize} \quad & f_0(x) \\
\text{such that} \quad & f_1(x) \le a_1, \ldots, f_l(x) \le a_l, \ x \in X,
\end{aligned}$$

where f_i, $i = 0, \ldots, l$, $l \ge 1$, are real valued convex continuous functions
defined on a Banach space X.

The data space $\mathcal{A}$ will be a subspace of the Cartesian product of $(l + 1)$
copies of $\mathcal{F}(X)$, endowed with the box metric:

$$d[(f_0, \ldots, f_l), (g_0, \ldots, g_l)] = \max_{i=0,\ldots,l} d(f_i, g_i),$$

(d is the usual metric inducing the uniform convergence on bounded sets).

Let $a = (f_0, f_1, \ldots, f_l) \in [\mathcal{F}(X)]^{l+1}$. The *feasible set* of the problem determined by a is the set

$$F(a) = \{x \in X : f_i(x) \le 0, \forall i = 1, \ldots, l\}.$$

The data space is the collection of all $a \in [\mathcal{F}(X)]^{l+1}$ for which $F(a) \neq \emptyset$ (with the inherited metric d), a natural choice. The function f_a associated with $a \in \mathcal{A}$ is defined in a standard way:

$$f(x) = \begin{cases} f_a(x) = f_0(x) & \text{if } x \in F(a), \\ \infty & \text{otherwise.} \end{cases}$$

Theorem 11.4.5 *Let $\mathcal{A}$ be the class of convex programming problems described above. Then the set of well-posed problems in $\mathcal{A}$ has a strong σ-porous complement in $(\mathcal{A}, d)$.*

Proof. Observe at first that $\mathcal{A} = \mathcal{F}(X) \times U$, where

$$U = \{(f_1, \ldots, f_l) : \exists x, \ f_i(x) \leq 0, \forall i = 1, \ldots, l\}.$$

Now, set

$$\mathcal{A}_0 = \{a = (f_0, f_1, \ldots, f_l) \in \mathcal{A} : \text{either } \inf f_0 = -\infty \text{ or }$$
$$f_0^{\inf f_0 + r} \text{ is unbounded } \forall r > 0\}.$$

Then, $\mathcal{A}_0 = \mathcal{A}_0' \times U$, where $\mathcal{A}_0'$ is the set from Theorem 11.4.4. Since the set $\mathcal{A}_0$ is strongly porous in $\mathcal{F}(X)$ we easily conclude that the set $\mathcal{A}_0$ is strongly porous in $\mathcal{A}$. Set further

$$\mathcal{A}_m = \{a \in \mathcal{A} \setminus \mathcal{A}_0 : f_a^{\inf f_a + r} \cap B(0; m) \neq \emptyset \ \forall r > 0\}, \quad m \geq 1.$$

It is seen that $\mathcal{A} = \bigcup_{m=0}^{\infty} \mathcal{A}_m$ and we will show that we can apply Corollary 11.4.2. Fix $m, k \in \mathbb{N}$, $m, k \geq 1$ and take $r_0 = 1$ and

$$\lambda = \frac{1}{2^{m+1}64ks},$$

where $s = \sum_{n=1}^{\infty} \frac{m+n}{2^n}$. As we saw in (11.14), with this choice of λ, the following estimate holds:

$$\|f - g\|_{m+1} \leq \frac{\varepsilon}{32ks},$$

for any two convex functions f, g satisfying $d(f, g) \leq \lambda r$, $r \leq 1$.

Let us now fix $a = (f_0, f_1, \ldots, f_l) \in \mathcal{A}_m$ and $0 < r \leq 1$. Then there exists $\bar{x} \in F(a)$ with $\|\bar{x}\| \leq m$ such that

$$f_0(\bar{x}) \leq \inf f_a + \frac{r}{32ks}.$$

Put

$$\bar{f}_0(x) = f(x) + \frac{r}{4s}\|x - \bar{x}\|, \quad x \in X,$$

$$\bar{f}_i(x) = f_i(x) + \frac{r}{4s}\|x - \bar{x}\| - \frac{r}{16ks}, \quad x \in X,$$

and set $\bar{a} = (\bar{f}_0, \bar{f}_1, \ldots, \bar{f}_l)$ and $\eta := \frac{r}{32ks}$. Suppose now $d(\bar{a}, b) < \lambda r$ where $b = (g, g_1, \ldots, g_l) \in \mathcal{A}$. First, it is easily seen that $d(a, b) < r$, showing the first part of condition $(\mathcal{P})$. Further, for every $i = 1, \ldots, l$, we have

$$g_i(\bar{x}) \leq \bar{f}_i(\bar{x}) + \frac{r}{32ks} = f_i(\bar{x}) - \frac{r}{16ks} + \frac{r}{32ks} \leq 0,$$

showing that $\bar{x} \in F(b)$.

Finally, suppose $x \in F(b)$, $\|x\| \leq m + 1$ and $\|x - \bar{x}\| > \frac{1}{2k}$. Then

$$f_i(x) + \frac{r}{8ks} - \frac{r}{16ks} \leq \bar{f}_i(x) \leq g_i(x) + \frac{r}{32ks} \leq \frac{r}{32ks},$$

yielding that $x \in F(a)$. Using this fact, exactly as in the proof of Theorem 11.4.4, we show that for such an x

$$f_b(x) = g(x) > g(\bar{x}) + \eta = f_b(\bar{x}) + \eta,$$

which completes the proof of the second part of $(\mathcal{P})$. □

Thus the only complication in the proof of the constrained problem is given by the fact that we have to manage the constraints in such a way that the pivot point $\bar{x}$ is in the feasible set of the functions g around $\bar{f}$ on one side, and that feasible points for the functions g close to $\bar{x}$ are also in the feasible set of the function under analysis.

We shall describe other results in convex programming in the next section. We shall deal, as we see, with more specific perturbations, and the tools to get our results are different.

11.4.3 Convex programming II

In this section we consider (almost) the same problem as in the previous section, but with a different point of view, by allowing different perturbations. In fact, we shall fix an objective function and we perturb it by means of linear terms, and on the constraint we allow perturbations only on the right-hand side. But let us see things in detail. Let X be a Banach space, let $f, f_1, \ldots, f_l \colon X \to (-\infty, \infty]$ be given convex functions. Let $p \in X^*$ and $a = (a_1, \ldots, a_l) \in \mathbb{R}^l$. They will serve as parameters. Here is the problem:

$$\mathcal{P}(p, a) \qquad \begin{array}{ll} \text{minimize} & f(x) - \langle p, x \rangle \\ \text{such that} & f_1(x) \leq a_1, \ldots, f_l(x) \leq a_l, x \in X. \end{array}$$

We shall write $g = (f_1, \ldots, f_l)$ and the inequalities defining the constraint set will be simply written $g(x) \leq a$. As usual, we denote by F the feasible set of the problem:

$$F = \{a \in \mathbb{R}^l : \exists x \in X \ g(x) \leq a\}.$$

F is a convex set with nonempty interior:

$$\operatorname{int} F = \bigcup_x \{a : \exists x,\ a \gg g(x)\}.$$

(Recall $a \gg g(x)$ means $a_i > g_i(x)$ for all i.) Set

$$v(p, a) = \inf\{f(x) - \langle p, x\rangle : g(x) \le a\},$$

with the standard convention that $\inf\{\emptyset\} = \infty$.

Now, set

$$F_a(x) = \begin{cases} f(x) & \text{if } g(x) \le a, \\ \infty & \text{otherwise.} \end{cases}$$

Thus the initial (constrained) minimum problem $\mathcal{P}(p, a)$ is equivalent to the (unconstrained) problem of minimizing $F_a(\cdot) - \langle p, \cdot\rangle$.

The following easy observation will be crucial throughout this section:

$$F_a^*(p) = \sup_x\{\langle p, x\rangle - F_{a,b}(x)\} = -v(p, a). \tag{11.15}$$

From (11.15), it is clear that v is concave in p, for every $a \in \mathbb{R}^l$. Moreover, it is easy to see that it is convex in a, for every $p \in X^*$. Thus, denoting by $S(p, a)$ the multifunction that to the given pair (p, a) associates the solution set of $\mathcal{P}(p, a)$, from (11.15) we have that $S(p, a) = \partial F_a^*(p)$. Now, for a given fixed $p \in X^*$, we can consider the convex programming problem with a as a parameter. We are exactly in the setting described on page 111. The solution set of the dual problem is then called the set of the Lagrange multipliers of the initial problem. Thus, denoting by $\Lambda(p, a)$ the Lagrange multifunction evaluated at (p, a), from Proposition 6.2.7 we get that $\Lambda(p, a) = \partial v^{**}(p, \cdot)(a)$. Thus the following fundamental formula holds, at the points where $v(p, \cdot)$ is lower semicontinuous:

$$S(p, a) \times \Lambda(p, a) = \partial v(p, a). \tag{11.16}$$

Now we make some assumptions in order to deal with meaningful problems. To take an example, suppose we consider $f = 0$, and no constraints. Clearly, every linear perturbation of f yields to an unbounded problem, so that it is nonsense to look for porosity or also genericity of well-posed problems. Furthermore, we need to consider only problems with nonempty feasible sets, and that are lower bounded. In any case, we must impose some restriction on the choice of the functions f, g with which we are dealing. It could be made a more general assumption, but let us agree on the following one:

$$\lim_{\|x\| \to \infty} \max\{f(x), f_1(x), \ldots, f_l(x)\} = \infty. \tag{11.17}$$

Thus, the data space is

$$\mathcal{A} = \{(p, a) \in X^* \times Y : |v(p, a)| < \infty \text{ and } (11.17) \text{ is satisfied}\}.$$

The following lemma and its subsequent corollary show that in this case the set $\mathcal{A}$ is big enough, i.e., contains an open set. This in particular implies that a null measure result is meaningful.

Lemma 11.4.6 *Let $\tilde{a} \in \operatorname{int} F$, and let p be such that $F_{\tilde{a}}(\cdot) - \langle p, \cdot \rangle$ is coercive. Then $v(\cdot, \cdot)$ is bounded on a neighborhood of $(p, \tilde{a})$.*

Proof. Fix $\hat{a} \gg \tilde{a}$. Since $F_{\tilde{a}}(\cdot) - \langle p, \cdot \rangle$ is coercive, then $p \in \operatorname{int}(\operatorname{dom} F_{\tilde{a}}^*)$ (See Exercise 5.1.11). This means that there are $\varepsilon > 0$ and $M > 0$ such that $F_{\hat{a}}^*(q) \leq M$ if $\|q - p\|_* \leq \varepsilon$. It follows that $v(q, \hat{a}) \geq -M$ if $\|q - p\|_* \leq \varepsilon$. Since $\hat{a} \gg \tilde{a}$, there is a neighborhood W of $\tilde{a}$ such that, for all $a \in W$, $a \leq \hat{a}$. Thus

$$v(p, a) \geq -M$$

if $\|p - q\|_* \leq \varepsilon$, and $a \in W$. As far as upper boundedness is concerned, simply observe that there exists x such that $g(x) \ll \tilde{a}$. Thus $v(p, a) \leq f(x) + \|p\|_* \|x\|$, for a in a suitable neighborhood of $\tilde{a}$ and for all p. $\qquad\square$

From Lemma 11.4.6 we get

Corollary 11.4.7 *There exists an open set $\Pi \subset X^*$ such that*

$$\operatorname{int} \mathcal{A} = \Pi \times \operatorname{int} F.$$

Proof. By Lemma 11.4.6, and because of assumption (11.17), v is bounded on a neighborhood N of $(0, \tilde{a})$, with $\tilde{a} \in \operatorname{int} F$. Thus $\mathcal{A}$ has nonempty interior. The conclusion now follows from Proposition 11.2.1. $\qquad\square$

From Lemma 11.4.6 it also follows that v is a concave/convex function, in particular locally Lipschitz around each point in the interior of $\mathcal{A}$.

We now give a new definition of well-posedness, suited to our setting. Since, in the convex programming problems we are considering, the Lagrange multipliers play an important role, this new definition of well-posedness should also take into account their behavior.

Definition 11.4.8 We say that the problem $\mathcal{P}(p, a)$ is *very well-posed* if

(i) $\mathcal{P}(p, a)$ is well-posed;
(ii) there is a unique Lagrange multiplier for $\mathcal{P}(p, a)$;
(iii) if $(p_n, a_n) \to (p, a)$ if $\lambda_n \in \Lambda(p_n, a_n)$, then $\lambda_n \to \lambda$.

In the language of multifunctions, the last condition amounts to saying that the Lagrange multiplier multifunction is upper semicontinuous and single valued at (p, a), as is easy to see.

The next result is the key to proving our porosity results.

Theorem 11.4.9 *Let X be a reflexive Banach space. Let $(p, a) \in \operatorname{int} \mathcal{A}$. Then $\mathcal{P}(p, a)$ is very well-posed if and only if $v(\cdot, a)$ is Fréchet differentiable at p and $v(p, \cdot)$ is Fréchet differentiable at a.*

Proof. The proof will show that actually Fréchet differentiability with respect to p is equivalent to well-posedness of the problem, while Fréchet differentiability with respect to a is related to the behavior of the Lagrange multiplier multifunction. Let us start by proving that, if $(p, a) \in \operatorname{int} \mathcal{A}$, and $v(\cdot, a)$ is Fréchet differentiable at $\bar{p}$, then $\mathcal{P}(p, a)$ is well-posed. We can suppose, without loss of generality, $(\bar{p}, \tilde{a}) = (0^*, 0)$. Remember that v is locally Lipschitz around $(0^*, 0)$. Call K one Lipschitz constant in a neighborhood of $(0, 0)$. For easy notation, we shall write F instead of F_0, and F_n instead of F_{a_n}, if $a_n \to 0$. Thus, by assumption, F^* is Fréchet differentiable at 0^*. Let $\bar{x}$ be the derivative. Then, by the Asplund–Rockafellar theorem (see Theorem 10.1.11), there is a forcing function c such that

$$F(x) - F(\bar{x}) \geq c(\|x - \bar{x}\|). \tag{11.18}$$

It follows that

$$F^*(p) \leq c^*(\|p\|) + \langle p, \bar{x} \rangle - F(\bar{x}). \tag{11.19}$$

Observe that, since c is forcing, its convolution with $\varepsilon \| \cdot \|$ is also forcing. The last one has a conjugate with effective domain contained in $[-\varepsilon, \varepsilon]$. Thus we can suppose, without loss of generality, $\operatorname{dom} c^* \subset [-\varepsilon, \varepsilon]$. Now, take $(p_n, a_n) \to (0^*, 0)$, and x_n such that $F_n(x_n) - v(p_n, a_n) \to 0$.

We then have

$$
\begin{aligned}
F_n(x_n) - F(\bar{x}) &\geq \sup_{\|p\| \leq \varepsilon} \{\langle p, x_n \rangle - F_n^*(p)\} - F(\bar{x}) \\
&\geq \sup_{\|p\| \leq \varepsilon} \{\langle p, x_n \rangle - F^*(p)\} - K\|a_n\| - F(\bar{x}) \\
&\geq \sup_{\|p\| \leq \varepsilon} \{\langle p, x \rangle - c^*(\|p\|) - \langle p, \bar{x} \rangle\} - K\|a_n\| \\
&\geq \sup_{p} \{\langle p, x_n - \bar{x} \rangle - c^*(\|p\|)\} - K\|(a_n)\| \\
&= c(\|x_n - \bar{x}\|) - K\|a_n\|.
\end{aligned}
$$

It follows that $c(\|x_n - \bar{x}\|) \to 0$, and thus $x_n \to \bar{x}$. We have shown that the problem $\mathcal{P}(p, a)$ is well-posed, provided $v(\cdot, a)$ is Fréchet differentiable at p. Now consider a point $(p, a) \in \operatorname{int} \mathcal{A}$ such that $v(p, \cdot)$ is Fréchet differentiable at a, with derivative λ. Without loss of generality we can again suppose $(p, a) = (0^*, 0)$. Fix $\varepsilon > 0$. Observe that

$$\lim_{R \to 0} \operatorname{diam} \partial v(0^*, \cdot) RB) = 0,$$

as $\partial v(0^*, \cdot)$ is (norm-norm) upper semicontinuous at 0, since $v(0^*, \cdot)$ is Fréchet differentiable at 0 (see Proposition 3.5.6). Thus, there is $r > 0$ such that

$$\operatorname{diam} \operatorname{co} \partial v(0^*, \cdot) 2rB) < \frac{\varepsilon}{2}.$$

Moreover, there are $H, K > 0$ such that

$$|v(p, a) - v(0^*, a)| \leq K\|p\|,$$

if $\|p\| \leq H$, $\|a,\| \leq H$. Take p such that $\|p\| \leq \frac{\varepsilon}{2k}$. We apply Lemma 3.6.4 to the convex functions $f(\cdot) = v(p, \cdot)$ and $g(\cdot) = v(0^*, \cdot)$, with $r \leq \frac{H}{2}$, $R = 2r$, $\delta = K\|p\|$, $\|a\| \leq r$. Let $\lambda_p \in \partial v(0^*, \cdot)(a)$. Then $d(\lambda_p, \operatorname{co} \partial v(0^*, \cdot))2rB) \leq \frac{\varepsilon}{2}$. It follows that

$$\|\lambda - \lambda_p\| < \varepsilon.$$

We have shown one implication. The proof of the converse is simpler. Condition (ii) in Definition 11.4.8 of a very well-posed problem is equivalent to saying that $v(p, \cdot)$ has a singleton as subdifferential at a. This implies, via Proposition 3.3.7, that actually $v(p, \cdot)$ is Fréchet differentiable at a. Moreover, by taking $a_n = a$, $p_n = p$ in the definition of well-posedness, we see that this implies the fact that $F_a(\cdot) - \langle p, \cdot \rangle$ is Tykhonov well-posed, and this in turn implies (see Theorem 10.1.11) that $F_a^*(\cdot) = -v(\cdot, a)$ is Fréchet differentiable at p. The proof is complete.   □

We finally have in our hands the tools to get the porosity result.

Theorem 11.4.10 *Let X be a reflexive Banach space with separable dual. Assume (11.17). Then the collection of $(p, a) \in \mathcal{A}$ such that $\mathcal{P}(p, a)$ is not very well-posed is σ-porous in $\mathcal{A}$.*

Proof. Clearly, it is enough to concentrate our attention on those (p, a) such that $(p, a) \in \operatorname{int} \mathcal{A}$. Then the claim immediately follows from Theorem 11.4.9 and Theorem 11.2.5.   □

If X is a finite-dimensional space, we can obtain another interesting result, i.e., not only are the majority of the problems very well-posed, but also the (solution, Lagrange multiplier) multifunction enjoys, for most problems, a Lipschitz stability property.

Theorem 11.4.11 *Let X be a Euclidean space and assume (11.17). Then the set of parameters (p, a) such that either the problem $\mathcal{P}(p, a)$ is not very well-posed or the (solution, Lagrange multiplier) multifunction $S(\cdot, \cdot) \times \Lambda(\cdot, \cdot)$ is not Lipschitz stable at (p, a) is a set of Lebesgue measure zero.*

Proof. Once again we use Theorem 11.4.9 together with a result by Mignot [Mi], asserting that, given a maximal monotone operator $A\colon X \to X^*$, where X is a Euclidean space, the set of the points where A is not Fréchet differentiable, is of null measure inside its domain. And of course Fréchet differentiability at a point implies Lipschitz stability at the point.   □

The results of this subsection are taken from [IL2], where also equality constraints are considered in the problem.

11.4.4 Quadratic programming

The result we shall illustrate in this subsection uses the variational principle established in Theorem 11.4.1, specifically in the form of its Corollary 11.4.2. Its proof is probably the most complicated, from a technical point of view, of the whole book. The main reason is that in the problem under consideration, there is neither convexity nor coercivity. This problem too, like that of the previous subsection, depends upon parameters ranging over a finite dimensional space, and thus we can also state a null-measure result. Let us introduce the setting of the problem. It is the quadratic programming in the N-dimensional Euclidean space $\mathbb{R}^N$. To be more precise, we consider problems of the form

$$\text{minimize} \quad \langle Q_0 x, x \rangle + \langle c_0, x \rangle$$
$$\text{such that} \quad \langle Q_1 x, x \rangle + \langle c_1, x \rangle \le \alpha_1, \ldots, \langle Q_l x, x \rangle + \langle c_l, x \rangle \le \alpha_l, \quad x \in \mathbb{R}^N,$$

where Q_i are $N \times N$ symmetric matrices, $c_i \in \mathbb{R}^N$, $\langle \cdot, \cdot \rangle$ is the usual scalar product in $\mathbb{R}^N$ and $\alpha_i \in \mathbb{R}$.

Every such problem is determined by the $3l + 2$-tuple

$$a = (Q_0, \ldots, Q_l, c_0, \ldots, c_l, \alpha_1, \ldots, \alpha_l).$$

The distance between two tuples, $a = (Q_0, \ldots, Q_l, c_0, \ldots, c_l, \alpha_1, \ldots, \alpha_l)$ and $b = (R_0, \ldots, R_l, d_0, \ldots, d_l, \beta_1, \ldots, \beta_l)$ is defined by

$$d(a, b) = \max_{0 \le i \le l} \{ \|Q_i - R_i\|, \|c_i - d_i\|, |\alpha_i - \beta_i| \},$$

where we set $\alpha_0 = \beta_0 = 0$. Here $\|Q\|$ and $\|x\|$ are the standard Euclidean norms of a matrix and a vector in the corresponding spaces. The following estimate holds (prove it) for $f_i(x) = \langle Q_i x, x \rangle + \langle c_i, x \rangle - \alpha_i$, $g_i(x) = \langle R_i x, x \rangle + \langle d_i, x \rangle - \beta_i$,

$$|f_i(x) - g_i(x)| \le 2(\|x\|^2 + 1) d(a, b), \ \forall x \in X. \tag{11.20}$$

This shows that the above defined metric d is compatible with the uniform convergence of f_i's on bounded sets.

As data space we shall take

$$\mathcal{A} = \big\{ a = (Q_0, \ldots, Q_l, c_0, \ldots, c_l, \alpha_1, \ldots, \alpha_l) : F(a) \ne \emptyset$$
$$\text{and} \ \max_{i=0,\ldots,l} \langle Q_i x, x \rangle \ge 0 \, \forall x \in \mathbb{R}^N \big\},$$

where, as in the previous section, $F(a)$ denotes the feasible set for the problem determined by a.

The additional requirement that the maximum of the quadratic forms be nonnegative is also quite natural. If for some a as above, there exists $\tilde{x} \in \mathbb{R}^N$ such that $\max_{i=0,\ldots,l} \langle Q_i \tilde{x}, \tilde{x} \rangle < 0$, then $t\tilde{x} \in F(a)$ for $t > 0$ large enough and hence, for all problems in a small ball around a, the corresponding objective function is unbounded below on the feasible set. Therefore, even generic well-posedness is not possible outside the above fixed class.

We begin our analysis by showing that certain "bad" sets of data are σ-porous in the data space.

Proposition 11.4.12 *The set*

$$\mathcal{Z} := \big\{ a = (Q_0, \ldots, Q_l, c_0, \ldots, c_l, \alpha_1, \ldots, \alpha_l) \in \mathcal{A} :$$
$$\exists 0 \neq x \in \mathbb{R}^N, \ \max_{i=0,\ldots,l} \langle Q_i x, x \rangle = 0 \big\}$$

is strongly σ-porous in $(\mathcal{A}, d)$.

Proof. Let
$$\mathcal{Z}_m := \{ a \in \mathcal{Z} : F(a) \cap B(0; m) \neq \emptyset \}.$$

Obviously, $\mathcal{Z} = \bigcup_{m \geq 1} \mathcal{Z}_m$ and we claim that each set $\mathcal{Z}_m$ is porous in $\mathcal{A}$. Let $\lambda = \frac{1}{4m^2}$ and $r_0 = 1$. Take $a \in \mathcal{Z}_m$, $r \in (0, r_0)$ and consider

$$\bar{a} := \Big(Q_0 + \frac{r}{2m^2} I, \ldots, Q_l + \frac{r}{2m^2} I, c_0, \ldots, c_l, \alpha_1 + \frac{r}{2}, \ldots, \alpha_l + \frac{r}{2} \Big),$$

where I is the $N \times N$ identity matrix. Now take $\bar{x} \in F(a)$ so that $\|\bar{x}\| \leq m$. Thus, for any $i = 1, \ldots, l$, we have

$$\langle Q_i \bar{x}, \bar{x} \rangle + \frac{r \|\bar{x}\|^2}{2m^2} + \langle c_i, \bar{x} \rangle \leq \alpha_i + \frac{r}{2},$$

showing that $\bar{x} \in F(\bar{a})$, i.e., $F(\bar{a}) \neq \emptyset$. Since the second condition in the definition of the class $\mathcal{A}$ is trivially fulfilled for $\bar{a}$, we obtain $\bar{a} \in \mathcal{A}$.

Take now any $b \in B(\bar{a}; \lambda r)$ with

$$b = (R_0, \ldots, R_l, d_0, \ldots, d_l, \beta_1, \ldots, \beta_l).$$

It is straightforward to see that $b \in B(a; r)$. Let us fix any $0 \neq x \in \mathbb{R}^N$. Then, for every $i = 0, 1, \ldots, l$, we have

$$\langle R_i x, x \rangle \geq \langle Q_i x, x \rangle + \frac{r}{2m^2} \|x\|^2 - \|R_i - Q_i - \frac{r}{2m^2} I\| \, \|x\|^2$$
$$\geq \langle Q_i x, x \rangle + (\frac{r}{2m^2} - \lambda r) \|x\|^2.$$

Since $a \in \mathcal{A}$, the latter together with the choice of λ, show that

$$\max_{i=0,\ldots,l} \langle R_i x, x \rangle > 0,$$

i.e., $b \notin \mathcal{Z}$, and this completes the proof. $\square$

Proposition 11.4.13 *The set*

$$\mathcal{E} := \{a \in \mathcal{A} : \exists b_n \to a, \ \inf f_{b_n} \to \infty\}$$

is strongly σ-porous in $(\mathcal{A}, d)$.

Proof. For $m = 1, 2, \ldots$, we set

$$\mathcal{E}_m := \{a \in \mathcal{E} : F(a) \cap B(0; m) \neq \emptyset\}$$

and observe that $\mathcal{E} = \bigcup_{m \geq 1} \mathcal{E}_m$. Thus, it is enough to prove that each $\mathcal{E}_m$ is porous in $\mathcal{A}$. To this end, set $\lambda = \frac{1}{4m^2}$, $r_0 = 1$ and let $a \in \mathcal{E}_m$ and $r \in (0, r_0)$. Set

$$\bar{a} := \left(Q_0, \ldots, Q_l, c_0, \ldots, c_l, \alpha_1 + \frac{3r}{4}, \ldots, \alpha_l + \frac{3r}{4}\right).$$

It is obvious that $\bar{a}$ is still in $\mathcal{A}$. Let $b \in B(\bar{a}; \lambda r)$ and

$$b = (R_0, \ldots, R_l, d_0, \ldots, d_l, \beta_1, \ldots, \beta_l).$$

We know that there is some $\bar{x} \in F(a)$ with $\|\bar{x}\| \leq m$. Then, if for every $i = 1, \ldots, l$ we put $\bar{\alpha}_i = \alpha_i + \frac{3r}{4}$, we have

$$\langle R_i \bar{x}, \bar{x} \rangle + \langle d_i, \bar{x} \rangle \leq \langle Q_i \bar{x}, \bar{x} \rangle + \|R_i - Q_i\| \, \|\bar{x}\|^2 + \langle c_i, \bar{x} \rangle + \|d_i - c_i\| \, \|\bar{x}\|$$
$$\leq \alpha_i + \lambda r m^2 + \lambda r m \leq \alpha_i + \frac{r}{4} + \frac{r}{4} = \bar{\alpha}_i - \frac{r}{4}$$
$$\leq \beta_i + \lambda r - \frac{r}{4} < \beta_i,$$

showing that $\bar{x} \in F(b)$. This gives $\inf f_b \leq f_b(\bar{x})$ and since by (11.20) we have $f_b(\bar{x}) < f_a(\bar{x}) + 1$ we see that $b \notin \mathcal{E}$. The proof is complete. $\qquad \square$

Now we are ready to prove the main result of this subsection.

Theorem 11.4.14 *Let $(\mathcal{A}, d)$ be the class of quadratic mathematical programming problems described above. Then the set of well-posed problems in $\mathcal{A}$ has a strongly σ-porous complement in $(\mathcal{A}, d)$.*

Proof. Put $\mathcal{A}_0 := \mathcal{Z} \cup \mathcal{E}$. By Propositions 3.2 and 3.3 $\mathcal{A}_0$ is σ-porous in $\mathcal{A}$. Next, we show that, denoting by $L_{b\beta}$ the level sets of the form $L_{b\beta} = \{x \in F(b) : f_b(x) \leq \beta\}$ if $a \in \mathcal{A} \setminus \mathcal{A}_0$, the following property holds:

$$\exists \bar{\beta} > 0, \ \forall \beta \geq \bar{\beta}, \ \exists m \geq 1, \ \forall b \in B_{\mathcal{A}}(a, \frac{1}{m}) \Longrightarrow \emptyset \neq L_{b\beta} \subset mB. \qquad (11.21)$$

Indeed, since $a \notin \mathcal{E}$, there is some $\bar{\beta} > 0$ so that the level sets $\{x \in F(b) : f_b(x) \leq \bar{\beta}\}$ are nonempty for b close to a. Now fix any $\beta \geq \bar{\beta}$ and suppose that there is a sequence $b_n \to a$ in $\mathcal{A}$ and a sequence $\{x_n\}$ with $x_n \in F(b_n)$ so that $f_{b_n}(x_n) \leq \beta$ for each n and $\|x_n\| \to \infty$. Let $y_n = \frac{x_n}{\|x_n\|}$ and, without loss of generality, $y_n \to \bar{y}$. Since $a \notin \mathcal{Z}$, at least one of the following two cases must hold:

(i) $\langle Q_0 \bar{y}, \bar{y} \rangle > 2\tau$ for some $\tau > 0$;

(ii)   there is some $i \geq 1$ with $\langle Q_i \bar{y}, \bar{y} \rangle > 2\tau$ for some $\tau > 0$.

In the first case we have that $\langle Q_0 y, y \rangle > 2\tau$ on some fixed ball B around $\bar{y}$ in $\mathbb{R}^N$. Thus $\langle Q_{0,n} y, y \rangle > \tau$ for any $y \in B$ and n large enough, yielding $\langle Q_{0,n} y_n, y_n \rangle > \tau$ eventually (here $Q_{0,n}$ are the corresponding matrices for b_n). But the latter implies $\langle Q_{0,n} x_n, x_n \rangle > \tau \|x_n\|^2$, contradicting $f_{b_n}(x_n) \leq \beta$ for every n. In the second case, as above $\langle Q_{i,n} x_n, x_n \rangle > \tau \|x_n\|^2$ for n large enough. This is a contradiction, because it means that, for n large enough, x_n will not satisfy the i-th constraint of b_n.

In both cases we arrived at a contradiction and thus (11.21) holds. Observe that, in particular, for any $a \in \mathcal{A} \setminus \mathcal{A}_0$, there exists $\bar{\beta} > 0$ such that, for b close to a, $\inf f_b$ is finite and $\inf f_b \leq \bar{\beta}$. Thus, applying (11.21) at first with $\beta = \bar{\beta}$ and then with $\beta = \bar{\beta} + 1$, we see that the sets

$$\mathcal{A}_m := \left\{ a \in \mathcal{A} \setminus \mathcal{A}_0 : d(a, b) < \frac{1}{m} \implies f_b^1 \subset B(0; m) \right\}, \quad m = 1, 2, \ldots,$$

provide a decomposition of $\mathcal{A} \setminus \mathcal{A}_0$, i.e., $\mathcal{A} \setminus \mathcal{A}_0 = \bigcup_{i=1}^{\infty} \mathcal{A}_m$.

We now show that Corollary 11.4.2 applies in order to get the conclusion of the theorem. To this end, let us fix $m \geq 1$ and $k \geq 1$ and set $r_0 = \frac{1}{m}$ and $\gamma = \lambda(2m^2 + 1)$, with positive λ so small that

$$3\gamma < \frac{1}{16k^2 m^2}.$$

With this choice, if we have $a = (Q_0, \ldots, Q_l, c_0, \ldots, c_l, \alpha_1, \ldots, \alpha_l)$ and $b = (R_0, \ldots, R_l, d_0, \ldots, d_l, \beta_1, \ldots, \beta_l)$ such that $d(a, b) < \lambda r$ for some $r > 0$, then (see (11.20)) for the data functions f_i of a and g_i of b, one has

$$\|f_i - g_i\|_m \leq \gamma r. \tag{11.22}$$

Fix $a \in \mathcal{A}_m$ and positive $r \leq r_0$ and choose $\bar{x} \in B(0; m)$ so that

$$f_a(\bar{x}) < \inf f_a + \gamma r. \tag{11.23}$$

Set

$$\bar{a} := (\bar{Q}_0, \ldots, \bar{Q}_l, \bar{c}_0, \ldots, \bar{c}_l, \bar{\alpha}_1, \ldots, \bar{\alpha}_l),$$

with

$$\bar{Q}_i := Q_i + \frac{rI}{2^2 m^2}, \quad i = 0, \ldots, l,$$

$$\bar{c}_i := c_i - \frac{r\bar{x}}{2m^2}, \quad i = 0, \ldots, l,$$

$$\bar{\alpha}_i := \alpha_i - \frac{r\|\bar{x}\|^2}{2^2 m^2} + \gamma r, \quad i = 1, \ldots, l.$$

Observe that for $i = 1, \ldots, l$,

$$\bar{f}_i(x) = f_i(x) + \frac{r}{4m^2}\|x - \bar{x}\|^2 - \gamma r, \tag{11.24}$$

while

$$\bar{f}_0(x) = f_0(x) + \frac{r}{4m^2}\|x - \bar{x}\|^2 - \frac{r\|\bar{x}\|^2}{4m^2}. \tag{11.25}$$

In particular, $\bar{x} \in F(\bar{a})$, which, together with the choice of $\bar{Q}_i$, shows that $\bar{a} \in \mathcal{A}$. Now let $b \in B_{\mathcal{A}}(\bar{a}; \lambda r)$. It is straightforward to check that $d(b, a) < r$. In particular, $d(b, a) < \frac{1}{m}$ and therefore $f_b^1 \subset B(0; m)$. We show the second condition of the property $(\mathcal{P})$.

First, let us see that $\bar{x} \in F(b)$. Denoting by g_i, $i = 0, \ldots, l$, the data functions corresponding to b, we have, according to (11.22) and (11.24), that for $i = 1, \ldots, l$,

$$g_i(\bar{x}) \leq \bar{f}_i(\bar{x}) + \gamma r = f_i(\bar{x}) \leq 0,$$

i.e., $\bar{x} \in F(b)$. Further, fix $x \in f_b^1$ and suppose that $\|x - \bar{x}\| > \frac{1}{2k}$. Observe that $\|x\| \leq m$. Moreover, we now show that x belongs to $F(a)$. Indeed, using successively (11.24) and (11.22), we have that, for $i = 1, \ldots, l$,

$$f_i(x) \leq \bar{f}_i(x) + \gamma r - \frac{r}{16m^2 k^2} \leq g_i(x) + r\left(2\gamma - \frac{1}{16m^2 k^2}\right) \leq 0,$$

the latter inequality being true because of the choice of γ.

Now, for the same x, having in mind (11.22), (11.23), (11.25) and the fact that $x \in F(a)$, we have

$$g_0(x) \geq \bar{f}_0(x) - \gamma r = f_0(x) + \frac{r\|\bar{x} - x\|^2}{4m^2} - \frac{r\|\bar{x}\|^2}{4m^2} - \gamma r$$

$$> f_0(\bar{x}) + \frac{r}{16k^2 m^2} - \frac{r\|\bar{x}\|^2}{4m^2} - 2\gamma r = \bar{f}_0(\bar{x}) + \frac{r}{16k^2 m^2} - 2\gamma r$$

$$\geq g_0(\bar{x}) + \frac{r}{16k^2 m^2} - 3\gamma r.$$

Summarizing,

$$g_0(x) > g_0(\bar{x}) + r\left(\frac{1}{16k^2 m^2} - 3\gamma\right).$$

Since $\frac{1}{16k^2 m^2} - 3\gamma > 0$, by choosing $\eta < r(\frac{1}{16k^2 m^2} - 3\gamma)$, we see that $\mathrm{diam} f_b^\eta \leq \frac{1}{k}$, and this ends the proof. $\qquad\square$

The class $\mathcal{A}$ can be seen as a subset of the finite dimensional space $\mathbb{R}^{(l+1)(N^2+N)+l}$ and the metric d is inherited by the Euclidean one. Since in finite dimensional spaces σ-porous sets are of Lebesgue measure zero, we have the following immediate corollary:

Corollary 11.4.15 *Let $\mathcal{A}$ be the class of quadratic mathematical programming problems introduced above. Then the set of all problems in $\mathcal{A}$ which are not well-posed is a set first category and of Lebesgue measure zero in $\mathbb{R}^{(l+1)(N^2+N)+l}$.*

The following example shows that the class $\mathcal{E}$ in the theorem above is nonempty.

Example 11.4.16 Consider the problem

$$\begin{aligned} \text{minimize} \quad & \inf x^2 \\ \text{such that} \quad & -x \le 0, 2x \le 0, \end{aligned}$$

and the approximating problems

$$\begin{aligned} \text{minimize} \quad & \inf x^2 \\ \text{such that} \quad & -x \le 0, -\frac{1}{n}x^2 + 2x \le -\frac{1}{n}. \end{aligned}$$

To conclude, I mention two other porosity principles (see [DR, Mar]), and some interesting papers dealing with porosity of "bad situations" in minimum problems (see [BMP, RZ, RZ2]).

12

More exercises

I believed myself to be a mathematician.
In these days I discovered that I am not even an amateur.

(R. Queneau, "Odile")

In this section we collect some more exercises, related to the whole content of the book.

Exercise 12.1 (About polar cones.) Let X be a reflexive Banach space, let $C \subset X$ be a closed convex cone. Then $C^{oo} = C$.

Hint. It is obvious that $C \subset C^{oo}$. Suppose now there is $x \in C^{oo} \setminus C$. Then there are $0^* \neq y^*$ and $a \in \mathbb{R}$ such that

$$\langle y^*, x \rangle > a \geq \langle y^*, c \rangle, \tag{12.1}$$

for all $c \in C$. Show that we can assume $a = 0$ in (12.1). It follows that $y^* \in C^o$ and thus, since $x \in C^{oo}$, we have that $\langle y^*, x \rangle \leq 0$.

Exercise 12.2 Let

$$f(x) = \begin{cases} -\sqrt{x} & \text{if } x \geq 0, \\ \infty & \text{elsewhere.} \end{cases}$$

Evaluate $f_k = f \nabla k \| \cdot \|$ for all k. Let $g(x) = f(-x)$. Find $\inf(f + g)$, $\inf(f_k + g_k)$ and their minimizers. Compare with the result of next exercise.

Exercise 12.3 With the notation of the previous exercise, suppose $f, g \in \Gamma(\mathbb{R}^n)$ and

$$\operatorname{ri} \operatorname{dom} f \cap \operatorname{ri} \operatorname{dom} g \neq \emptyset.$$

Then, for all large k, we have

$$\inf(f + g) = \inf(f_k + g_k)$$

and

$$\operatorname{Min}(f + g) = \operatorname{Min}(f_k + g_k).$$

Hint. Prove that $\inf(f + g) \leq \inf(f_k + g_k)$. There is $y \in \mathbb{R}^n$ such that

$$- \inf(f + g) = f^*(y) + g^*(-y).$$

Take $k > \|y\|$. Then

$$
\begin{aligned}
- \inf(f + g) &= f^*(y) + g^*(-y) = (f^* + I_{kB})(y) + (g^* + I_{kB})(-y) \\
&= (f_k)^*(y) + (g_k)^*(-y) \geq \inf_{z \in \mathbb{R}^n} ((f_k)^*(z) + (g_k)^*(-z)) \\
&= - \inf(f_k + g_k) \geq - \inf(f + g).
\end{aligned}
$$

Observe that the above calculation also shows that y as above is optimal for the problem of minimizing $(f_k)^*(\cdot) + (g_k)^*(- \cdot)$ on $\mathbb{R}^n$.

Now, using $k > \|y\|$,

$$
\begin{aligned}
x \in \mathrm{Min}(f + g) &\Leftrightarrow f(x) + g(x) = -f^*(y) - g^*(-y) \\
&\Leftrightarrow x \in \partial f^*(y) \cap \partial g^*(-y) \\
&\Leftrightarrow x \in \partial(f^* + I_{kB})(y) \cap \partial(g^* + I_{kB})(-y) \\
&\Leftrightarrow x \in \partial(f_k)^*(y) \cap \partial(g_k)^*(-y) \\
&\Leftrightarrow x \in \mathrm{Min}(f_k + g_k).
\end{aligned}
$$

Exercise 12.4 Let $\{x_n^i\}$, $i = 1, \ldots, k$ be k sequences in a Euclidean space, and suppose $x_n^i \to x^i$ for all i. Prove that $\mathrm{co} \bigcup x_n^i$ converges in the Hausdorff sense to $\mathrm{co} \bigcup x^i$.

Exercise 12.5 Let X be a Banach space and suppose $f, g \in \Gamma(X)$, $f \geq -g$, $f(0) = -g(0)$. Then

$$\{y^* : f^*(y^*) + g^*(-y^*) \leq 0\} = \partial f(0) \cap -\partial g(0).$$

Exercise 12.6 Let X be a Banach space, let $f \in \Gamma(X)$ be Fréchet differentiable, and let $\sigma > 0$. Set

$$S_\sigma := \{x \in X : f(x) \leq f(y) + \sigma \|y - x\|, \forall y \in X\},$$

and

$$T_\sigma := \{x \in X : \|\nabla f(x)\|_* \leq \sigma\}.$$

Prove that $S_\sigma = T_\sigma$ are closed sets. Which relation holds between the two sets if f is not assumed to be convex?

Exercise 12.7 In the setting of Exercise 12.6, prove that f is Tykhonov well-posed if and only if $S_\sigma \neq \emptyset$ for all $\sigma > 0$ and $\mathrm{diam}\, S_\sigma \to 0$ as $\sigma \to 0$. Deduce an equivalence when f is also Fréchet differentiable. Is convexity needed in both implications? Give an example when the equivalence fails if f is not convex.

Hint. Suppose f is Tykhonov well-posed. Clearly, $S_\sigma \neq \emptyset$ for all σ. Without loss of generality, suppose $f(0) = 0 = \inf f$. Suppose $\operatorname{diam} S_\sigma \geq 2a$, for some $a > 0$, and let $0 < m = \inf_{\|x\|=a} f(x)$. There is $x_n \in S_{\frac{1}{n}}$ such that $\|x_n\| \geq a$. Show that this leads to a contradiction. Conversely, show that $\bigcap_{\sigma>0}$ is a singleton and the set of the minimizers of f. From the Ekeland variational principle deduce that

$$f^{\inf f + a^2} \subset B_a(S_a)$$

and use the Furi Vignoli characterization of Tykhonov well-posedness. As an example, consider $f(x) = \arctan x^2$.

Variational convergences are expressed in terms of set convergences of epigraphs. On the other hand, not only is the behavior of the epigraphs important. How the level sets move under convergence of epigraphs is an important issue. Thus, the next exercises provide gap and excess calculus with level sets and epigraphs. In the space $X \times \mathbb{R}$ we shall consider the box norm.

Exercise 12.8 Let X be a metric space, let $f \colon X \to (-\infty, \infty]$ be lower semicontinuous, let $C \in c(X)$ Prove that

(i) $D(C, f^a) = d$ implies $D(C \times \{a - d\}, \operatorname{epi} f) = d$.
(ii) $\forall b \in \mathbb{R}$ and $\forall a \geq b$ such that $f^a \neq \emptyset$,

$$D(C \times \{b\}, \operatorname{epi} f) \geq \min\{D(C, f^a), a - b\}.$$

(iii) $\forall b \in \mathbb{R}$ and $\forall a \geq b$ such that $f^a \neq \emptyset$, $D(C \times \{b\}, \operatorname{epi} f) = d$ implies $b + d \geq \inf f$.
(iv) $D(C \times \{b\}, \operatorname{epi} f) = d$ implies $D(C, f^{b+d+\varepsilon}) \leq d$, for all $\varepsilon > 0$.
(v) $D(C \times \{b\}, \operatorname{epi} f) = d$ implies $D(C, f^{b+d-\varepsilon}) \geq d$, for all $\varepsilon > 0$.

Exercise 12.9 Let X be a metric space, let $f \colon X \to (-\infty, \infty]$ be lower semicontinuous, let $C \in c(X)$ Prove that

(i) $e(C, f^a) = d$ implies $e(C \times \{a - d\}, \operatorname{epi} f) = d$.
(ii) $\forall b \in \mathbb{R}$ and $\forall a \geq b$ such that $f^a \neq \emptyset$,

$$e(C \times \{b\}, \operatorname{epi} f) \leq \max\{e(C, f^a), a - b\}.$$

(iii) $\forall b \in \mathbb{R}$ and $\forall a \geq b$ such that $f^a \neq \emptyset$, $e(C \times \{b\}, \operatorname{epi} f) = d$ implies $b + d \geq \inf f$.
(iv) $e(C \times \{b\}, \operatorname{epi} f) = d$ implies $e(C, f^{b+d+\varepsilon}) \leq d$, for all $\varepsilon > 0$.
(v) $e(C \times \{b\}, \operatorname{epi} f) = d$ implies $e(C, f^{b+d-\varepsilon}) \geq d$, for all $\varepsilon > 0$.

Exercise 12.10 Let X be an E-space and $f \in \Gamma(X)$. Then, setting $f_n(x) = f(x) + \frac{1}{n}\|x\|^2$, prove that $f_n \to f$ for the Attouch–Wets convergence and that $f_n(\cdot) - \langle p, \cdot \rangle$ is Tykhonov well-posed for all n and for all $p \in X^*$.

Exercise 12.11 Let X be a reflexive Banach space, and $f \in \Gamma(X)$. Find a sequence $\{f_n\}$ such that $f_n \in \Gamma(X)$ are Tykhonov well-posed, everywhere Fréchet differentiable, and $f_n \to f$ for the Attouch–Wets convergence.

Hint. Take an equivalent norm $\| \cdot \|$ in X such that both X and X^* are now E-spaces. From Exercise 12.10 we know that $f^* + \frac{1}{n} \| \cdot \|_*^2 - \langle p, \cdot \rangle$ is Tykhonov well-posed for all n. Thus $(f^* + \frac{1}{n} \|x\|_*^2)^*$ is everywhere Fréchet differentiable for all n. It follows that $g_n(x) = (f^* + \frac{1}{n} \| \cdot \|_*^2)^*(x) + \frac{1}{n} \|x\|^2$ is Fréchet differentiable and Tykhonov well-posed for all n. Prove that $g_n \to f$ for the Attouch–Wets convergence.

Exercise 12.12 Consider the following game. Rosa and Alex must say, at the same time, a number between 1 and 4 (inclusive). The one saying the highest number gets from the other what was said. There is one exception for otherwise the game is silly. If Alex says n and Rosa $n - 1$, then Rosa wins n, and conversely. Write down the matrix associated with the game, and find its value and its saddle points.

Hint. Observe that it is a fair game, and use Exercise 7.2.6.

We make one comment on the previous exercise. The proposed game (or maybe an equivalent variant of it) was invented by a rather famous person, with the intention of creating a computer program able to *learn* from the behavior of an opponent, in order to be able to understand its psychology and to beat it after several repetitions of the game. Unfortunately, he had a student with some knowledge of game theory, proposing to him the use of the optimal strategy, whose existence is guaranteed by the theorem of von Neumann. Thus, when telling the computer to play this strategy over and over, no clever idea could do better than a tie (on average) with resulting great disappointment for the famous person. I like this story, since it shows well how challenging game theory can be from the point of view of psychology.

Exercise 12.13 Consider the following game. Emanuele and Alberto must show each other one or two fingers and say a number, at the same time. If both are right or wrong, they get zero. If one is wrong and the other one is right, the one who is right gets the number he said. Determine what they should play, knowing that both are very smart. Do the same if the winner always gets 1, instead of the number he said.

Hint. The following matrix should tell you something.

$$\begin{pmatrix} 0 & 2 & -3 & 0 \\ -2 & 0 & 0 & 3 \\ 3 & 0 & 0 & -4 \\ 0 & -3 & 4 & 0 \end{pmatrix}.$$

Ask yourself if the result of Exercise 7.2.6 can be used. My answer (but you should check) is that they always say "three" and play 1 with probability x, 2 with probability $1 - x$, where $\frac{4}{7} \le x \le \frac{3}{5}$.

Exercise 12.14 Let $f \colon \mathbb{R}^2 \to \mathbb{R}$ be continuous convex, and suppose $\lim_{|x| \to \infty} f(x, mx) = \infty$ for all $m \in \mathbb{R}$. Prove that f is Tykhonov well-posed in the generalized sense. Does the same hold in infinite dimensions?

Hint. Consider a separable Hilbert space with basis $\{e_n : n \in \mathbb{N}\}$, and the function

$$f(x) = \sum_{n=1}^{\infty} \frac{\langle x, e_n \rangle^2}{n^2} - \langle x^*, x \rangle,$$

where $x^* = \sum \frac{1}{n} e_n$. Then show f is not even lower bounded.

Exercise 12.15 This is a cute example taken from T. Rockafellar's book *Convex Analysis*, i.e., the example of a function $f \colon \mathbb{R}^2 \to \mathbb{R}$ continuous convex, assuming a minimum on each line, and not assuming a minimum on $\mathbb{R}^2$. Let C be the epigraph of the function $g(x) = x^2$ and consider the function $f(x,y) = d^2[(x,y), C] - x$. Prove that f fulfills the above property, and prove also that f is $C^1(\mathbb{R}^2)$.

Exercise 12.16 Let $f \in \Gamma(\mathbb{R}^n)$. The following are equivalent:

- f is lower bounded and $\mathrm{Min}\, f = \emptyset$;
- $0 \in \mathrm{dom}\, f^*$ and there is y such that $(f^*)'(0, ; y) = -\infty$.

Hint. Remember that $f^*(0) = -\inf f$ and that $\mathrm{Min}\, f = \partial f^*(0)$. Prove that $\partial f(x) = \emptyset$ if and only if there exists a direction y such that $f'(x; y) = -\infty$ (remember that $f'(x; \cdot)$ is sublinear).

Exercise 12.17 Prove that $\mathrm{cl\, cone\, dom}\, f = \left(0^+((f^*)^a)\right)^\circ$, for $a > -f(0)$.

Hint. Observe that $(f^*)^a \neq \emptyset$. $(f^*)^a = \{x^* : \langle x^*, x \rangle - f(x) \le a, \forall x \in \mathrm{dom}\, f\}$. Thus $z^* \in (0^+((f^*)^a))^\circ$ if and only if $\langle z^*, x \rangle \le 0$ for all $x \in \mathrm{dom}\, f$, if and only if $\langle z^*, y \rangle \le 0$ for all $y \in \mathrm{cl\, cone\, dom}\, f$.

Exercise 12.18 This is much more than an exercise. Here I want to introduce the idea of "minimizing" a function which is not real valued, but rather takes values in a Euclidean space. This subject is known under the name of *vector optimization* (also Pareto optimization, multicriteria optimization) and it is a very important aspect of the general field of optimization. Minimizing a function often has the meaning of having to minimize some cost. However, it can happen that one must take into account several cost functions at the same time, not just one. Thus it is important to give a meaning to the idea of minimizing a function $f = (f_1, \ldots, f_n)$, where each f_i is a scalar function. And this can be generalized by assuming that f takes values on a general space, ordered in some way (to give a meaning to the idea of minimizing). Here I want to talk a little about this. I will consider very special cases, in order to avoid any technicalities. What I will say can be deeply generalized. The interested reader could consult the book by Luc [Luc] to get a more complete idea of the subject.

So, let $P \subset \mathbb{R}^l$ be a pointed (i.e., $P \cap -P = \{0\}$) closed and convex cone with nonempty interior. The cone P induces on $\mathbb{R}^l$ the order relation $\le_P$ defined as follows: for every $y_1, y_2 \in \mathbb{R}^l$,

$$y_1 \leq_P y_2 \overset{\text{def}}{\iff} y_2 \in y_1 + P.$$

Here are some examples of cones: in $\mathbb{R}^n$, $P = \{x = (x_1, \ldots, x_n) : x_i \geq 0, \forall i\}$; in $\mathbb{R}^2$, $P = \{x = (x, y) :$ either $x > 0$ or $x = 0$ and $y \geq 0\}$: this cone, which is *not* closed, induces the so called *lexicographic order*. In l^2, let $P = \{x = (x_1, \ldots, x_n, \ldots) : x_i \geq 0, \ \forall i\}$: this cone has empty interior, in l^∞ let $P = \{x = (x_1, \ldots, x_n, \ldots) : x_i \geq 0, \ \forall i\}$: this cone has nonempty interior.

Given C, a nonempty subset of $\mathbb{R}^l$, we denote by $\text{Min}\, C$ the set

$$\text{Min}\, C \overset{\text{def}}{=} \{y \in C : C \cap (y - P) = \{y\}\}.$$

The elements of the set $\text{Min}\, C$ are called the *minimal points* of C (with respect to the order induced by the cone P).

This is not the only notion of minimality one can think of. For instance, the above notion of minimality can be strengthened by introducing the notion of *proper minimality*. A point $y \in C$ is a *properly minimal point* of C if there exists a convex cone P_0 such that $P \setminus \{0\} \subset \text{int}\, P_0$ and y is a minimal point of C with respect to the order given by the cone P_0. We denote the set of the properly minimal points of C by $\text{Pr Min}\, C$.

The concept of minimal point can also be weakened. Define the set

$$\text{Wmin}\, C \overset{\text{def}}{=} \{y \in C : C \cap (y - \text{int}\, P) = \emptyset\}$$

of the weakly minimal points of the set C. Clearly

$$\text{Pr Min}\, C \subset \text{Min}\, C \subset W\, \text{Min}\, C.$$

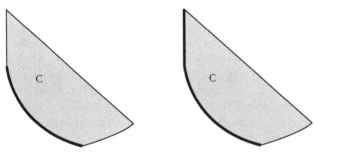

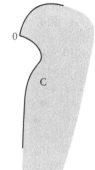

Efficient points of C.

Weakly efficient points of C.

0 is not properly efficient for C.

Figure 12.1.

Let us now consider a function $f\colon \mathbb{R}^k \to \mathbb{R}^l$. Let A be a subset of $\mathbb{R}^k$. The set of the efficient points of A is

$$\text{Eff}(A, f) \overset{\text{def}}{=} \{x \in A : f(x) \in \text{Min } f(A)\}.$$

In the same way we can introduce the sets $\text{WEff}(A, f)$ and $\text{PrEff}(A, f)$.

And it is clearly possible and interesting to define a notion of convexity for vector valued functions. Here it is.

Let $A \subset \mathbb{R}^k$ be a convex set, and $f\colon A \subset \mathbb{R}^k \to \mathbb{R}^l$. Then f is said to be a *P-convex* (or simply convex, when it is clear which is the cone P inducing the order relation) function on A if for every $x_1, x_2 \in A$ and for every $\lambda \in [0, 1]$,

$$f(\lambda x_1 + (1 - \lambda)x_2) \in \lambda f(x_1) + (1 - \lambda)f(x_2) - P,$$

and it is said to be a *strictly P-convex* function if for every $x_1, x_2 \in A$, $x_1 \neq x_2$ and for every $\lambda \in (0, 1)$,

$$f(\lambda x_1 + (1 - \lambda)x_2) \in \lambda f(x_1) + (1 - \lambda)f(x_2) \setminus \text{int } P.$$

Now I only suggest some results, focusing essentially on some aspects of convexity, and stability. I leave the proofs as exercises, and sometimes outline the main ideas of the proofs. The first is an existence result, which is stated in a very particular case.

Proposition 12.19 *Under the setting previously described, let $A \subset \mathbb{R}^k$ be nonempty, closed and such that there exists $x \in \mathbb{R}^k$ such that $A \subset x + P$. Then $\text{Min } A$ is nonempty.*

Proof. (Outline) Without loss of generality, suppose $x = 0$. Prove that there exists $x^* \in \mathbb{R}^k$ such that $\langle x^*, p \rangle > 0$ for all $p \in P$, $p \neq 0$ (the origin can be separated from $\text{co}(A \cap \partial B)$, since the cone P is pointed). Prove that $\lim_{c \in C, \|c\| \to \infty} \langle x^*, c \rangle = \infty$ (arguing by contradiction). Then $g(a) = \langle x^*, a \rangle$ assumes minimum on A. Prove that if $\bar{x}$ minimizes g on A then $\bar{x} \in \text{Min } A$.

With a little more effort one could prove that under the previous assumptions $\text{Pr Min } A$ is actually nonempty. $\square$

We now see some properties of the convex functions.

Proposition 12.20 *Let $A \subset \mathbb{R}^k$ be a convex set and let $f\colon \mathbb{R}^k \to \mathbb{R}^l$ be a P-convex function. Then*

(i) $f(A) + P$ *is a convex subset of* $\mathbb{R}^l$.
(ii) f *is continuous.*
(iii)   *If f is strictly $P-$ convex then $\text{WEff}(A, f) = \text{Eff}(A, f)$.*
(iv)   *Defining in the obvious way the level sets of f, prove that, for all $a, b \in \mathbb{R}^k$ such that $f^a \neq \emptyset$, $f^b \neq \emptyset$, it holds $0^+(f^a) = 0^+(f^b)$.*
(v)   *Calling H the common recession cone of the level sets of f, show that, if $0^+(A) \cap H = \{0\}$, then $f(A) + P$ is closed.*

We turn now our attention to convergence issues. Prove the following.

Proposition 12.21 *Let C_n be closed convex subsets of $\mathbb{R}^l$. Suppose $C_n \xrightarrow{K} C$. Then*

(i) $\operatorname{Li} \operatorname{Min} C_n \supset \operatorname{Min} C$;
(ii) $\operatorname{Li} Pr \operatorname{Min} C_n \supset Pr \operatorname{Min} C$;
(iii) $\operatorname{Ls} W \!\min C_n \subset \operatorname{Min} C$.

Proof. (Outline) For (i), it is enough to prove that for every $c \in C$ and for every $\varepsilon > 0$ there exists $y_n \in \operatorname{Min} C_n$ such that $d(y_n, c) < \varepsilon$. There exists a sequence $\{c_n\}$ such that $c_n \in C_n$ for all n and $c_n \to c$. Show that $D_n := (c_n - P) \cap C_n \subset B(c; \varepsilon)$ eventually. Since $\operatorname{Min} D_n$ is nonempty and $\operatorname{Min} D_n \subset \operatorname{Min} C_n$, the conclusion of (i) follows. The proof of (ii) relies on the fact that the proper minimal points are, under our assumptions, a dense subset of the minimal points. The proof of (iii) is straightforward. □

Thus the minimal and properly minimal sets enjoy a property of lower convergence, while the weakly minimal sets enjoy a property of upper convergence. Easy examples show that opposite relations do not hold in general. However it should be noticed that, if $\operatorname{Min} A = W \operatorname{Min} A$, then actually from (i) and (iii) above we can trivially conclude that $\operatorname{Min} C_n$ converges to $\operatorname{Min} C$ in Kuratowski sense.

Theorem 12.22 *Let $A_n \subset \mathbb{R}^k$ be closed convex sets, let f_n and f be P-convex functions. Suppose*

(i) $0^+(A) \cap H_f = \{0\}$;
(ii) $A_n \xrightarrow{K} A$;
(iii) $f_n \to f$ *with respect to the continuous convergence (i.e., $x_n \to x$ implies $f_n(x_n) \to f(x)$).*

Then

$$f_n(A_n) + P \xrightarrow{K} f(A) + P.$$

Theorem 12.23 *Under the same assumptions as the previous theorem we have*

(i) $\operatorname{Min} f(A) \subset \operatorname{Li} \operatorname{Min} f_n(A_n)$.
(ii)   *If moreover f is strictly convex,*

$$\operatorname{Min} f_n(A_n) \xrightarrow{K} \operatorname{Min} f(A) \quad and \quad \operatorname{Eff}(A_n, f_n) \xrightarrow{K} \operatorname{Eff}(A, f).$$

If anyone is really interested in having the proofs of the previous exercises, he can send me an e-mail and I will send back the paper.

A

Functional analysis

A.1 Hahn–Banach theorems

Recall that a *topological real vector space* is a real vector space X, endowed with a Hausdorff topology making continuous the operations of sum and multiplication by a real number. We shall indicate by X' the space of the continuous linear functionals from X to $\mathbb{R}$.

A fundamental theorem in this setting is the following analytic form of the Hahn–Banach theorem.

Theorem A.1.1 *Let X be a vector space and let $p: X \to \mathbb{R}$ be a sublinear functional. Let $E \subset X$ be a subspace and $l: E \to \mathbb{R}$ a linear form such that $l(e) \leq p(e)\ \forall e \in E$. Then there exists a linear functional $L: X \to \mathbb{R}$ extending l and satisfying $L(x) \leq p(x)\ \forall x \in X$.*

The previous theorem does not involve topological structures, but it is not difficult to get results for linear continuous functionals from it. Here are two examples.

Corollary A.1.2 *Let X be a topological vector space and let $p: X \to \mathbb{R}$ be a sublinear continuous functional. Then there exists a linear bounded functional $L: X \to \mathbb{R}$ satisfying $L(x) \leq p(x)\ \forall x \in X$.*

Proof. Take $E = \{0\}$ in the previous theorem. Boundedness of L follows from the fact that p is bounded above in a neighborhood of the origin. □

From Corollary A.1.2 it follows in particular that in a Euclidean space, given a real valued sublinear functional p, it is possible to find a linear functional minorizing p.

Corollary A.1.3 *Let X be a Banach space and let $E \subset X$ be a subspace. Let $l: E \to \mathbb{R}$ be a linear bounded functional. Then there exists a linear bounded functional $L: X \to \mathbb{R}$ extending l and having the same norm of l.*

Proof. It is enough to apply Theorem A.1.1, with $p(x) = \|x\|\|l\|$. □

We are particularly interested in the so called geometric forms of the Hahn–Banach theorem. To introduce them, let $0 \neq x^* \in X'$ and $c \in \mathbb{R}$. We shall use the familiar notation $\langle x^*, x \rangle$ rather than $x^*(x)$, and as we have throughout the book, we shall call a set H of the form

$$\{x \in X : \langle x^*, x \rangle = c\}$$

a hyperplane. We say that H (strictly) separates two sets A, B if each of the half spaces determined by H contains one of the sets. In formula, $(A \subset \{x : \langle x^*, x \rangle < c\})$ $A \subset \{x : \langle x^*, x \rangle \leq c\}$, $(B \subset \{x : \langle x^*, x \rangle > c\})$, $B \subset \{x : \langle x^*, x \rangle \geq c\}$. We are interested in conditions guaranteeing that convex sets can be separated. It is worth noticing that it is not always possible to separate two disjoint convex sets:

Example A.1.4 In $L^2[0, 1]$ the sets

$$A := \{f \colon [0, 1] \to \mathbb{R} : f \text{ is continuous and } f(0) = a\},$$

$$B := \{f \colon [0, 1] \to \mathbb{R} : f \text{ is continuous and } f(0) = b\},$$

(with $a \neq b$) are dense hyperplanes and no linear bounded functional can separate them.

From the previous Hahn–Banach theorem, it is possible to get:

Theorem A.1.5 *Let X be a (real) topological vector space, let A be a nonempty open convex set, and B a nonempty convex set such that $A \cap B = \emptyset$. Then there exists a hyperplane separating A from B.*

Proof. (Outline) To begin with, let us suppose B is a singleton, say $\{b\}$. Without loss of generality, we can suppose $0 \in A$. Consider the Minkowski functional m_A associated to A: $m_A(x) = \inf\{\lambda > 0 : x \in \lambda A\}$ (see Exercise 1.2.15), the linear space Y generated by b and the linear functional l, defined on Y and such that $l(b) = m_A(b)$. It is easy to verify that $l(y) \leq m_A(y), \forall y \in Y$. We appeal to Theorem A.1.1 to claim the existence of a linear functional x^*, defined on all of X, extending l, and such that $\langle x^*, x \rangle \leq m_A(x), \forall x \in X$. Moreover, $\langle x^*, a \rangle \leq 1$ $\forall a \in A$. Setting $W = A \cap -A$, then W is a *symmetric* open convex set contained in A. Given $\varepsilon > 0$, we have that

$$|\langle x^*, x \rangle| \leq \varepsilon,$$

for all $x \in \varepsilon W$. This shows that $x^* \in X'$. Now, for $a \in A$ we have

$$\langle x^*, a \rangle \leq 1 \leq m_A(b) = l(b) = \langle x^*, b \rangle.$$

This establishes the theorem when B is a singleton. In the general case, as in the previous step, separate the set $M = A - B$ (which is open and convex) from the origin. We have existence of $x^* \in X'$ such that, $\forall m \in M$,

$$\langle x^*, m \rangle \leq \langle x^*, 0 \rangle = 0,$$

i.e.,

$$\langle x^*, a \rangle \leq \langle x^*, b \rangle,$$

for all $a \in A, b \in B$. Setting $c = \sup\{\langle x^*, a \rangle : a \in A\}$, we then have

$$\langle x^*, a \rangle \leq c \leq \langle x^*, b \rangle,$$

for all $a \in A, b \in B$. This allows us to conclude. Observe also that due to openness of A, we actually have the more precise information $\langle x^*, a \rangle < c$ for all $a \in A$. □

Remember that a topological vector space is said to be *locally convex* if the origin has a fundamental system of neighborhoods made by convex sets.

Theorem A.1.6 *Let X be a (real) topological locally convex vector space. Let A be a nonempty compact convex set and B a nonempty closed convex set such that $A \cap B = \emptyset$. Then there exists a hyperplane strictly separating A from B.*

Proof. (Outline) Since A is a compact set, there is an open convex symmetric neighborhood N of the origin such that $A + N \cap B = \emptyset$. Now apply the previous result to $A + N$ and B, to get existence of $x^* \neq 0$ and $c \in \mathbb{R}$ such that

$$A + N \subset \{x : \langle x^*, x \rangle \leq c\}, \quad B \subset \{x : \langle x^*, x \rangle \geq c\}.$$

This implies that there is $a > 0$ such that $A \subset \{x : \langle x^*, x \rangle \leq c - a\}$. The proof is complete. □

Remark A.1.7 The previous theorems are used in this book mainly with X a Banach space and $X' = X^*$ its topological dual, but also with $X = Y^*$ and Y a Banach space. In this case X is endowed with the weak* topology, and so X' is (isomorphic to) Y. In other words the bounded linear functional giving rise to the separation is an element of the space Y.

In the first theorem, in general it is not possible to get a strict separation. Think of an open convex set and one of its boundary points.

An interesting application of the first theorem is the following: given a nonempty convex set A and a point $x \in A$, x is said to be a supporting point for A if there exists a closed hyperplane, of the form $\{x \in X : \langle x^*, x \rangle = c\}$, containing x and leaving A on one of the half spaces determined by the hyperplane (so that x is a minimum, or a maximum, point on A for x^*). Then, if A is a closed convex set with nonempty interior, each of its boundary points is a supporting point. This property fails if we do not assume that A has interior points:

Exercise A.1.8 Let A be the following subset of l^2:

$$A := \{x = (x_1, x_2, \ldots, x_n, \ldots) : x_i \geq 0 \, \forall i \text{ and } \|x\| \leq 1\}.$$

Verify that, if $x \in A$ is such that $x_i > 0 \, \forall i$ and $\|x\| < 1$, then x is not a supporting point for A.

A consequence of the second theorem is:

Corollary A.1.9 *If A is a closed convex set, then it is the intersection of the closed half spaces containing it.*

Corollary A.1.10 *Let X be a Banach space, let A, B be closed convex sets such that $D(A, B) > 0$ (remember, $D(A, B) = \inf\{d(a, b) : a \in A, b \in B\}$). Then there exists a hyperplane strictly separating them.*

Proof. Take $a > 0$ so small that $D(A, B) > 2a$. Then $S_a[A]$ and B can be separated, by Theorem A.1.5.   □

We now provide an finite dimensional version of the Hahn–Banach theorem. I believe it is useful, and it is not a direct consequence of the infinite-dimensional case.

Let us begin by proving some auxiliary, yet interesting results.

Theorem A.1.11 *Let C be a convex subset of the Euclidean space $\mathbb{R}^l$, let $\bar{x} \in \overline{C^c}$. Then there are an element $0 \neq x^* \in \mathbb{R}^l$ and a real k such that*

$$\langle x^*, c \rangle \geq k \geq \langle x^*, \bar{x} \rangle,$$

$\forall c \in C$.

Proof. At first, suppose $\bar{x} \notin \overline{C}$. The we can project $\bar{x}$ on $\overline{C}$. Call p its projection. Then

$$\langle p - \bar{x}, c - p \rangle \geq 0,$$

$\forall c \in C$. Setting $x^* = p - \bar{x}$, the above inequality can be written

$$\langle x^*, c - \bar{x} \rangle \geq \|x^*\|^2,$$

i.e.,

$$\langle x^*, c \rangle \geq \|x^*\|^2 + \langle x^*, \bar{x} \rangle$$

$\forall c \in C$ and this shows the claim in the particular case $\bar{x} \notin \overline{C}$. Now, if $\bar{x} \in \overline{C} \backslash C$, take a sequence $\{x_n\} \subset C^c$ such that $x_n \to \bar{x}$. From the first step of the proof, find x_n^* and k_n such that

$$\langle x_n^*, c \rangle \geq k_n > \langle x^*, x_n \rangle,$$

$\forall c \in C$. Observe that, without loss of generality, we can suppose $\|x_n^*\| = 1$. Moreover, it is $\{k_n\}$ bounded (with this choice of x_n^*). Thus, possibly passing to a subsequence, we can suppose $x_n^* \to x^*$, $k_n \to k$. Now we can take the limit in the above string of inequalities, to get

$$\langle x^*, c \rangle \geq k \geq \langle x^*, \bar{x} \rangle,$$

$\forall c \in C$.   □

Remembering the definition of a supporting hyperplane (see Definition 3.2.2), we get from Theorem A.1.11 the following corollary (compare it with the infinite-dimensional situation described in Exercise A.1.8).

Corollary A.1.12 *Let C be a closed convex subset of $\mathbb{R}^l$ and let x be in the boundary of C. Then there is a hyperplane supporting C at x.*

Theorem A.1.13 *Let A, C be closed convex subsets of $\mathbb{R}^l$ such that $\operatorname{ri} A \cap \operatorname{ri} C = \emptyset$. Then there is $0 \neq x^*$ such that*

$$\langle x^*, a \rangle \geq \langle x^*, c \rangle,$$

$\forall a \in A, \ \forall c \in C.$

Proof. Since $0 \in (\operatorname{ri} A - \operatorname{ri} C)^c$, we can apply Theorem 4.2.16 to find $0 \neq x^*$ such that

$$\langle x^*, x \rangle > 0,$$

$\forall x \in \operatorname{ri} A - \operatorname{ri} C$. This amounts to saying that

$$\langle x^*, a \rangle \geq \langle x^*, c \rangle,$$

$\forall a \in \operatorname{cl} \operatorname{ri} A = A, \ \forall c \in \operatorname{cl} \operatorname{ri} C = C.$ $\qquad\qquad\square$

A.2 The Banach–Dieudonné–Krein–Smulian theorem

When X is a reflexive Banach space, the weak* topology in X^*, of course, agrees with the weak topology. In particular, a closed convex set is weakly closed. We are interested in seeing a general weak* closedness criterion, without assuming reflexivity. The theorem we want to prove here is the following:

Theorem A.2.1 *Let X be a Banach space with topological dual X^*. Suppose A is a convex subset of X^* such that $A \cap rB^*$ is weak* compact for all $r > 0$. Then A is weak* closed.*

Proof. The proof relies on some intermediate results. First of all, observe that A is norm closed. This easily follows from the fact that if a sequence $\{a_n\} \subset A$ is convergent to a, then it is bounded; thus it belongs to $A \cap rB^*$ for some $r > 0$ and by compactness, $a \in A$.

We next consider, given a set S in X, its *polar set*:

$$S^\circ := \{x^* \in X^* : |\langle x^*, x \rangle| \leq 1 \text{ for all } x \in S\}.$$

Fix $r > 0$ and denote by $\mathcal{F}$ the collection of all *finite* subsets of $\frac{1}{r}B$. Then we have

Lemma A.2.2

$$\bigcap_{S \in \mathcal{F}} S^\circ = rB^*.$$

Proof. Call $F = \bigcap_{S \in \mathcal{F}} S^\circ$. Then $S \subset \frac{1}{r}B$ implies $S^\circ \supset \left(\frac{1}{r}B\right)^\circ = rB^*$. Thus

$$F \supset rB^*.$$

Conversely, we show that $(rB^*)^c \subset F^c$. Take x^* with $||x^*|| > r$. Then there exists $x \in X$ such that $||x|| = 1$ and $\langle x^*, x \rangle > r$. Set $S = \{\frac{x}{r}\} \subset \mathcal{F}$. Thus $x^* \notin S^\circ$ and thus $x^* \notin F$.   □

Proposition A.2.3 *Suppose K is a convex subset of X^* such that $K \cap rB^*$ is weak* compact for all $r > 0$. Suppose moreover $K \cap B^* = \emptyset$. Then there is $x \in X$ such that $\langle x^*, x \rangle \geq 1$ for all $x^* \in K$.*

Proof. Set $S_0 = \{0_X\}$. Suppose we have found finite sets $S_0, \ldots, S_{k-1}$ so that $jS_j \subset B$ and

$$S_0^\circ \cap \cdots \cap S_{k-1}^\circ \cap K \cap kB^* = \emptyset.$$

For $k = 1$ the previous formula is true. Set

$$Q = S_0^\circ \cap \cdots \cap S_{k-1}^\circ \cap K \cap (k+1)B^*.$$

Suppose $S^\circ \cap Q \neq \emptyset$ for every finite set $S \subset \frac{1}{k}B$. As Q is weak* compact, this would mean that $Q \cap \bigcap_{S \in \mathcal{F}} S^\circ \neq \emptyset$ and appealing to the Lemma A.2.2, we would finally have $Q \cap kB^* \neq \emptyset$, a contradiction. Thus there must be a finite set S_k such that $kS_k \subset B$ and satisfying

$$S_0^\circ \cap \cdots \cap S_k^\circ \cap K \cap (k+1)B^* = \emptyset.$$

As a result, we find a sequence of finite sets S_k such that

$$K \cap \bigcap_{k=1}^{\infty} S_k^\circ = \emptyset.$$

The set $\{\bigcup_{k \in \mathbb{N}} S_k\}$ is countable; thus we can arrange it in a sequence $\{x_n\}$. Clearly $x_n \to 0$ (remember that $kS_k \subset B$). Now we consider the linear bounded operator T from X^* to the Banach space c_0 of the sequences convergent to 0 (the norm of an element $r = (r_1, r_2, \ldots, r_n, \ldots) \in c_0$ being $||r|| = \sup_{n \in \mathbb{N}} |r_n|$):

$$T : X^* \to c_0 \quad Tx^* = \{\langle x^*, x_n \rangle\}.$$

$T(K)$ is a convex subset of c_0. From $K \cap \bigcap_{k=1}^{\infty} S_k^\circ = \emptyset$ we get

$$||Tx^*|| := \sup_{n \in \mathbb{N}} |\langle x^*, x_n \rangle| \geq 1,$$

for all $x^* \in K$. This means that the set $T(K)$ does not intersect the unit (open) ball of c_0. Appealing to the first Hahn–Banach theorem gives us existence of a nonzero element of l^1, call it y, and a constant $c \in \mathbb{R}$, such that

$$\langle y, z \rangle \leq c \leq \langle y, u \rangle,$$

$\forall z \in B_{c_0}$, $\forall u \in T(K)$. Since $c > 0$, we can suppose $c \geq 1$. Thus, the element $x = \sum_{n \in \mathbb{N}} y_n x_n$ fulfills the required properties. □

We are now able to prove the theorem. Suppose $x^* \notin A$. Then $0 \notin A - \{x^*\}$. As this last set is norm closed, there is $r > 0$ such that $rB^* \cap A - \{x^*\} = \emptyset$. Thus

$$B^* \cap \frac{1}{r}(A - \{x^*\}) = \emptyset.$$

The set $K := \frac{1}{r}(A - \{x^*\})$ fulfills the assumptions of Proposition A.2.3. Thus, there exists $x \in X$ such that $\langle y^*, x \rangle \geq 1$ for all $y^* \in \frac{1}{r}(A - \{x^*\})$. It follows that for all $a^* \in A$,

$$\langle a^*, x \rangle \geq \langle x^*, x \rangle + r.$$

In other words, the weak* open set

$$O := \left\{ z^* \in X^* : \langle z^*, x \rangle < \langle x^*, x \rangle + \frac{r}{2} \right\}$$

is such that $O \cap A = \emptyset$ and $x^* \in O$. Thus A^c is a weak* open set and this concludes the proof. □

Exercise A.2.4 Prove that the dual space of c_0, the Banach space defined in the previous proof, is the space l^1 of the elements $y = (y_1, \ldots, y_n, \ldots)$ such that $\sum_{n \in \mathbb{N}} |y_n| < \infty$, equipped with the norm $\|y\| = \sum_{n \in \mathbb{N}} |y_n|$.

B

Topology

In this appendix we provide some topological results. The first one is the Baire theorem, the others are related to hypertopologies. In this setting, we start by proving a necessary condition for metrizability of the hyperspace, endowed with the topologies having the lower Vietoris topology as lower part. Then we take a look at the convergence of nets for the Kuratowski and Mosco convergences. In particular, we see when the convergence obtained by substituting sequences with nets in their definitions is topological, i.e., is a convergence of nets for a given topology. The result with Kuratowski convergence is classical; the other is surely less well known. Finally, we present a unified approach to the study of the hypertopologies. I believe that this approach is not necessary to a first understanding of these topologies; this is the reason why it has been moved into an appendix. But I also believe that this point of view is interesting, and worth mentioning somewhere in this book.

B.1 The Baire theorem

Definition B.1.1 A topological space is said to be a *Baire space* if any countable union of open dense sets is nonempty.

Proposition B.1.2 *A complete metric space is a Baire space.*

Proof. Let $x \in X$, $r > 0$, and let A_n, $n \in \mathbb{N}$, be a countable family of open and dense set; we shall prove that

$$B[x; r] \cap \bigcap O_n \neq \emptyset,$$

so showing something more than what was claimed in the proposition. Let $x_1 \in X$, $r_1 > 0$ be such that

$$B[x_1; r_1] \subset B(x; r).$$

Let $x_2 \in X$, $r_2 > 0$ be such that

(i) $B[x_2; r_2] \subset A_1 \cap B(x_1; r_1)$;

(ii) $r_2 < \frac{1}{2}$.

And by induction find $x_n \in X$, $r_n > 0$ such that

(iii) $B[x_n,; r_n] \subset A_{n-1} \cap B(x_{n-1}; r_{n-1})$;

(iv) $r_n < \frac{1}{n}$.

It follows from (iii) and (iv) that $d(x_n, x_m) < \frac{1}{n}$ if $m > n$, and thus $\{x_n\}$ is a Cauchy sequence. Let $x_0 = \lim x_n$. Since $x_n \in B[x_m; r_m]$ for $n \geq m \geq 1$, then $x_0 \in B[x_m; r_m] \subset A_m$. Thus $x_0 \in \bigcap A_n$. Also $x_0 \in B[x_1; r_1] \subset B(x; r)$, and this concludes the proof.   □

Exercise B.1.3 Show that in a Baire space X, if a countable family F_n of closed subsets is such that $X = \bigcup_{n \in \mathbb{N}} F_n$, then at least one of the sets F_n has nonempty interior.

B.2 First countability of hypertopologies

Proposition B.2.1 *Suppose $\{X\}$ has a countable neighborhood system for the lower Vietoris topology. Then X must be separable.*

Proof. Suppose $(\mathcal{V}_n)_{n \in \mathbb{N}}$ is a countable neighborhood system for $\{X\}$. Each $\mathcal{V}_n$ must contain an open basic set of the form $(V_1)^- \cap \cdots \cap (V_{m(n)})^-$. It follows that there must be a neighborhood system of the form

$$\mathcal{W}_{n,k} = \left\{ B\left(p_{n1}; \frac{1}{k}\right)^- \cap \cdots \cap B\left(p_{nm(n)}; \frac{1}{k}\right)^- : n, k, \in \mathbb{N} \right\},$$

for suitable points p_{ni}, $i = 1, \ldots, m(n)$. Let $I = \{p_{ni} : n \in \mathbb{N}, i \leq m(n)\}$, a countable set. Now suppose X is not separable. Then there must be a point p and $a > 0$ such that $B(p; a) \cap I^c = \emptyset$. It follows that $B(p, ; a)^-$ is a neighborhood of $\{X\}$ not containing any of the $\mathcal{W}_{n,k}$.   □

It follows that for the Fell, Wijsman, and Vietoris topologies, a necessary condition for metrizability is that X is separable. The same argument (with obvious modifications), holds on the space $C(X)$ with the same topologies, and with the Mosco topology.

B.3 Convergence of nets

In this section we study the topological nature of the Kuratowski and Mosco convergences of nets. In order to do this, we start by giving some definitions. A set $(T, \leq)$ is said to be a *directed set* if $\leq$ is a preorder (this means a symmetric and transitive relation) with the property that, if $u, s \in T$, there exists $t \in T$ such that $t \geq u$, $t \geq s$. Then $S \subset T$ is said to be *cofinal* to T if for each $t \in T$ there exists $s \in S$ such that $s \geq t$. It is said to be *residual* to

T if it is of the form $\{s \in T : s \geq t\}$ for some $t \in T$. A net in a topological space Y is a function $f : (T, \leq) \to Y$, usually denoted by (y_t), $t \in T$. The definitions of converging nets, in the Kuratowski sense, are the following:

Definition B.3.1 Given a net (A_t), $t \in T$, T a directed set, $A_t \in c(X)$, define

$$\operatorname{Ls} A_t = \big\{x \in X : \text{ each open set } O \text{ containing } x \text{ has nonempty}$$
$$\text{intersection with } A_s, \ s \in S, S \text{ a set cofinal to } T\big\},$$

$$\operatorname{Li} A_t = \big\{x \in X : \text{ each open set } O \text{ containing } x \text{ has nonempty}$$
$$\text{intersection with } A_s, s \in S, S \text{ a residual set of } T\big\}.$$

Then $(A_t), t \in T$ converges in the Kuratowski sense to A if

$$\operatorname{Ls} A_t \subset A \subset \operatorname{Li} A_t.$$

The Mosco convergence is defined in the same way, but using the weak topology in the definition of Ls.

The main results we want to analyze here deal with the question whether these convergences arise from some topologies. As far as the convergence of sequences, we have already seen that this is always true, and the answer is given by the Fell and Mosco topologies. However these are not always 1-countable, i.e., it is not in general possible to describe them by using sequences. To tackle the problem, we need some more notation.

So, let $(T, \leq)$ be a directed set, and suppose that for each $t \in T$ another directed set $(E_t, \leq_t)$ is defined. Let us now consider yet another directed set $(D, \preceq)$, where $D = T \times (\times E_t)$, ordered in the pointwise fashion: for $(u, \alpha), (s, \beta) \in T \times (\times E_t)$

$$(u, \alpha) \preceq (s, \beta) \text{ if } u \leq s \text{ and } \alpha_t \leq_t \beta_t \text{ for all } t \in T.$$

Suppose for any $t \in T$ and $\gamma \in E_t$ an element $x_{t\gamma}$ is given. Then we can consider the *iterated limit*:

$$\lim_{t \in T} \lim_{\gamma \in E_t} x_{t\gamma}.$$

We can also consider the *diagonal limit*:

$$\lim_{(t, \alpha) \in D} x_{t\alpha_t}.$$

Then a necessary condition for a convergence to be topological is: if $x = \lim_{t \in T} \lim_{\gamma \in E_t} x_{t\gamma}$, then $x = \lim_{(t, \alpha) \in D} x_{t\alpha_t}$.

A hint to understanding why the previous condition is necessary is that, given a convergence of nets, one can define a closure operator cl aimed at defining the closed sets of the space. The previous condition ensures that $\mathrm{cl}(\mathrm{cl}\,A) = \mathrm{cl}\,A$, for every subset A.

The following result holds.

Theorem B.3.2 *If X is not a locally compact space, then the upper Kuratowski convergence is not topological.*

Proof. Let x_0 be a point in X without compact neighborhoods and let $x \neq x_0$. Let $\mathcal{U} = \{U_n : n \in \mathbb{R}\}$ be a (countable) basis of neighborhoods of x_0, such that $U_n \supset U_{n+1}$ for all n. Since for each U_n, $\mathrm{cl}\,U_n$ is not a compact set, there exists a sequence $\{x_{nm}\}_{m \in \mathbb{N}}$, without limit points. Set $E_t = \mathbb{N}$ for all $t \in T$, and let D be the directed set, ordered pointwise. Set, for each $n, m \in \mathbb{N}$, $A_{nm} = \{x, x_{nm}\}$. For a fixed n,

$$\lim_{m \in \mathbb{N}} A_{nm} = \{x\},$$

in the Kuratowski sense, as the sequence $\{x_{nm}\}$ does not have limit points. Hence

$$\{x\} = \lim_{n \in \mathbb{N}} \lim_{m \in \mathbb{N}} A_{nm},$$

in the Kuratowski sense. On the other hand, $x_0 \in \lim_{(n,\alpha) \in D} A_{U_{\alpha U}}$, in the lower, and so upper, Kuratowski topology. For, if we take an open set A containing x_0, then there exists $\hat{n}$ such that $A \supset U_{\hat{n}}$. Fix arbitrary $\hat{\alpha} \in \times_n \mathbb{N}_n$. Then $A_{n\alpha_n} \cap A \neq \emptyset$ for each $(n, \alpha) \succeq (\hat{n}, \hat{\alpha})$. This implies that x_0 must belong to *any* set A which is $\lim_{(U,\alpha) \in D} A_{U_{\alpha U}}$ in the upper Kuratowki sense. Thus the iterated upper Kuratowski limit $\{x\}$ is not a diagonal upper Kuratowski limit and this concludes the proof.   □

The previous result is a classical one, the next one is less well known.

Theorem B.3.3 *If X is a reflexive, infinite-dimensional Banach space, then the upper Mosco convergence (on $C(X)$) is not topological.*

Proof. Let us take a closed hyperplane H and a norm one element x_0 such that $X = H \oplus \mathrm{sp}\,\{x_0\}$. Define $T = \{y_1^*, \ldots, y_n^*\}$, where $n \in \mathbb{N}$, $y_i^* \in X^* \setminus \{0^*\}$, and make it a directed set by inclusion. To each $t \in T$, $t = \{y_1^*, \ldots, y_n^*\}$, we associate an element $a_t \in H$ such that $\|a_t\| = 1$, $\langle y_i^*, a_t \rangle = 0$ for $i = 1, \ldots, n$. Now set $E_t = \mathbb{N}$ and D as above. For each $t \in T$, we consider the sequence

$$A_{tn} = \{x : x = naa_t + bx_0, 0 \leq a \leq 1, b \geq -a\}.$$

Then, for $t \in T$ the sequence (A_{tn}) Mosco converges to A_t, where

$$A_t = \{x : x = aa_t + bx_0, a \geq 0, b \geq 0\}.$$

Now, let us show that $\lim_{t \in T} A_t = A$ in Mosco's sense, where A is the set

$$A = \{x : x = bx_0, b \geq 0\}.$$

Clearly, $A \subset \text{Li}\, A_t$. Now, suppose $z \notin A$. Then there exists $0^* \neq y^* \in X^*$ such that $\langle y^*, z \rangle > 0 = \sup\{\langle y^*, x \rangle : x \in A\}$. Consider the following weak neighborhood $\mathcal{W}$ of z:

$$\mathcal{W} = \left\{ x \in X : \langle y^*, x \rangle > \frac{\langle y^*, z \rangle}{2} \right\}.$$

If $t > t_0 = \{y^*\}$, then $\forall x_t \in A_t$, $\langle y^*, x_t \rangle = \langle y^*, bx_0 \rangle$, for some $b \geq 0$, whence $\langle y^*, x_t \rangle \leq 0$, showing that $A_t \cap \mathcal{W} = \emptyset$ for $t > t_0$. To conclude, let us show that $-x_0 \in \text{Ls}\, A_{t\alpha_t}$, with $(t, \alpha) \in D$. So, let us fix a weak neighborhood $\mathcal{W}$ of x_0. We can suppose $\mathcal{W}$ is of the form

$$\mathcal{W} = \{ x \in X : |\langle y_i^*, x + x_o \rangle| < \varepsilon \},$$

for some $\varepsilon > 0$, $y_1^*, \ldots, y_n^* \in X^*$. Set $\hat{t} = \{y_1^*, \ldots, y_n^*\}$ and let $\hat{\alpha}$ be arbitrary. As $\alpha_t a_t - x_0 \in A_{t\alpha_t}$ for all (t, α), then, for $(t, \alpha) \succeq (\hat{t}, \hat{\alpha})$ we have

$$\langle y_i^*, \alpha_t a_t - x_0 \rangle = \langle y_i^*, -x_0 \rangle.$$

Thus $\alpha_t a_t - x_0 \in \mathcal{W}$ and so $\mathcal{W} \cap A_{t\alpha_t} \neq \emptyset$, and this concludes the proof. $\square$

To conclude, let us observe that the following proposition holds:

Proposition B.3.4 *The following are equivalent:*

(i) $A = F^+ \lim A_t$.

(ii)   *For each $a_s \in A_s$, $s \in S$, S cofinal set to T, such that a_s is contained in a compact set K and $\lim a_s = a$, then $a \in A$.*

The previous proposition shows that in a non locally compact space, $\text{Ls}\, A_t$ can be a bigger set of an F^+-lim A_t, and so the upper Kuratowski convergence is finer than the upper Fell.

Let us summarize the previous results concerning the Fell and Kuratowski convergences.

Corollary B.3.5 *If X is locally compact, then the upper Kuratowski convergence of nets is topological; a compatible topology is the upper Fell topology. If X is not locally compact, the upper Kuratowski (and so the Kuratowski) convergence of nets is not topological.*

B.4 A more sophisticated look at hypertopologies

The Wijsman, Hausdorff and Attouch–Wets topologies can be defined in terms of continuity of certain geometric functionals (distances, excesses). For instance, a sequence $\{A_n\}$ in $c(X)$ converges in the sense of Wijsman if $d(\cdot, A_n) \to d(\cdot, A)$, which is to say that the Wijsman topology is connected with continuity of the family of functions $\{d(x, \cdot) : x \in X\}$. It is possible, and useful, to extend this idea to characterize other topologies. Remember that for two closed sets A, F, the *gap* between them is defined as

$$D(A, F) := \inf_{a \in A, f \in F} d(a, f) = \inf_{a \in A} d(a, F) = \inf_{f \in F} d(f, A).$$

Observe that for every compact set K,

$$\{F \in c(X) : F \in (K^c)^+\} = \{F \in c(X) : D(F, K) > 0\}.$$

So that $(K^c)^+$ is an *open* set in every topology over $c(X)$ such that the family of functions

$$\{c(X) \ni F \mapsto D(K, F) : K \text{ is compact}\}$$

is lower semicontinuous. This shows that the upper Fell topology is related to lower semicontinuity of the gap functional. It is then natural to ask whether it is possible to describe the hypertopologies as the weakest ones making continuous (or semicontinuous) families of geometric functionals of the form

$$\{f(A, \cdot) : c(X) \to \mathbb{R} : A \in \Omega\},$$

where Ω is a given family of subsets of X, and f is a geometrical functional to be specified.

This approach is useful for several reasons. Just to cite one of them, topologies defined in this way (and called *initial* topologies) all share good topological properties (for instance, they are completely regular, and metrizable under general conditions). Moreover, once we have described the topologies as initial ones, it will be easy to make comparisons among them. It is clear that if $\Omega_1 \subset \Omega_2$, the topology generated by Ω_1 is coarser than the topology generated by Ω_2. Finally, this approach can suggest, as we shall see, how to introduce new topologies, some of them also very useful for applications.

In the sequel to this section we describe the main ideas of this approach, paying attention mainly to the topologies defined in terms of gap functionals.

Suppose we are given a family Ω of closed subsets of X always fulfilling the following property:

- Ω contains the singletons of X: $\{x\} \in \Omega \; \forall x \in X$.

We shall consider two types of topologies:

$$\tau^-_{\Omega, f(A, \cdot)},$$

which is the weakest topology making *upper semicontinuous* the functionals of the family

$$\{C \mapsto f(A, C), A \in \Omega\},$$

and

$$\tau^+_{\Omega, f(A, \cdot)},$$

which is the weakest topology making *lower semicontinuous* the functionals of the family

$$\{C \mapsto f(A, C), A \in \Omega\},$$

with the choice of either $f(A, \cdot) = D(A, \cdot)$, or $f(A, \cdot) = e(A, \cdot)$, $A \in \Omega$. It is not difficult to verify that the topologies we have labeled with a $-$ sign are lower topologies, and those with a $+$ sign are upper topologies. We now intend to see what kind of connections there are between $\tau^-_{\Omega, f(A, \cdot)}$, $\tau^+_{\Omega, f(A, \cdot)}$ and the topologies introduced in the previous sections.

For the convenience of the reader, we start by collecting, in the next exercise, some elementary facts which will be useful later.

Exercise B.4.1 An *arbitrary* family of subsets of a set Y, whose union is Y, is a *subbasis* for a topology on Y, the coarsest topology containing the subsets. The collection of all finite intersections of the elements of the family are a basis for the topology.

Show that given a topological space (Z, σ) and a family of functions, indexed by $i \in I$,

$$f_i \colon Y \to (Z, \sigma),$$

then the weakest topology τ on Y making the functions f_i continuous, has a subbasis

$$\{y \in Y : f_i(y) \in \mathcal{O}, i \in I, \mathcal{O} \text{ open in } Z\}.$$

Moreover, show that $y_n \overset{\tau}{\to} y$ if and only if $f_i(y_n) \overset{\sigma}{\to} f_i(y) \ \forall i \in I$. Suppose now we have a family of functions

$$g_i \colon Y \times Y \to [0, \infty),$$

such that $g_i(y, y) = 0$, $\forall i \in I, \forall y \in Y$, and we define a convergence c of sequences in the following way:

$$y_n \overset{c}{\to} y \text{ if } g_i(y_n, y) \to 0, \forall i \in I.$$

Then there is a topology τ in Y such that

$$y_n \overset{c}{\to} y \iff y_n \overset{\tau}{\to} y.$$

A local subbasis at $\bar{y} \in Y$ is provided by

$$\{y \in Y : g_i(y, \bar{y}) < \varepsilon, \ i \in I, \varepsilon > 0\}.$$

We start by considering the lower topologies. The first result shows that when using gap functionals, different choices of Ω actually do *not* provide different topologies.

Proposition B.4.2 *Let Ω and Ω_s be the following classes of sets:*

$$\Omega = c(X);$$
$$\Omega_s = \{\{x\} : x \in X\}.$$

Then the following topologies coincide on $c(X)$:

(i) V^-, *the lower Vietoris topology;*
(ii) $\tau_{\Omega,D}^-$;
(iii) $\tau_{\Omega_s,D}^-$.

Proof. Noticing that $\tau_{\Omega_s,D}^-$ is nothing else than the lower Wijsman topology, the equivalence of conditions (i) and (iii) is shown in Proposition 8.2.3. Moreover, since the topology $\tau_{\Omega_s,D}^-$ is coarser than $\tau_{\Omega,D}^-$, since $\Omega_s \subset \Omega$, the proof will be completed once we show that the lower Vietoris topology is finer than $\tau_{\Omega,D}^-$. To do this, let us fix $A \in \Omega$ and prove that the function

$$D(A,\,\cdot\,)\colon (c(X),V^-) \to [0,\infty)$$

is upper semicontinuous. Equivalently, let us show that

$$\{C \in c(X) : D(A,C) < r\}$$

is open for each $r \geq 0$. So let $F \in \{C \in c(X) : D(A,C) < r\}$. Since $D(A,F) < r$, there are $a \in A$, $\bar{x} \in F$ such that $d(a,\bar{x}) < r$. Setting $V = \{x \in X : d(a,x) < r\}$, then $F \in V^-$ and if $C \in V^-$, then obviously $D(A,C) < r$.   □

The consequence of the above result is clear. It is enough to know that a topology τ on $c(X)$ is such that for $x \in X$ the function

$$C \mapsto d(x,C)\colon (c(X),\tau) \to [0,\infty),$$

is upper semicontinuous, to conclude that the richer family of functions

$$C \mapsto D(F,C)\colon (c(X),\tau) \to [0,\infty), \quad F \in c(X),$$

is upper semicontinuous.

A different situation arises when *lower* semicontinuity of gap functionals is considered. In this case different choices of the family Ω can indeed produce different topologies. In a certain sense, to get lower semicontinuity of an inf function (as the gap function is) requires some form of compactness, while upper semicontinuity does not. And dually, upper semicontinuity of excess functions (which are sup functions) is useful to produce different lower hypertopologies.

To introduce the next results, let us recall a way to describe the lower Attouch–Wets and Hausdorff topologies. A local subbasis at $A \in c(X)$, in the lower Attouch–Wets topology, is given by

$$\{C \in c(X) : e(A \cap B(x_0;r),C) < \varepsilon\},$$

where r, ε range over the positive real numbers, while a local basis at $A \in c(X)$, in the lower Hausdorff topology is given by

$$\{C \in c(X) : e(A,C) < \varepsilon\},$$

where ε ranges over the positive real numbers (See Exercise B.4.1).
Then the following results hold.

Proposition B.4.3 *Let Ω be the family in $c(X)$ of the bounded sets $B \subset X$. Then the following two topologies agree on $c(X)$:*

(i) AW^-: *the lower Attouch–Wets topology;*

(ii) $\tau^-_{\Omega, e(B, \cdot)}$.

Proof. Let us start by showing that AW^- is finer than $\tau^-_{\Omega, e(B, \cdot)}$. Let B be a bounded set, let $r > 0$ and let $F \in c(X)$ be such that $e(B, F) < r$. Let us seek an AW^- neighborhood $\mathcal{I}$ of F such that if $C \in \mathcal{I}$, then $e(B, C) < r$. This will show the upper semicontinuity of the function

$$e(B, \cdot): (c(X), AW^-) \rightarrow [0, \infty),$$

and will allow us to complete the proof. So, let $\varepsilon > 0$ be such that $e(B, F) < r - 2\varepsilon$ and let $\alpha > 0$ be such that $B(x_0; \alpha) \supset B_r[B]$. Define

$$\mathcal{I} = \{C \in c(X) : e(F \cap B(x_0; \alpha), C) < \varepsilon\}.$$

Let $C \in \mathcal{I}$. As $e(B, F) < r - 2\varepsilon$, for every $b \in B$ there exists $x \in F$ such that $d(x, b) < r - \frac{3\varepsilon}{2}$. Hence $x \in B(x_0; \alpha) \cap F$. As $C \in \mathcal{I}$, there exists $c \in C$ such that $d(x, c) < \varepsilon$, whence $d(b, c) < r - \frac{\varepsilon}{2}$ showing that $e(B, C) < r$ and this ends the first part of the proof. To conclude, simply observe that a basic open neighborhood of a set A in the lower AW topology is of the form

$$\mathcal{I} = \{F \in c(X) : e(A \cap B(x_0; \alpha), F) < \varepsilon\},$$

and that this set is open in the $\tau^-_{\Omega, e(B, \cdot)}$ topology, as $A \cap B(x_0; \alpha)$ is a bounded set.   □

Proposition B.4.4 *Let $\Omega = c(X)$ be the family of all closed subsets $F \subset X$. Then the two following topologies agree on $c(X)$:*

(i) H^-: *the lower Hausdorff topology;*

(ii) $\tau^-_{\Omega, e(F, \cdot)}$.

Proof. The proof is similar to the proof of the previous proposition and is left as an exercise.   □

We have seen that the lower Attouch–Wets topology is the weakest topology τ making upper semicontinuous the functions

$$F \mapsto e(B, F): (c(X), \tau) \rightarrow [0, \infty),$$

where B is any bounded set. An analogous result holds for the Hausdorff metric topology, but considering B an arbitrary (closed) set.

To conclude this part on the lower topologies, we recall that upper semicontinuity of gap functions characterize the lower Vietoris topology, while upper semicontinuity of excesses functions are exploited to characterize the

Attouch–Wets and Hausdorff topologies. It seems that there is not much room to imagine new lower topologies (with very few exceptions). The situation is quite different with upper topologies. In this case, we shall see that we can produce a much richer variety of topologies by exploiting lower semicontinuity of gap functionals. Let us see how.

Given the family Ω, we shall denote by τ_Ω^{++} the topology having the following family as a subbasis:

$$\{C \in c(X) : D(A,C) > 0, A \in \Omega\}.$$

Moreover, we shall say that the family Ω is *stable*, if $\forall A \in \Omega$, $\forall r > 0$, $B_r[A] \in \Omega$. Here is the first result.

Proposition B.4.5 *Let Ω be a stable family. Then the two following topologies agree on $c(X)$:*

(i) τ_Ω^{++};

(ii) $\tau_{\Omega, D(A, \cdot)}^+$.

Proof. Clearly the topology $\tau_{\Omega, D(A, \cdot)}^+$ is finer; this follows from the definition and does not depend on the fact that Ω is stable. So, it remains to show that, $\forall B \in \Omega$,

$$D(B, \cdot) \colon (c(X), \tau_\Omega^{++}) \to \mathbb{R}$$

is a lower semicontinuous function or, equivalently,

$$\mathcal{O} = \{C \in c(X) : D(B,C) > r\}$$

is an open set for all $r \geq 0$. Let $C \in \mathcal{O}$ and let $\varepsilon > 0$ be such that $D(B,C) > r + 2\varepsilon$. Then, $\forall b \in B$, $\forall c \in C$, $d(b,c) > r + 2\varepsilon$. Now, if $d(x,B) \leq r + \varepsilon$, then, $\forall b \in B$,

$$d(x,c) \geq d(b,c) - d(b,x) > \varepsilon.$$

This implies $D(B_{r+\varepsilon}[B], C) > 0$, and, as Ω is stable and $B \in \Omega$, it follows that $B_{r+\varepsilon}[B] \in \Omega$, hence

$$C \in \{F \in c(X) : D(B_{r+\varepsilon}[B], F) > 0\} \subset \mathcal{O}.$$

We have found a τ_Ω^{++} open neighborhood of C which is contained in $\mathcal{O}$. This ends the proof.   □

The previous proposition was proved in $c(X)$. But the same proof holds if we substitute $c(X)$ with some meaningful subset, such as the set $C(X)$ of the closed convex subsets of a Banach space X, or the set of the weakly closed subsets of X.

Merging Propositions B.4.2 and B.4.5, we get the following result.

Theorem B.4.6 *Let Ω be a stable family. Then the topology having as a subbasis sets of the form*

$$\{C \in c(X) : C \cap V \neq \emptyset, V \text{ open in } X\},$$

(for the lower part) and of the form

$$\{C \in c(X) : D(A, C) > 0, A \in \Omega\},$$

(for the upper part), is the weakest topology making continuous all functionals of the family

$$\{C \mapsto D(C, F) : F \in \Omega\}.$$

We now can state some useful corollaries.

Corollary B.4.7 *Let (X, d) be such that every closed enlargement of a compact set is still a compact set. Then the Fell topology on $c(X)$ is the weakest topology making continuous all functionals of the family*

$$\{A \mapsto D(A, K) : K \subset X \text{ is a compact set}\}.$$

Corollary B.4.8 *Let X be a reflexive Banach space. Then the Mosco topology on $C(X)$ is the weakest topology making continuous all functionals of the family*

$$\{A \mapsto D(A, wK) : wK \subset X \text{ is a weakly compact set}\}.$$

Exercise B.4.9 Prove that the upper Hausdorff topology agrees with the topology τ_Ω^{++}, where $\Omega = c(X)$.

Having in mind the Vietoris and Hausdorff topologies, we can construct another topology, weaker than both, by considering the lower part of the Vietoris, and the upper part of the Hausdorff. This is the proximal topology. In view of the above exercise, we have the following:

Lemma B.4.10 *The proximal topology is the weakest topology making continuous all functionals of the family*

$$\{A \mapsto D(A, F) : F \subset X \text{ is a closed set}\}.$$

Exercise B.4.9, and the definition of proximal topology can induce the idea of asking what happens when considering the topology τ_Ω^{++}, when Ω is the family of the bounded subsets of X. In other words, which upper topology generates the basis

$$\{F \in c(X) : D(B, F) > 0, B \text{ bounded}\}?$$

The rather expected answer is in the next proposition.

Proposition B.4.11 *Let Ω be the subset of $c(X)$ of the bounded subsets of X. Then the following topologies on $c(X)$ are equivalent:*

(i)   AW$^+$;
(ii) τ_Ω^{++}.

Proof. To prove that AW$^+$ is finer than τ_Ω^{++}, let us show that

$$\mathcal{I} = \{F \in c(X) : D(B, F) > 0\}$$

is open in the AW$^+$ topology, for every B bounded set. So, let $C \in \mathcal{I}$. There is $\varepsilon > 0$ such that $D(B, C) > 2\varepsilon$. Let $\alpha > 0$ be such that $B(x_0; \alpha) \supset B_\varepsilon[B]$. Consider the AW$^+$ neighborhood of C:

$$\mathcal{N} = \{F \in c(X) : e(F \cap B(x_0; \alpha), C) < \varepsilon\}.$$

Then $\mathcal{N} \subset \mathcal{I}$. For, let $F \in \mathcal{N}$. Suppose $D(B, F) = 0$. Then there are $b \in B$, $x \in F$ such that $d(x, b) < \varepsilon$. Then $x \in F \cap B_\varepsilon[B]$ and there is $c \in C$ such that $d(x, c) < \varepsilon$. It follows that

$$D(B, C) \le d(b, c) \le d(b, x) + d(x, c) < 2\varepsilon,$$

which is impossible. Conversely, suppose

$$D(B, \cdot) : (c(X), \tau) \to [0, \infty)$$

is lower semicontinuous for each bounded set B, and let us show that τ is finer than AW$^+$. Let $C \in c(X)$, $\mathcal{N}$ of the form

$$\mathcal{N} = \{F \in c(X) : e(F \cap B(x_0; \alpha), C) < \varepsilon\},$$

and seek for a bounded set B and for $\delta > 0$ such that the set

$$\mathcal{I} = \{A \in c(X) : D(B, A) > \delta\}$$

fulfills the conditions $C \in \mathcal{I} \subset \mathcal{N}$. Without loss of generality, we can suppose that C does not contain $B(x_0; \alpha)$. For, taking possibly a larger α, the neighborhood $\mathcal{N}$ becomes smaller. Let $0 < \sigma < \varepsilon$ be such that

$$B = (S_\sigma[C])^c \cap B(x_0; \alpha)$$

is nonempty. Given $0 < \delta < \sigma$, then

$$D(B, C) \ge D((B_\sigma[C])^c, C) \ge \sigma > \delta.$$

Moreover, if $D(B, F) > \delta$ and if $x \in F \cap B(x_0; \alpha)$, then necessarily $d(x, C) < \sigma < \varepsilon$, showing that $e(F \cap B(x_0; \alpha), C) < \varepsilon$, and this ends the proof.   □

    In view of the previous results, one can think of other different families Ω, in order to introduce new topologies. The following examples are very natural.

- $\Omega = \{$closed balls of $X\}$.
- If X is a Banach space, $\Omega = \{$closed convex subsets of $X\}$.

- If X is a Banach space, $\Omega = \{$closed bounded convex subsets of $X\}$.

Of course, we have to check that different families generate different topologies, but this is not difficult. Moreover, not all topologies have the same importance, especially for the applications. The three above are used in some problems. The first one is called ball-proximal, while the family Ω of all closed convex sets generates on $C(X)$ the so called linear topology. More important for the applications are the topologies generated by the following two classes Ω.

Definition B.4.12 Let (X, d) be a metric space. Define on $c(X)$ the *bounded proximal* topology as the weakest one on $c(X)$ making continuous the family of functionals

$$\{A \mapsto D(A, F) : F \subset X \text{ is a closed bounded set}\}.$$

Definition B.4.13 Let X be a normed space. Define on $C(X)$ the *slice* topology as the weakest one on $C(X)$ making continuous the family of functionals

$$\{A \mapsto D(A, C) : C \subset X \text{ is closed convex and bounded}\}.$$

Let X^* be a dual space. Define on $C(X^*)$ the *slice** topology as the weakest one on $C(X)$ making continuous the family of functionals

$$\{A \mapsto D(A, C) : C \subset X \text{ is a weak}^* \text{ closed, convex and bounded}\}.$$

These topologies have interesting properties for optimization problems. In particular, the slice topology is a natural extension of the Mosco topology in nonreflexive spaces. Their coincidence in a reflexive setting is obvious by observing that the family of the weakly compact convex sets coincides with that one of the closed bounded convex sets. Moreover, the slice topology is clearly finer than the Mosco topology when X is not reflexive, and it is coarser than the AW. The bounded proximal topology is coarser than the AW, in particular as far as the lower part is concerned, as the upper parts coincide. On the other hand, for several problems to require having AW$^-$ as lower part is restrictive and not useful. The lower Vietoris will be enough.

We just mention that the Vietoris topology can also be characterized as an initial topology. Here is the idea. Given (X, d) metric space, we have already observed that the Vietoris topology in $c(X)$ is not affected by changing d in an equivalent way. Thus the Vietoris topology is finer than *all* different Wijsman topologies generated by distances equivalent to d. It can be shown that the Vietoris topology is actually the supremum of all these Wijsman topologies, and it can thus be characterized as a weak topology as well.

Let us finally summarize the results of this section. We have seen that it is possible to characterize several hypertopologies as initial ones, which means as the weakest topologies making continuous certain families of geometric functionals, such as gap and excess functionals. More precisely, lower

semicontinuity of gap functionals allows characterizing several upper hypertopologies, while upper semicontinuity of excess functionals is related to lower
hypertopologies. As already mentioned, this approach is useful. Even without
mentioning the advantage of having a unified theory highlighting several interesting aspects of hypertopologies, it is a fact that this approach suggested
to scholars how to define new topologies. This is not merely an intellectual
exercise, or a simple way to produce a number of well-published papers. There
is much more, and the section dedicated to stability should clarify this. Having different topologies allows establishing stability results for several *different*
classes of functions, and this is without a doubt very useful. The most com-
pelling example is, in my opinion, the bounded proximal topology. It shares,
especially in problems without constraints, several good properties with the
much finer Attouch–Wets topology. This means that having introduced the
bounded proximal topology allows for stability results for much broader classes
of problems.

Exercise B.4.14 Prove that the Wijsman topology is the weakest topology
making continuous all functionals of the family

$$\{A \mapsto D(A, K) : K \subset X \text{ is a compact set}\}.$$

Prove that in any normed space the Wijsman topology is the weakest topology
making continuous all functionals of the family

$$\{A \mapsto D(A, B) : B \subset X \text{ is a closed ball}\}.$$

The Exercise B.4.14 provides another way to get the already proved fact
that the Wijsman and Fell topologies coincide when X is locally compact.
Moreover, in the normed spaces, we see that the Wijsman topology is generated by a stable family Ω.

C

More game theory

This appendix deals with some aspects of noncooperative game theory. Why include it in a book like this, aside from the fact that I know something about it? Well, the mathematics of this book is optimization. But, as I said elsewhere, optimization does not deal only with minimizing or maximizing scalar functions. Thus, even if most of the material presented here is related to real valued functions to be minimized, I also like to give some small insight into other aspects of optimization. And surely, game theory is a major aspect of optimization.

There are situations in which an optimizer is not alone. His final result will depend not only upon his choices, but also upon the choices of other agents. Studying this as a typical optimization problem is exactly a matter of game theory. So, I believe that many of the aspects that this book deals with, and that are studied in the setting of scalar optimization, could be as well considered in game theory. Thus this appendix can be a very short introduction to some topics in game theory for those readers who are not familiar with this theory, in the hope to convince some of them that the subject is worth knowing.

We already have considered some aspects of the finite, two player zero sum games. These games are the starting point of the theory, but it is quite clear that in many situations two agents could *both* benefit from acting in a certain way rather than in another, so that the zero sum property is lost. Thus there is a need to go beyond zero sum games. Let us quickly recall what we already know about them. First of all, a natural concept of solution arises from analyzing games with a simple structure. A solution is a pair which is a saddle point for the payment function of the first player. (Remember, the payment function of the second player is the opposite of that of the first one. What one gains in any circumstances is what the other one loses.) We also have learned interesting things from Theorem 7.2.2. Essentially, the two players must take into account that there is a competitor, but to solve the game they behave as if they are alone, because they need to either maximize or minimize a given function (it is *only* in this function that the presence of another player is important, but each agent can find it by himself). So they independently solve a problem (and these problems are in duality), and each pair of solutions they find is a solution for the game. This means that there is no need to coordinate their actions when implementing the game. Moreover, at each equilibrium point they receive the same amount of utility, and this makes them indifferent as to which equilibrium will be actually implemented, exactly as when once we find two or more minima of a function we are not interested in which one to use, given that they all have the same value, which is what matters. Unfortunately, or better *fortunately*, nothing similar happens with nonzero sum games. We shall spend some time arguing that what is a solution of a game is not so obvious, and that properties like those we mentioned above can utterly fail. After this, we quickly formalize what is intended for a game in normal or strategic form, and we arrive at an existence theorem for the equilibrium.

The philosophy underlying optimization is that a decision maker is "rational and intelligent". If he has some cost function on his hands, he will try to minimize; if he has some utility function, he will try to get the maximum. (A word of caution, to get the maximum does not mean necessarily to win, to get more money, and so on, but to maximize the satisfaction. You might be happier to losing a card game with your little child, rather than seeing him frustrated). We can try to apply this philosophy in game theory. But how should we formalize it? Remember that in some sense in the zero sum setting the problem was hidden, since the idea of saddle point arose naturally from the analysis of a game (see Example 7.2.3). But when the game is not zero sum? I believe we could all agree on the (weak) form of rationality given in the following definition:

A player will not make a certain choice if he has another, better choice, no matter what the other players do.

The above rationality axiom can be called the elimination of dominated strategies. A consequence of it is that if a player has a dominant strategy, i.e., a strategy allowing him to do better than any other one, *no matter what the other players do*, he will select that strategy.

All of this is quite natural, and we are ready to accept it without any doubt. Now we have to make a choice. And, as usual in game theory, we want to be optimistic and decide that we will maximize gains, rather than minimize pains. So, look at the following example, thinking here and for the rest of the section that we prefer more to less, and let us try to see if the above principle helps in finding a solution.

Example C.1 The game is described by the following bimatrix:

$$\begin{pmatrix} (10, 10) & (0, 15) \\ (15, 0) & (5, 5) \end{pmatrix}.$$

It is clear how to read it, taking into account what we did with a matrix in a zero sum game. There the entry of position ij was what the player selecting the row i gets from the player selecting the column j. Here the entry ij is a pair, and the first item is what the first player (choosing rows) gets, the second one what the second player gets. Let us see if we can use the above rationality axiom to find a solution for the game. What does the first player do? She looks at her payments, since she is interested in them, and not in what the second one will get. Thus, she looks at the following matrix, representing her utilities:

$$\begin{pmatrix} 10 & 0 \\ 15 & 5 \end{pmatrix}.$$

The analysis of the situation is very easy! It is clear that the second row is better than the first one, since $15 > 10$ and $5 > 0$. We know what the first player will do. Now, the second step is to realize that the game is clearly symmetric. The two players face the same situations, and evaluate them in the same way. Thus the second player will choose the right column. We have found the solution of the game, which is $(5, 5)$. They will both get 5.

Nothing strange? Maybe you will be surprised if I claim that this is the most famous example in game theory, but this is the truth. Because these four numbers, used in a smart way, serve at the same time as a model of very common, very familiar situations, and as an example of the puzzling situations one immediately meets when dealing with multiple agents. If you look again at the bimatrix, you will immediately realize that the players could both get more by selecting the first row and the first column. In other words, rationality imposes a behavior which provides a very poor result for both. This, as I said, is not academic. This example seems to model well very many situations of real life. By the way, it was illustrated the first time to a conference of psychologists, with a story that will be described later in an exercise. This is not surprising, since it is clear that it proposes a strong intellectual challenge: cooperation can be worthwhile for both, but at the same time it is not individually rational.

Let us continue in this analysis of paradoxical situations immediately arising when getting out of the zero sum case. There is something which is absolutely clear when a decision maker is alone. Suppose he has a utility function $f(\cdot)$ to maximize. Suppose then that for some reasons his utility function changes. Denote by g the new one. Finally suppose that $f(x) \leq g(x)$ for all x. At the end, will he do better in the first model or in the second? Well, if $f(x) \leq g(x)$ for all x, there is no doubt that $\max_x f(x) \leq \max_x g(x)$. I do not need to be a Fields medalist to understand this. Does the same thing happen in game theory? Look at the following bimatrix:

$$\begin{pmatrix} (8, 8) & (-1, 0) \\ (0, -1) & (-2, -2) \end{pmatrix}.$$

It is not difficult to check that in this case too there is a dominant strategy for both players, which will provide 8 to both. Nice, no dilemma now, 8 is the maximum they could get and fortunately they are able to get it. But make a comparison now with the bimatrix of the Example C.1. In the second game both players gain less than in the first game *in any circumstances*. As a result, they are better off in the second.

Let us now consider a fairly common situation. Before going to work, Daniela and Franco discuss what they will do in the evening. Daniela would like to go to a concert, but Franco wants to see someone's pictures of his last trip to Himalaya. But, in any case, both (surprising?) prefer better to stay together. A bimatrix describing the situation could be the following:

$$\begin{pmatrix} (10,0) & (-1,-1) \\ (-5,-5) & (0,10) \end{pmatrix}.$$

Observe that the chosen numbers are rather arbitrary, but not the ordering relations between them. For instance it is clear that the pair $(-5,-5)$ represents the sad situation where Daniela is watching uninteresting (for her, and also for me) pictures, while Franco is sleeping on a seat at the concert hall. Well, here we cannot proceed by eliminating dominated strategies, as one can easily verify. However I think that everybody will agree that a good suggestion to them is to stay together (not really very smart, they told us they like better to stay together!). At the same time, it seems to be impossible, given the model, to distinguish between the two different situations. Going together to the concert, versus going together to see pictures. What I want to point out with this example, which is indeed almost as famous as that of Example C.1, is the following. There are two equilibria, and, in contrast to the case of the theory of decisions, this causes problems, since the two players are *not* indifferent as to which equilibrium will be implemented. That is not all. In the zero sum case, as we have observed at the beginning of the section, each player must take into account that he is not alone in the world, but does not need to coordinate with the others to arrive to a solution. Here, suppose Daniela and Franco go to work planning to decide later with a phone call, and suppose for any reason they cannot be in touch for the rest of the day. Where should they go? Every choice they have makes sense. One can decide to be generous and go where the partner prefers; but if they both do this, it is a nightmare. Or else, they can decide that it is better not to risk and to go where they like, but in this case they will be alone; in other words, even if some solution of the game is available, it is necessary for them to coordinate in order to arrive to a solution. And this could be made difficult by the fact that they like different solutions.

One more game, a voting game, illustrates another interesting situation.

Example C.2 Suppose there are three people, and that they have to choose between three different alternatives, say A, B, C. They vote, and if there is an alternative getting at least two votes, this will be the decision. Otherwise the alternative voted by the first player will win. The (strict) preferences of the three players are as follows:

$$A \succ B \succ C,$$

$$B \succ C \succ A,$$

$$C \succ A \succ B.$$

We can use the procedure to eliminate dominated strategies, and to implement dominant strategies. Somebody could try to make all calculations, but it is not surprising that with this method the first player will vote A, which is his preferred outcome, while the two other players will eliminate A and B respectively, which are the worst choices for them. At this point the first player becomes "dummy", and the other two face the following situation (verify it):

$$\begin{pmatrix} A \ A \\ C \ A \end{pmatrix},$$

where the second player chooses the rows, the first one representing the choice of B, while the first column represents the choice of C for the third player. At this point the solution is clear. Since both like C better than A, the final result will be C. What is interesting in this result is that the game has a stronger player, which is the first one, and the final result, obtained by a certain (reasonable) procedure is what he dislikes the most. We shall see that his preference, result A, can be supported by an idea of rationality. What is interesting to point out here is that even a procedure like implementing dominant strategies can be dangerous. More precisely, for him the strategy to announce A is only weakly dominant, i.e., it is not always *strictly* better for him to play it. And this can cause the problem we have just noticed.

Notwithstanding all of these problems, a theory can be developed. First of all, it is necessary to have a model for a game, and a new idea of equilibrium, since the procedure of eliminating dominated strategies can stop, or in certain games cannot even start.

After the contributions of von Neumann to the theory of zero sum games, the next step was the famous book by von Neumann–Morgestern, *The Theory of Games and Economic Behavior*, whose publication was later taken as the official date of the birth of game theory. There, in order to include in the theory the situations which cannot be modeled as zero sum games, a cooperative approach was developed. It was an attempt to study the mechanisms of interaction between agents having different, but not necessarily opposite, interests. At the beginning of the 1950s, J. F. Nash proposed a different model, and a new idea of equilibrium, which nowadays is considered better suited for the theory. Here it is.

Definition C.3 A *two player noncooperative game in strategic (or normal) form* is a quadruplet $(X, Y, f: X \times Y \to \mathbb{R}, g: X \times Y \to \mathbb{R})$. A Nash *equilibrium* for the game is a pair $(\bar{x}, \bar{y}) \in X \times Y$ such that

- $f(\bar{x}, \bar{y}) \geq f(x, \bar{y})$ for all $x \in X$;
- $g(\bar{x}, \bar{y}) \geq f(\bar{x}, y)$ for all $y \in Y$.

Let us make a quick comment on this definition: X and Y are the strategy spaces of player one and two, respectively. Every pair (x, y), when implemented, gives rise to a result which provides utility $f(x, y)$ to the first player, and $g(x, y)$ to the second one. And an equilibrium point is a pair with the following feature: suppose somebody proposes the pair $(\bar{x}, \bar{y})$ to the players. Can we expect that they will object to it? The answer is negative, because each one, *taking for granted that the other one will play what was suggested to him*, has no incentive to deviate from the proposed strategy. A simple idea, but worth a Nobel Prize.

What does the rational player do, once he knows (or believes) that the second player plays a given strategy y? Clearly, he maximizes his utility function $x \mapsto f(x, y)$, i.e., he will choose a strategy x belonging to $\text{Max}\{f(\cdot, y)\}$. Denote by BR_1 the following multifunction:

$$\text{BR}_1: Y \to X, \quad \text{BR}_1(y) = \text{Max}\{f(\cdot, y)\}$$

(BR stands for "best reaction"). Define BR_2 similarly for the second player and finally define

$$\text{BR}: X \times Y \to X \times Y, \quad \text{BR}(x, y) = (BR_1(y), BR_2(x)).$$

Then it is clear that a Nash equilibrium for a game is nothing else than a fixed point for BR: $(\bar{x}, \bar{y})$ is a Nash equilibrium for the game if and only if

$$(\bar{x}, \bar{y}) \in \text{BR}(\bar{x}, \bar{y}).$$

Thus a fixed point theorem will provide an existence theorem for a Nash equilibrium. Here convexity plays a role. Remember that Kakutani's fixed point theorem states that if a multifunction $F: Z \to Z$, where Z is a compact convex subset of a Euclidean space which is nonempty closed convex valued and has closed graph, then F has a fixed point.

Thus, the following theorem holds:

Theorem C.4 *Given the game* $(X, Y, f: X \times Y \to \mathbb{R}, g: X \times Y \to \mathbb{R})$, *suppose* f, g *continuous and*

- $x \mapsto f(x, y)$ *is quasi concave for all* $y \in Y$;
- $y \mapsto g(x, y)$ *is quasi concave for all* $x \in X$.

Then the game has an equilibrium.

Proof. Remember that quasi concavity of a function h is by definition convexity of the a level sets of h, $\{z : h(z) \geq a\}$. Thus the assumptions guarantee that BR is nonempty closed convex valued. I leave as an exercise to show that it has closed graph.   □

The above proof was the first one published by Nash. In subsequent articles he published two different proofs of the same theorem. He was not happy about using Kakutani's theorem; he was more interested in using Browder's fixed point theorem, and this is rather curious.

Observe that the zero sum games fit in this theory. It is enough to set $g = -f$. Moreover, in a zero sum game a pair is a saddle point if and only if it is a Nash equilibrium. Also, observe that von Neumann's theorem on the existence of equilibria in mixed strategies for finite games can be derived as a consequence of Nash's theorem.

We have so far considered games in *normal* (or *strategic*) *form*, which means, roughly speaking, that we take the available strategies of the players as primitive objects. But in practice this is usually not the case when you have to play a game. So, let me spend few words on how a simple game, with a finite number of moves, can be analyzed in a fruitful way. Since I do not intend to enter into the theory, to make things as simple as possible I will just consider an example.

Example C.5 There are three politicians who must vote whether or not to increase their salaries. The first one publicly declares his vote (Yes or No), then it is the turn of the second one, finally the third one declares his move. The salary will be increased if at least two vote Y. They all have the same preferences, in increasing order, vote Y and do not get more salary (a nightmare, no money and a lot of criticism by the electors), vote N and do not get more money, vote Y and get the money, vote N and get the money (very nice, look altruistic and get the money!) Let us say that their level of satisfaction is d, c, b, a, respectively, with then $d < c < b < a$. If you are one of the politicians, would you prefer to be the first, the second, or the third one to vote?

First of all, let us emphasize that, notwithstanding that the three politicians have the same moves to play (they must say Y or N), their strategy sets are quite different; think about it. Next, a very efficient and complete way to describe such a game is to build up the associated *game tree*. Instead of wasting too many words, let us see a self explanatory picture (Fig. C.1):

It should be clear what we shall call nodes and branches. Also, it is not difficult to understand how to find "the solution" of such a game (and of all games written in the same form). An effective method is the so called *backward induction*. It means that the game must be analyzed starting from the end, and not the beginning. Let us try to understand what will happen at every terminal node, i.e., a node such that all branches going out from it lead to

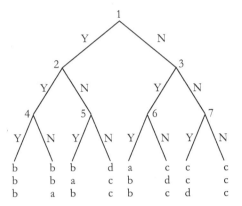

Figure C.1.

a final situation. In our particular game all these nodes are attached to the
third voter. And they are labeled by the digits 4, 5, 6, 7. At node 4, the player
observes that if he chooses Y he gets b, otherwise he gets a. Since he likes
a better than b, he will choose the branch N. It is easy to check what will
happen at all other nodes. What is really important is that the third player
knows what to do at every node where he is called upon to make a decision,
and the other players know what he will do. At this point, the second voter is
able to decide for his best, at nodes 2 and 3. For instance, you can check that
at node 2 he will say N. It is clear that, doing this, we are able to arrive at
the top, and to know the exit of every game of this type, if played by rational
players. In our example, you can check that the first voter will vote against
the increase of the salaries, while the other ones will vote in favor. Think
a little about it. It is an easy intuition to think that the final result of the
process will be that the three politicians will be richer: more money is always
better. And, with a little more thought, we can understand that it is logical
that the first one will vote against the proposal, because in this case he will
force the two other to vote in favor of it. But if you ask a group of people
to say what they think will happen in such a situation, several of them will
probably answer they would like to be the last voter, thinking, erroneously,
that the first two will vote in favor because they want the money. Thus, such
a simple example shows that games which can be described by such a tree
are always solvable. True, we are systematically applying a rationality axiom
(when called upon to decide, everyone makes the decision which is best for
himself), and we are able to arrive at a conclusion. So, the next question
is: what kind of games can be modeled in the above form? It is clear that
such games must be finite, i.e., they must end after a finite number of moves.
Moreover, each player knows all possible developments of the game, and the
whole past history once he is called upon to decide. They are called *finite
games with perfect recall.* It is clear that there are games which cannot be

described in this way. Most of the games played with cards are fun exactly because no player has complete information about the situation. However, very popular games fit in the above description, think of chess and checkers, for instance. But now a natural question arises. Why are games which are determined (in the sense that we know how to find a solution, using backward induction), interesting to play? I mean, if both players agree on what the final outcome will be, why play the game? Here it clearly appears what is so challenging in such games, and why game theory, even when it provides a satisfactory (theoretical) answer to the problem of identifying a solution, still has a lot of work to do. The key point is that even for very simple games, with few moves and easy rules, it is out of question to be able to explicitly write down the tree of the game. Actually, what is so interesting in studying these games is exactly the fact that the good player has at least an intuition that certain branches must not be explored, since it is very likely that they will cause trouble to the player who must select a branch. The domain of artificial intelligence is of course deeply involved in such questions. The fact that IBM spent a lot of money to create a computer and programs able to beat a human being in a series of chess games, is perfectly rational.

A last observation: the solution determined by the above procedure of backward induction in finite games of perfect recall, is of course a Nash equilibrium of the game. Even more, it can be shown that, when translating into normal form such types of games, it can happen that other Nash equilibria arise, prescribing for some player a choice which is not optimal at a certain branch (it should be noticed that such a branch is *never* reached when effectively implementing the Nash equilibrium). Thus the equilibria provided by the backward induction procedure are particular Nash equilibria, which are called *subgame perfect*. Thus, a natural question is whether these subgame perfect equilibria avoid some bad situations such as the one described in Example C.1. An answer is given by the following famous example.

Example C.6 (The centipedes) Andrea and Stefano play the following game. They are rational, they know what to do, but the final result leaves them very disappointed.

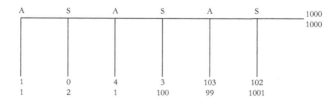

Figure C.2.

To conclude, some more exercises, just for fun.

Exercise C.7 Two men are suspected of a serious crime. The judge makes the following proposal to them: if one confesses that both are guilty, while the other one does not, the one who confesses will be free, as a precious collaborator of justice, while the other one will be condemned to ten years. If both confess, they will be condemned to five years. On the other hand, if they do not confess, the judge does not have evidence that they are guilty, so that they will be condemned to one year of jail, for a minor crime. Prove that this game fits the model of Example C.1. Observe how smart the judge is. It is likely that without the proposal, the two men will not confess.

Exercise C.8 Prove that in Example C.2 the result of A is the outcome of a Nash equilibrium. Is it the same with B?

Exercise C.9 Paola, player number one, is with her son Tommaso, player number two, at the shopping center. Tommaso wants an ice cream. Paola does not like to buy it, since she thinks that too many ice creams will increase the dentist's bill. Tommaso knows that he could cry to be more convincing, even if he does not like to do it. Thus the first player has two moves, to buy or not to buy the ice cream. Tommaso can decide, after his mother's decision, whether to cry or not. Let us quantify their utility functions. If Paola buys the ice cream, her utility is 0, while the utility of Tommaso is 20. If she announces that she will not buy the ice cream, Tommaso can cry, and the utilities are, in this case, -10 for Paola and -1 for Tommaso, while if he does not cry, the utilities are 1 for both. Build up the tree of the game, find the solution with the backward induction, write it in normal form, and find all Nash equilibria.

Exercise C.10 Analyze the following game. There are two groups of matches on the table. One player, when he moves, can take as many matches as he wants from a group, or the same amount of matches from both. The player clearing the table is the winner. Try to list the strategies of the players when the matches are 3 and 4. Try to prove that the second player wins if the initial situation is a pair (x, y) $(x < y)$ of matches, such that there is n such that $x = [nt]$, $y = [nt] + n$, where $[a]$ stands for the integer part of the positive number a, and $t = \frac{1+\sqrt{5}}{2}$.

Hint. A winning set W for the second player is a set of a pair of nonnegative integers (x, y) such that

(i) $(0, 0) \in W$;
(ii) if $(x, y) \in W$, then every possible move (u, v) immediately following (x, y) is such that $(u, v) \notin W$;
(iii) if $(u, v) \notin W$, there is $(x, y) \in W$ immediately following (u, v).

The meaning of this is clear. Starting from a pair $(x, y) \in W$, the first player is obliged by the rules to select something outside W, then the second player has the possibility to get in W again, and so on. It is not difficult to construct

W. Start from $(0,0)$, then successively put $(1,2)$, $(3,5)$, $(4,7)$, $(6,10)$, etc. Observe that at each step the difference between the two digits increases by one unit, and that each natural number must be in the list in one and only one pair. The key property of t above is that it is irrational, that $\frac{1}{t} + \frac{1}{t+1} = 1$ and that each natural number m can be written in one and only one way either in the form $m = [na]$, or $m = [nb]$, if $\frac{1}{a} + \frac{1}{b} = 1$, and $a, b > 1$ are irrational. Thus, t being as above, each natural number m can be written as either $m = [nt]$ or $n = [nt] + n$. Now, prove that if a pair belongs to W, any possible move from this pair takes the player going out of W. Finally, the interesting part is how to enter W from a pair outside it, since this describes the winning strategy. Let $(x,y) \notin W$, $x < y$. If $x = [nt] + n$, leave $[nt]$ on the group where there were y matches. If $x = [nt]$, there are two possibilities. Either $y > [nt] + n$ and in such a case make them become $[nt] + n$ or $y < [nt] + n$. Now it is up to you to conclude!

Exercise C.11 (The Nash bargaining problem.) This exercise quickly proposes the Nash solution to a bargaining problem between two players, a situation which is intermediate between noncooperative and cooperative games. A bargaining problem is modeled as a pair (C,x), where $C \subset \mathbb{R}^2$ and $x \in C$. The meaning is the following: a vector in C represents a possible distribution of utilities among the two players. They get x if they do not agree on a distribution (x is called the *disagreement* point). In order to have a true bargaining situation, let us suppose that there is an element of C whose coordinates are both greater than the coordinates of x; if this does not happen at least one player is not interested in bargaining. Convexity can be justified by making some reasonable assumption on the utility functions, which we will not discuss here. Consider the set $\mathcal{B}$ of all bargaining problems. A *solution* of the bargaining problem is a function f assigning to each pair (C,x) in $\mathcal{B}$ an element of C. Observe, if we have a solution of the bargaining problem we have a rule for solving every bargaining situation! What kind of properties should a solution f have? Nash proposed the following list of properties:

(i) Suppose (C,x) and (A,y) are two bargaining situations connected by the following property: there are $a, b > 0$ and $k_1, k_2 \in \mathbb{R}$ such that $z = (z_1, z_2) \in C$ if and only if $w = (az_1 + k_1, bz_2 + k_2) \in A$; Moreover $y = (ax_1 + k_1, bx_2 + k_2)$. Then $f(A,y) = (af_1[(C,x)] + k_1, bf_2[(C,x)] + k_2)$ (invariance with respect to admissible transformation of utility functions).

(ii) Suppose (C,x) is such that $(u,v) \in C$ implies $(v,u) \in C$, and suppose $x = (a,a)$ for some $a \in \mathbb{R}$. Then $f(C,x)$ must be of the form (b,b) (the two players are in a perfectly symmetric situation, so that the result should be the same for both).

(iii) Given the two problems (A,x) and (C,x) (observe, same disagreement point), if $A \supset C$, and if $f[(A,x)] \in C$, then $f[(C,x)] = f[(A,x)]$ (the alternatives in A which are not in C are *irrelevant* alternatives).

(iv) Given (C, x), if $y \in C$ and there is $u \in C$ such that $u_1 > y_1$ and $u_2 > y_2$, then $f(C, x) \neq y$ (efficiency, which is perfectly justified since we are not in a noncooperative setting).

Here is the Nash theorem:

There is one and only one f satisfying the above properties. Precisely, if $(C, x) \in \mathcal{B}$, $f(C, x)$ is the point maximizing on C the function $g(u, v) = (u - x_1)(v - x_2)$.

In other words, the players must maximize the product of their utilities.

Prove the Nash theorem.

Hint. f is well defined: the point maximizing g on C exists and is unique. It is easy to show that f satisfies the above list of properties. Less simple is the proof of uniqueness. Call h another solution. First of all, observe that properties (ii) and (iv) imply $h = f$ on the subclass of the symmetric games. Then take a general problem (C, x) and, by means of a transformation as in property (i), send x to the origin and the point $f(C, x)$ to $(1, 1)$. Observe that the set $\hat{C}$ obtained in this way is contained in the set $A = \{(u, v) : u, v \geq 0, u + v \leq 2\}$. Then $(A, 0)$ is a symmetric game, so that $f(A, 0) = g(A, 0) = (1, 1)$. The independence of irrelevant alternatives provides $h(\hat{C}, 0) = (1, 1)$. Now via the first property go back to the original bargaining situation, and conclude from this.

Exercise C.12 Two men, one rich and one poor, must decide how to divide 500 Euros between them. If they do not agree, they will get nothing. The rich man, when receiving the amount l of money will get a satisfaction $u_1(l) = cl$, where $c > 0$. The utility function of the poor man is instead $u_2(l) = \ln(1 + \frac{l}{100})$. Find what Nash proposes to the players.

D

Symbols, notations, definitions and important theorems

D.1 Sets

- In a metric space X, $B(x; r)$ ($B[x; r]$) is the open (closed) ball centered at x with radius r; in a normed linear space X, either B or B_X is the unit ball, rB is the ball centered at the origin and with radius r.
- If X is a Banach space, X^* is the continuous dual space, i.e., the space of all linear continuous functionals on X, $\langle x^*, x \rangle$ is the pairing between $x \in X$ and $x^* \in X^*$.
- For a set $A \subset X$,

  (1) $A^c = X \setminus A$ is the complement of A, the set of the elements belonging to X but not to A;

  (2) cl A (or $\bar{A}$) is the closure of A;

  (3) int A is the interior of A;

  (4) diam$A$ is the diameter of A : diam$A = \sup\{d(x, y) : x, y \in A\}$;

  (5) $B_r[A] = \{x \in X : d(x, A) \le a\}$.

  If X is a linear space,

  (1) co A is the convex hull of A: the smallest convex set containing A;

  (2) cone A is the conic hull of A: the smallest cone containing A;

  (3) aff A is the affine hull of A: the smallest affine space containing A;

  (4) ri A is the relative interior of A: the interior points of A inside the space aff A;

  (5) $0^+(A)$ is the recession cone of A:

  $$0^+(A) = \{x : x + a \in A \quad \forall a \in A\};$$

  (6) The indicator function of A is the function $I_A(\cdot)$ valued zero inside A, ∞ outside A.

- The simplex in $\mathbb{R}^m$ is $\{x \in \mathbb{R}^m : x_i \ge 0, \sum x_i = 1\}$.
- The projection $p_A(x)$ of a point x on the set A is the set of the points of A nearest to x.

- The distance of a point x to a set A is

$$d(x, A) = \inf\{d(x, a) : a \in A\}.$$

- The excess of a set A over a set C is

$$e(A, C) = \sup\{d(a, C) : a \in A\}.$$

D.2 Functions

- The epigraph of f is

$$\operatorname{epi} f := \{(x, r) \in X \times \mathbb{R} : f(x) \leq r\} \subset X \times \mathbb{R}.$$

- The strict epigraph of f is

$$\text{s-epi}\, f := \{(x, r) \in X \times \mathbb{R} : f(x) < r\}.$$

- The effective domain of f is

$$\operatorname{dom} f := \{x \in X : f(x) < \infty\}.$$

- The level set at height $a \in \mathbb{R}$ of f is

$$f^a := \{x \in X : f(x) \leq a\}.$$

- $\operatorname{Min} f = \{x : f(x) = \inf f\} = f^{\inf f}.$
- The set $\mathcal{F}(X)$ is

$$\mathcal{F}(X) := \{f : X \to [-\infty, \infty] : f \text{ is proper and convex}\}.$$

- The set $\Gamma(X)$ is

$$\Gamma(X) := \{f \in \mathcal{F}(X) : f \text{ is lower semicontinuous}\}.$$

- Inf-convolution or epi-sum of f and g is

$$\begin{aligned}(f \nabla g)(x) &:= \inf\{f(x_1) + g(x_2) : x_1 + x_2 = x\} \\ &= \inf\{f(y) + g(x - y) : y \in X\}.\end{aligned}$$

- The lower semicontinuous regularization of f is $\bar{f}$ is

$$\operatorname{epi} \bar{f} := \operatorname{cl} \operatorname{epi} f.$$

- The lower semicontinuous convex regularization of f is $\hat{f}$ is

$$\operatorname{epi} \hat{f} := \operatorname{cl} \operatorname{co} \operatorname{epi} f.$$

- The directional derivative of f at x along the vector d is

$$f'(x; d) := \lim_{t \to 0^+} \frac{f(x + td) - f(x)}{t}.$$

- A subgradient $x^* \in X^*$ of f at the point $x_0 \in \operatorname{dom} f$ satisfies $\forall x \in X$,

$$f(x) \geq f(x_0) + \langle x^*, x - x_0 \rangle;$$

  $\partial f(x_0)$ is the set of the subgradients of f at x_0.
- An ε-subgradient $x^* \in X^*$ of f at the point $x_0 \in \operatorname{dom} f$ satisfies $\forall x \in X$,

$$f(x) \geq f(x_0) + \langle x^*, x - x_0 \rangle - \varepsilon.$$

- The Fenchel conjugate of f is

$$f^*(x^*) = \sup_{x \in X} \{ \langle x^*, x \rangle - f(x) \}.$$

- The strong slope $|\nabla f|(x)$ of f at x is

$$|\nabla f|(x) = \begin{cases} \limsup_{y \to x} \frac{f(x) - f(y)}{d(x,y)} & \text{if } x \text{ is not a local minimum,} \\ 0 & \text{if } x \text{ is a local minimum.} \end{cases}$$

- Given $f \colon X \to (-\infty, \infty]$ and $A \subset X$: a minimizing sequence $\{x_n\}$ for $f \colon f(x_n) \to \inf f$; a Levitin–Polyak minimizing sequence $\{x_n\}$ for (A, f): $\lim f(x_n) = \inf_A f$ and $d(x_n, A) \to 0$; a strongly minimizing sequence $\{x_n\}$ for (A, f): $\limsup f(x_n) \leq \inf_A f$ and $d(x_n, A) \to 0$.
- The problem f is Tykhonov well-posed if every minimizing sequence converges to the minimum point of f; the problem (A, f) is Levitin–Polyak (strongly) well-posed if every Levitin–Polyak minimizing (strongly minimizing) sequence converges to the minimum point of f over A.
- Well-posedness in the generalized sense means every minimizing sequence (in the appropriate sense) has a subsequence converging to a minimum point.

D.3 Spaces of sets

- $c(X)$ is the set of the closed subsets of a metric space X; $C(X)$ is the set of all closed convex subsets of a normed space X.
- Given sets $G \subset X$, $A \subset X$:

$$V^- := \{ A \in c(X) : A \cap V \neq \emptyset \},$$
$$G^+ := \{ A \in c(X) : A \subset G \}.$$

- V^-, the lower Vietoris topology on $c(X)$, is the topology having as a subbasis of open sets the family

$$\{V^- : V \text{ is open in } X\};$$

  V^+, the upper Vietoris topology on $c(X)$, is the topology having as a basis of open sets the family:

$$\{G^+ : G \text{ is open}\};$$

  $V = V^-$ and V^+, the Vietoris topology, has as a basis of open sets

$$G^+ \cap V_1^- \cap \cdots \cap V_n^-,$$

  with $G, V_1, \ldots, V_n$ open in X and $n \in \mathbb{N}$.
- $F^- = V^-$ is the lower Fell topology on $c(X)$; F^+, the upper Fell topology on $c(X)$, is the topology having as a basis of open sets the family:

$$\{(K^c)^+ : K \text{ is compact}\};$$

  $F = F^-$ and F^+ is the Fell topology; a basis for it is given by the family of sets

$$(K^c)^+ \cap V_1^- \cap \cdots \cap V_n^-,$$

  with $V_1, \ldots, V_n$ open, K compact and $n \in \mathbb{N}$.
- The Hausdorff (extended) distance between the closed sets A, C is

$$h(A, C) := \max\{e(A, C), e(C, A)\}.$$

- The inferior and superior limits in the Kuratowski sense of a sequence of sets are

$$\text{Li } A_n := \{x \in X : x = \lim x_n, x_n \in A_n \text{ eventually}\}$$

  and

$$\text{Ls } A_n := \big\{x \in X : x = \lim x_k, x_k \in A_{n_k},$$
$$n_k \text{ a subsequence of the integers}\big\}.$$

  The Kuratowski limit A of a sequence $\{A_n\}$ of sets is

$$\text{Ls } A_n \subset A \subset \text{Li } A_n.$$

- The Wijsman limit A of a sequence $\{A_n\}$ of sets is

$$\lim d(x, A_n) = d(x, A), \forall x \in X.$$

- The Attouch–Wets limit A of a sequence $\{A_n\}$ of sets: let $x_0 \in X$, where X is a metric space. If A, C are nonempty sets, define

$$e_j(A, C) := e(A \cap B(x_0; j), C) \in [0, \infty),$$
$$h_j(A, C) := \max\{e_j(A, C), e_j(C, A)\}.$$

If C is empty and $A \cap B(x_0; j)$ nonempty, set $e_j(A, C) = \infty$. Then the sequence $\{A_n\}$ converges to A if

$$\lim_{n \to \infty} h_j(A_n, A) = 0 \text{ for all large } j.$$

- The Mosco limit A of a sequence $\{A_n\}$ of convex sets is

$$\text{w-Ls } A_n \subset A \subset \text{Li } A_n.$$

- The bounded proximal topology satisfies $A_n \to A$ if and only if

$$D(A_n, F) \to D(A, F),$$

for every $F \subset X$ which is a closed bounded set.
- Let X be a normed space. The slice topology on $C(X)$: $A_n \to A$ if and only if

$$D(A_n, C) \to D(A, C)$$

for every $C \subset X$ which is a closed convex bounded set.
- Let X^* be a dual space. The slice* topology on $C(X^*)$: $A_n \to A$ if and only if

$$D(A_n, C) \to D(A, C),$$

for every $C \subset X$ which is a weak* closed convex bounded set.

D.4 Definitions

- affine set: Definition 1.1.10.
- approximate subdifferential: Definition 3.7.1.
- Attouch–Wets convergence: Definition 8.2.13.
- Baire space: Definition B.1.1.
- bounded proximal topology: Definition 8.5.2.
- converging net in Kuratowski sense: Definition B.3.1.
- convex combination: Definition 1.1.4.
- convex function: Definition 1.2.1.
- convex function (classical): Definition 1.2.3.
- convex hull: Definition 1.1.6.
- convex lower semicontinuous regularization: Definition 5.2.2.
- convex set: Definition 1.1.1.
- cooperative game: Definition 7.4.1.

- core of a cooperative game: Definition 7.4.3.
- directional derivative: Definition 3.1.1.
- E-space: Definition 10.4.5.
- excess of A over B: Definition 8.1.7.
- extreme point: Definition 1.1.8.
- Fell topology: Definition 8.1.4.
- Fenchel conjugate: Definition 5.1.1.
- forcing function: Definition 10.1.8.
- Fréchet differentiable function: Definition 3.3.1.
- Fréchet differentiable subgradient: Definition 3.6.2.
- Gâteaux differentiable function: Definition 3.3.1.
- Hausdorff metric topology 8.1.
- Kuratowski convergence: Definition 8.1.15.
- inf-convolution (Episum): Definition 1.2.20.
- Lagrangean: Definition 6.5.1.
- Levitin–Polyak minimizing sequence: Definition 10.1.12.
- Levitin–Polyak well-posed problem: Definition 10.1.13.
- Lipschitz stable multifunction: Definition 3.6.1.
- linear topology: Definition 8.5.3.
- lower semicontinuous function: Definition 2.2.1.
- lower semicontinuous regularization: Definition 2.2.3.
- maximal monotone: Definition 3.5.13.
- monotone operator: Definition 3.5.11.
- Mosco convergence: Definition 8.3.1.
- nowhere dense set: Definition 11.1.3.
- outer density point: Definition 11.1.5.
- porous set: Definition 11.1.1.
- proper function: Definition 1.2.16.
- proximal topology: Definition 8.5.1.
- recession cone: Definition 1.1.15.
- regular problem: Definition 6.4.1.
- relative interior: Definition 3.2.9.
- saddle point: Definition 6.5.3.
- slice topology: Definition 8.5.4.
- strong slope: Definition 4.2.1.
- strongly minimizing sequence: Definition 10.1.12.
- strongly porous set: Definition 11.1.7.
- strongly smooth space: Definition 10.4.3.
- strongly well-posed problem: Definition 10.1.13.
- subdifferential of a concave/convex function: Definition 11.2.3.
- subgradient: Definition 3.2.1.
- sublinear function: Definition 1.2.14.
- supporting functional: Definition 3.2.2.
- two player noncooperative game in strategic form C.3.
- two player noncooperative game: Definition C.3.

D.5 Important theorems

- Theorem 2.2.8: A convex lower semicontinuous function on a Banach space is continuous at the interior points of its effective domain.
- Theorem 2.2.21: A function $f \in \Gamma(X)$ is the pointwise supremum of the affine functions minorizing it.
- Theorem 4.1.1: The Weierstrass theorem on existence of minima.
- Theorem 4.2.5: The Ekeland variational principle.
- Corollary 4.2.13: For $f \in \Gamma(X), \partial f$ is nonempty on a dense subset of dom f.
- Theorem 4.2.17: Let $f \in \Gamma(X)$. Then, for all $x \in X$

$$f(x) = \sup\{f(y) + \langle y^*, x - y \rangle : (y, y^*) \in \partial f\}.$$

- Theorem 5.2.8 The Fenchel conjugation is a bijection between $\Gamma(X)$ and $\Gamma^*(X^*)$.
- Theorem 5.4.2: The Attouch–Brézis theorem on the conjugate of the sum.
- Theorem 7.1.1 The duality between two linear programming problems.
- Theorem 7.2.5: The theorem of von Neumann on zero sum games: A two player, finite, zero sum game has equilibrium in mixed strategies.
- Theorem 7.3.6: On two feasible linear programming problems in duality.
- Theorem 7.4.8: Nonemptiness of the core via balanced coalitions.
- Theorem 8.4.1: Completeness of the hyperspace endowed with the Hausdorff metric topology.
- Theorem 8.4.3: Topological completeness of the hyperspace of convex sets endowed with the Mosco topology.
- Theorem 8.4.4: Compactness of the hyperspace with the Fell topology.
- Theorem 8.6.3 and Theorem 8.6.4: Characterization of Kuratowski (resp. Mosco) convergence of a sequence of lower semicontinuous (resp. lower semicontinuous convex) functions.
- Theorem 8.6.6 The first general stability result.
- Theorem 9.1.2, Theorem 9.1.4, Theorem 9.1.6: On the continuity of the conjugation with respect to the Mosco, slice and Attouch–Wets convergences.

- Theorem 9.2.5: On the continuity of the sum with respect to the Attouch–Wets convergence.
- Theorem 9.3.1: On Mosco convergence of functions and lower convergence of associated differential operators.
- Theorem 10.1.11: Tykhonov well-posedness of a function and Fréchet differentiability of its Fenchel conjugate.
- Theorem 10.2.14: The basic result on the connections between stability, with Mosco convergence, and Tykhonov well-posedness.
- Theorem 10.2.24, Theorem 10.2.25: The basic results on the connections between stability, with Attouch–Wets convergence, and Tykhonov well-posedness.
- Theorem 10.4.6: Equivalent conditions to the Tykhonov well-posedness of the best approximation problem.
- Theorem 10.4.15: The subdifferential of the distance function on a general Banach space.
- Theorem 11.2.5: On the σ porosity of the set of points of non Fréchet differentiability of a concave/convex function.
- Theorem 11.3.8: The Ioffe–Zaslavski principle.
- Theorem 11.4.1: The porosity principle.
- Theorem 11.4.5, Theorem 11.4.10: The σ-porosity of the set of the non well-posed problems in convex programming.
- Theorem 11.4.14: The σ-porosity of the set of the non well-posed problems in quadratic programming.
- Theorem A.1.1, Theorem A.1.5, Theorem A.1.6: The Hahn–Banach theorems.
- Theorem A.2.1: The Banach–Dieudonné–Krein–Smulian theorem.
- Theorem B.4.6: On a characterization of hypertopologies as initial ones.
- Theorem C.4: The theorem of Nash on the existence of equilibria in non-cooperative games.

References

[AR]   E. ASPLUND AND R. T. ROCKAFELLAR, Gradients of convex functions, *Trans. Amer. Math. Soc.* **139** (1969), 443–467.

[AB]   H. ATTOUCH AND H. BRÉZIS, Duality for the sum of convex functions in general Banach spaces, *Aspects of Mathematics and its Applications*, J. A. Barroso, ed. Elsevier Science Publishers (1986), pp. 125–133.

[Be]   G. BEER, A Polish topology for the closed subsets of a Polish space, *Proc. Amer. Math. Soc.* **113** (1991), 1123–1133.

[Be2]   G. BEER, *Topology on Closed and Closed Convex Sets,* Mathematics and Its Applications, Vol. 268, Kluwer Academic Publishers, 1993.

[BL]   G. BEER AND R. LUCCHETTI, Convex optimization and the epidistance topology, *Trans. Amer. Math. Soc.* **327** (1991), 795–813.

[BT]   G. BEER AND M. THÉRA, Attouch–Wets convergence and a differential operator for convex functions, *Proc. Amer. Math. Soc.* **122** (1994), 851–858.

[BF]   J. M. BORWEIN AND S. FITZPATRICK, Mosco convergence and the Kadeč property, *Proc. Amer. Math. Soc.* **106** (1989), 843–852.

[BoLe]   J. M. BORWEIN AND A. S. LEWIS, *Convex Analysis and Nonlinear Optimization*, Springer, New York, 2000.

[BP]   J. M. BORWEIN AND D. PREISS, A smooth variational principle with applications to subdifferentiability and to differentiability of convex functions, *Trans. Amer. Math. Soc.* **303** (1987), 517–527.

[BMP]   S. DE BLASI, J. MYJAK, AND P. L. PAPINI, Porous sets in best approximation theory, *J. London Math. Soc.* **44** (1991), 135–142.

[DGZ]   R. DEVILLE, G. GODEFROY, AND V. ZIZLER, *Smoothness and Renormings in Banach Spaces*, Pitman Monographs and Surveys in Pure and Appl. Math., Longman Scientific & Technical, 1993.

[DR]   R. DEVILLE AND J. P. REVALSKI, Porosity of ill-posed problems, *Proc. Amer. Math. Soc.* **128** (2000), 1117–1124.

[DZ]   A. L. DONTCHEV AND T. ZOLEZZI, *Well-Posed Optimization Problems*, Lectures Notes in Mathematics 1543, Springer-Verlag, Berlin, 1993.

[ET]   I. EKELAND AND R. TEMAM, *Convex Analysis and Variational Problems*, North-Holland, Amsterdam, 1976.

[HUL]   J. B. HIRIART-URRUTY AND C. LEMARÉCHAL, *Convex Analysis and Minimization Algorithms I*, Springer-Verlag, Berlin, 1993.

[IL]   A. IOFFE AND R. LUCCHETTI,*Generic existence, uniqueness and stability in optimization*, Nonlinear Optimization and Related Topics, G. Di Pillo and F. Giannessi eds., Kluwer Academic Publishers, Dordrecht, 2000, pp. 169–182.

[IL2]   A. IOFFE AND R. LUCCHETTI, Typical convex program is very well-posed, to appear in Math. Program.

[ILR]   A. D. IOFFE, R. LUCCHETTI, AND J. P. REVALSKI, A variational principle for problems with functional constraints, *SIAM J. Optim.* **12** (2001), 461–478.

[ILR2]   A. D. IOFFE, R. LUCCHETTI, AND J. P. REVALSKI, Almost every convex or quadratic programming problem is well posed, *Math. Oper. Res.* **29** (2004), 369–382.

[IZ]   A. D. IOFFE AND A. J. ZASLAVSKI, Variational principles and well-posedness in optimization and calculus of variations. *SIAM J. Control Optim.* **38** (2000), 566–581.

[LL]   A. LECHICKI AND S. LEVI, Wijsman convergence in the hyperspace of a metric space, *Boll. Unione Mat. Ital. B* **1** (1987), 435–451.

[LeL]   A. LEWIS AND R. LUCCHETTI, Nonsmooth duality, sandwich and squeeze theorems, *Siam J. Control Optim.* **38** (2000), 613–626.

[Luc]   D. T. LUC, *Theory of Vector Optimization*, Lecture Notes in Economics and Mathematical Systems, Springer-Verlag, Berlin, 1989.

[LSS]   R. LUCCHETTI, P. SHUNMUGARAJI, AND Y. SONNTAG, Recent hyper-topologies and continuity of the value function and of the constrained level sets, *Numer. Funct. Anal. Optim.* **14** (1993), 103–115.

[LT]   R. LUCCHETTI AND A. TORRE, Classical set convergences and topologies, *Set-Valued Anal.* **2** (1994), 219–240.

[Mar]   E. MARCHINI, Porosity and variational principles. *Serdica Math. J.* **28** (2002), 37–46.

[Mi]   F. MIGNOT, Contrôle dans les inéquations variationnelles elliptiques, *J. Funct. Anal.* **22** (1976), 130–185.

[Ow]   G. OWEN, *Game Theory*, Second edition, Academic Press, Orlando, 1982.

[Ph]   R. R. PHELPS, *Convex Functions, Monotone Operators and Differentiability*, Lecture Notes in Mathematics 1364, Springer-Verlag, Berlin Heidelberg, 1993.

[PZ]   D. PREISS AND L. ZAJÍČEK, Fréchet differentiation of convex functions in a Banach space with separable dual, *Math. Oper. Res.* **4** (1979), 425–430.

[RZ]   S. REICH AND A. J. ZASLAVSKI, The set of divergent descent methods in a Banach space is sigma-porous, *SIAM J. Optim.* **11** (2001), 1003–1018.

[RZ2]   S. REICH AND A. J. ZASLAVSKI, Well-posedness and porosity in best approximation problems, *Topol. Methods Nonlinear Anal.* **18** (2001), 395–408.

[Ro]   R. T. ROCKAFELLAR, *Convex Analysis*, Princeton University Press, Princeton, New Jersey, 1970.

[Si]   S. SIMONS, *Minimax and Monotonicity*, Lecture Notes in Mathematics 1693, Springer-Verlag, Berlin, 1998.

[Si2]   S. SIMONS, A new version of the Hahn–Banach theorem, *Arch. Math. (Basel)* **80** (2003), 630–646.

[SRo]   J. E. SPINGARN AND R. T. ROCKAFELLAR, The generic nature of optimality conditions in nonlinear programming, *Math. Oper. Res.* **4** (1979), 425–430.

[St]   Ts. STOYANOV, A measure on the space of compact subsets in $\mathbb{R}^n$ and its application to some classes of optimization problems, *C. R. Acad. Bulgare Sci.* **42** (1989), 29–31.

[Zaj]   L. ZAJÍČEK, Porosity and σ-porosity, *Real Anal. Exchange* **13** (1987/88), 314–350.

[Za]   C. ZALINESCU, *Convex Analysis in General Vector Spaces*, World Scientific, 2002.

[Zo]   T. ZOLEZZI, Extended well-posedness of optimization problems, *J. Optim. Theory Appl.* **91** (1996), 257–268.

Index